Ferrous Materials

Hans Berns · Werner Theisen

Ferrous Materials

Steel and Cast Iron

Translated by Gillian Scheibelein, B.Sc. (Hons.)

 Springer

Professor em. Dr.-Ing. Hans Berns
Professor Dr.-Ing. Werner Theisen
Ruhr-Universität Bochum
Lehrstuhl Werkstofftechnik
44780 Bochum
berns@wtech.rub.de
wth@wtech.rub.de

ISBN 978-3-540-71847-5   e-ISBN 978-3-540-71848-2

DOI 10.1007/978-3-540-71848-2

Library of Congress Control Number: 2008933410

© 2008 Springer-Verlag Berlin Heidelberg

This work is subject to copyright. All rights are reserved, whether the whole or part of the material is concerned, specifically the rights of translation, reprinting, reuse of illustrations, recitation, broadcasting, reproduction on microfilm or in any other way, and storage in data banks. Duplication of this publication or parts thereof is permitted only under the provisions of the German Copyright Law of September 9, 1965, in its current version, and permission for use must always be obtained from Springer. Violations are liable for prosecution under the German Copyright Law.

The use of general descriptive names, registered names, trademarks, etc. in this publication does not imply, even in the absence of a specific statement, that such names are exempt from the relevant protective laws and regulations and therefore free for general use.

The authors do not warrant the actuality, correctness, completeness or quality of the provided information. The authors disclaim reliability for material as well as ideal damages that result from the use or disuse of the provided information or the use of deficient or incomplete information, as far as they are not demonstrably the result of an intended or wantonly negligent fault committed by the authors.

Production: le-tex publishing services oHG, Leipzig, Germany
Cover design: eStudioCalamar S.L., F. Steinen-Broo, Girona, Spain

Printed on acid-free paper

9 8 7 6 5 4 3 2 1

springer.com

Preface

The annual global production of ferrous materials has risen by more than 60 % in the last ten years to a total of about 1300 million tons. This exceeds the sum of all other metallic materials by more than one order of magnitude. The importance of ferrous materials is based on their variety of different properties that can be customised by alloying and processing. These relationships are discussed for steel and cast iron conjointly. Part A discusses the fundamental principles in the Chapters *Constitution, Microstructure, Heat treatment* and *Properties*. The much larger Part B discusses processing and applications of European standard materials as well as recent developments. It deals with *unalloyed and high-strength materials, materials for surface layer treatment* and for *tools*, as well as *chemically resistant, creep-resistant* and *functional materials*.

This book is intended for engineers working with ferrous materials who wish to deepen their understanding or who are looking for advice. The necessary practical relevance arises from the authors' first-hand knowledge gained during many years of industrial experience. Students are also avid readers of the three German editions. Harmonisation of European standards has enabled publication of an English edition.

We extend our thanks to Ms Gillian Scheibelein B.Sc. for translating the German text and to SCHMOLZ + BICKENBACH for financial support of this task. We thank Dr.-Ing. Markus Karlsohn from the Chair of Materials Technology for editing and preparing the printable LaTeX version. Our appreciation also goes to further contributors: Dipl.-Ing. Stephan Huth, Dipl.-Ing. André Oppenkowski, Dipl.-Ing. Tanja Macher and cand. ing. Marius Weber.

Bochum, Spring 2008

Hans Berns *Werner Theisen*

Note: Unless stated otherwise, the alloy contents given in % are mass percentages.

Contents

A Fundamentals of ferrous materials   **3**
H. BERNS
- A.1 Constitution 3
  - A.1.1 Pure iron 5
  - A.1.2 Iron-carbon 9
    - A.1.2.1 The iron-cementite system 11
    - A.1.2.2 The iron-graphite system 13
  - A.1.3 Alloyed iron 14
- A.2 Microstructure 21
  - A.2.1 Near-equilibrium microstructure 26
    - A.2.1.1 Steel 26
    - A.2.1.2 Cast iron 31
  - A.2.2 Non-equilibrium microstructure 36
    - A.2.2.1 Shaping 37
    - A.2.2.2 Austenite transformation 38
    - A.2.2.3 Post-quenching morphology 42
    - A.2.2.4 Reheating of quenched microstructures . . . 48
  - A.2.3 Morphology of cementite and graphite 53
- A.3 Heat treatment 57
  - A.3.1 Annealing processes 58
    - A.3.1.1 Baking 58
    - A.3.1.2 Stress-relief annealing 58
    - A.3.1.3 Soft annealing of steel 59
    - A.3.1.4 Soft annealing of cast iron 62
    - A.3.1.5 Normalising 62
    - A.3.1.6 Temper annealing of cast iron 63
    - A.3.1.7 Solution annealing 63
    - A.3.1.8 Homogenising 63
  - A.3.2 Hardening and related processes 64
    - A.3.2.1 Hardening 64
    - A.3.2.2 Tempering 68
    - A.3.2.3 QT treatment 69
    - A.3.2.4 Transformation in the bainite range 70
  - A.3.3 Surface layer treatment/Coating 71

Contents

- A.3.4 Side-effects ... 72
  - A.3.4.1 Thermal side-effects ... 73
  - A.3.4.2 Thermochemical side-effects ... 74
- A.4 Properties ... 79
  - A.4.1 Mechanical properties ... 79
    - A.4.1.1 Loading ... 79
    - A.4.1.2 Behaviour of steel ... 82
    - A.4.1.3 Behaviour of grey cast iron ... 99
    - A.4.1.4 Behaviour of white cast iron ... 102
  - A.4.2 Tribological properties ... 103
    - A.4.2.1 Friction ... 104
    - A.4.2.2 Wear ... 106
  - A.4.3 Chemical properties ... 109
    - A.4.3.1 Wet corrosion ... 109
    - A.4.3.2 High-temperature corrosion ... 115
  - A.4.4 Special physical properties ... 117
    - A.4.4.1 Magnetic properties ... 117
    - A.4.4.2 Thermal expansion ... 120
    - A.4.4.3 Conductivity ... 121

B Ferrous materials and their applications ... 125
- B.1 Materials for general applications ... 125
  - B.1.1 Unalloyed structural steels ... 125
    H. BERNS
    - B.1.1.1 Properties ... 126
    - B.1.1.2 Grades and applications ... 134
  - B.1.2 Cast iron ... 144
    W. THEISEN
    - B.1.2.1 Composition of grey cast iron ... 144
    - B.1.2.2 Cast iron with flake graphite ... 147
    - B.1.2.3 Cast iron with spheroidal graphite ... 150
    - B.1.2.4 Cast iron with vermicular graphite ... 152
    - B.1.2.5 Malleable cast iron ... 154
    - B.1.2.6 Processing and applications of cast iron ... 157
- B.2 High-strength materials ... 165
  - B.2.1 Weldable rolled steels ... 165
    H. BERNS
    - B.2.1.1 Fine-grain steels ... 165
    - B.2.1.2 Multi-phase steels ... 168
    - B.2.1.3 Applications of weldable steels ... 175
    - B.2.1.4 Lightweight steels ... 182
    - B.2.1.5 Pearlitic rolled steels ... 184
  - B.2.2 Steels treated from the forging temperature ... 184
    - B.2.2.1 Martensitic steels ... 185
    - B.2.2.2 Ferritic-pearlitic steels ... 188

B.2.3	Structural steels for full heat treatment		190
	B.2.3.1	QT steels	190
	B.2.3.2	Ultrahigh-strength steels	198
	B.2.3.3	Hard steels	203
B.2.4	Cast iron for full heat treatment		207

W. THEISEN

B.2.4.1	Quenching and tempering	207
B.2.4.2	Transformation in the bainite range / ADI	208

B.3 Materials for surface layer treatments 217

H. BERNS

B.3.1	Materials for surface-hardening		217
	B.3.1.1	Process engineering aspects of surfacehardening	217
	B.3.1.2	Materials and the surface layer	220
	B.3.1.3	Applications	223
B.3.2	Nitriding steels		224
	B.3.2.1	Process engineering aspects of nitriding	224
	B.3.2.2	Materials and the surface layer	228
	B.3.2.3	Applications	232
B.3.3	Case hardening steels		234
	B.3.3.1	Process engineering aspects of case hardening	234
	B.3.3.2	Materials and the surface layer	241
	B.3.3.3	Applications	246

B.4 Tools for processing minerals 251

W. THEISEN

B.4.1	Loading and material concepts		251
	B.4.1.1	Hard phases	252
	B.4.1.2	Metal matrix	254
B.4.2	Tools made of hot-formed steel		256
B.4.3	Cast tools		259
	B.4.3.1	Pearlitic white cast iron	259
	B.4.3.2	Martensitic nickel white cast iron	260
	B.4.3.3	Martensitic chromium white cast iron	261
B.4.4	Coated tools		264
	B.4.4.1	Hard-facing	264
	B.4.4.2	Powder metallurgical coatings	267
	B.4.4.3	Composite casting	269

B.5 Tools for processing materials 273

W. THEISEN

B.5.1	Cold-work tools		274
	B.5.1.1	Properties	277
	B.5.1.2	Coated tools	283
	B.5.1.3	Applications of cold-work tools	288
B.5.2	Tools for processing plastics		291
B.5.3	Hot-work tools		293
	B.5.3.1	Properties	294

Contents

- B.5.3.2 Applications 296
- B.5.4 Tools for machining applications 299
  - B.5.4.1 Properties 300
  - B.5.4.2 Applications 303
- B.6 Chemically resistant materials 309
  H. BERNS
  - B.6.1 General information 309
    - B.6.1.1 Alloying concept 309
    - B.6.1.2 Matrix properties 311
  - B.6.2 Stainless steels 317
    - B.6.2.1 Properties 319
    - B.6.2.2 Applications 331
  - B.6.3 Heat-resistant steels 336
    - B.6.3.1 Properties 336
    - B.6.3.2 Applications 339
  - B.6.4 Cast iron 341
    - B.6.4.1 Ferritic cast iron 341
    - B.6.4.2 Austenitic cast iron 342
    - B.6.4.3 White cast iron / carbide-rich steels 344
- B.7 Creep-resistant materials 349
  H. BERNS
  - B.7.1 Properties 352
    - B.7.1.1 Normalised as well as QT steels 352
    - B.7.1.2 Austenitic steels 357
    - B.7.1.3 Cast iron 360
  - B.7.2 Applications 360
    - B.7.2.1 Steam power plants 360
    - B.7.2.2 Gas turbines 362
    - B.7.2.3 Estimation of the service life 363
    - B.7.2.4 Petrochemistry 365
    - B.7.2.5 Valves 365
- B.8 Functional materials 369
  H. BERNS
  - B.8.1 Magnetically soft materials 369
  - B.8.2 Magnetically hard materials 373
  - B.8.3 Non-magnetisable materials 375
  - B.8.4 Materials with a special thermal expansion 376
  - B.8.5 Materials with a shape memory 378
  - B.8.6 Electrical resistance heating alloys 380

C Appendix 383

- C.1 Designation systems for steel and cast iron 383
  W. THEISEN
  - C.1.1 Standardisation 383
  - C.1.2 Designations for steels and cast steels 384

		Unalloyed steels 386
		Alloyed steels 386
		High-alloy steels 387
		High-speed tool steels 387
	C.1.3	Designation of cast irons 388
C.2	A brief discourse on the history of iron 392
H. BERNS
	C.2.1	From a bloomery to a shaft furnace 392
	C.2.2	The spread of iron-making 395
	C.2.3	Cast iron and the fining process 395
	C.2.4	Mild steel 397
	C.2.5	Ferrous materials 398
C.3	Bibliography for figures and tables 400

Keyword Index   **403**

List of alloying and tramp elements   **416**

Ni
substituted
Cr
C interstitial
Fe
X5CrNi18-10

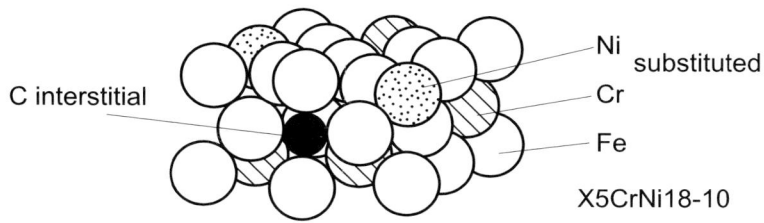

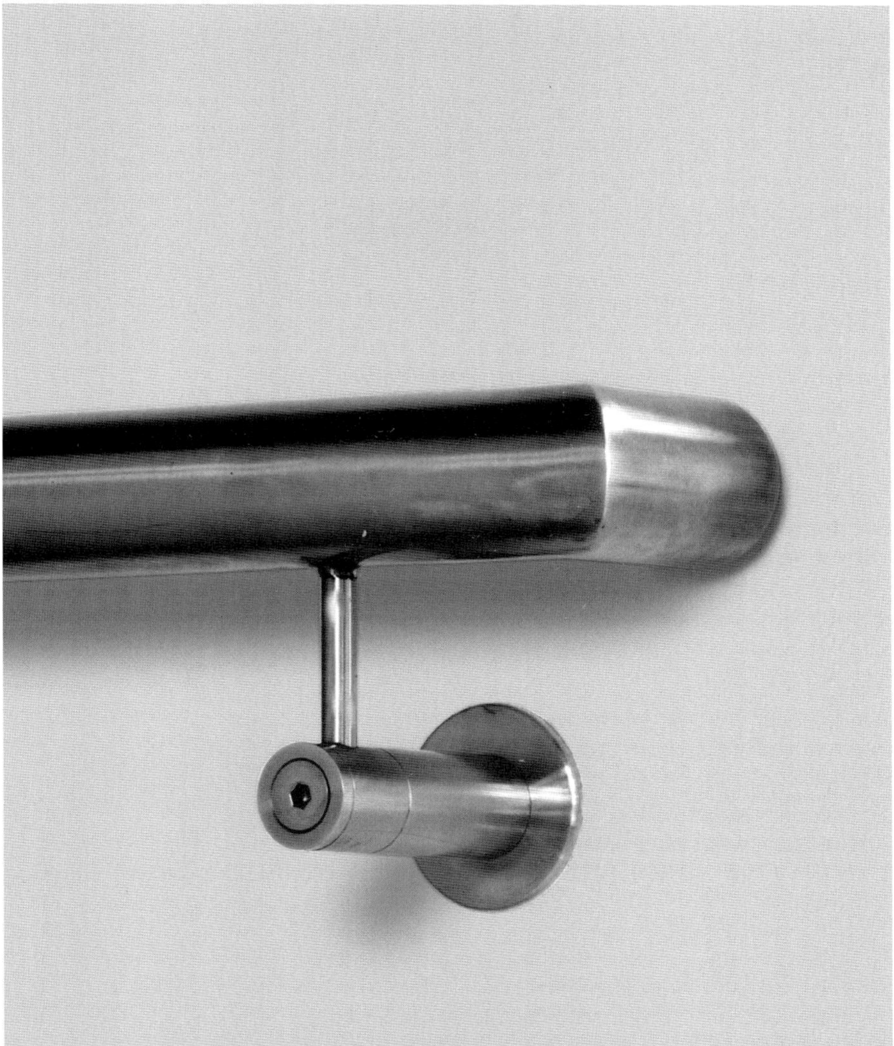

Handrail of stainless austenitic steel
(SCHMOLZ + BICKENBACH Distributions GmbH, Düsseldorf, Germany)

A

Fundamentals of ferrous materials

A.1 Constitution

This chapter describes the different constitutions or states of ferrous materials with respect to the ordering of the atoms. Above the boiling point, the atoms are in the gaseous state and they fly through space completely disordered and well apart. Between the boiling point and the solidification point, they move around in the liquid state and are still disordered, but close together. Below the solidification point, they are in the solid state and are ordered into crystals. Their movements are now limited to oscillating about fixed points in the crystal lattice. The atoms can have more than one type of arrangement in the crystalline state. A region that has the same ordering is known as a phase. Several phases may be present in a ferrous material, e.g. during a liquid/solid transition or as crystals with different types of ordering. The state is said to be homogeneous if only one phase is present and heterogeneous if there is more than one.

Which phases form ultimately depends on three state parameters: temperature, pressure and concentration. In contrast to the above-mentioned effect of temperature, the effect of pressure is not immediately obvious because incompressible liquid and solid phases only respond with a change of state at a high pressure; however, most production processes and applications take place at atmospheric pressure, i.e. approximately 1 bar. Nevertheless, if we look more closely, we can see that phase transformations are frequently associated with a change in volume, and this may lead to large pressure changes on a local scale. A pressure increase favours the denser phase and thus stabilises it. The compressible gas phase reacts to even a slight increase in pressure, e.g. with an increase in the boiling point and thus stabilisation of the denser liquid phase. Inversely, a very low pressure is used to deposit vapours on surfaces or to degas melts. In spite of these examples illustrating the influence of pressure on the state of ferrous materials, the assumption of constant atmospheric pressure is usually justified in practice. On the other hand, the concentration is a variable state parameter like the temperature. It is primarily responsible for

A.1 Constitution

the diversity of solid phases. In ferrous materials, concentration refers to the amounts of tramp and alloying elements. Tramp elements are unintentional and originate during production. Alloying elements are intentionally added to bring about certain ordered states, thus creating the prerequisites for certain desired properties. If we disregard minor contaminants, such as oxides and sulphides, production starts with a homogeneous melt consisting of iron and other dissolved elements that is solidified by casting into ingots, bars and shaped pieces or by atomisation into powders and semi-finished products. At a given alloy concentration and constant pressure, any changes of state during solidification and subsequent cooling or reheating depend solely on the temperature.

The basic premise for these discussions on the constitution of ferrous materials is an extremely slow rate of temperature change that allows a thermodynamic equilibrium to be established between the different phases. This gives the atoms enough time to diffuse and accumulate to form the various phases, thus maintaining the balance between formation and dispersal of phases that belong to a given temperature. In addition, a certain amount of time is required to add or dissipate the latent heat associated with a phase transition. Because gaseous atoms have a higher kinetic energy, the heat of condensation is greater than the heat of solidification. Nevertheless, the transformation of one solid phase into another is generally accompanied by evolution or absorption of heat. State and phase equilibria are usually only approached during

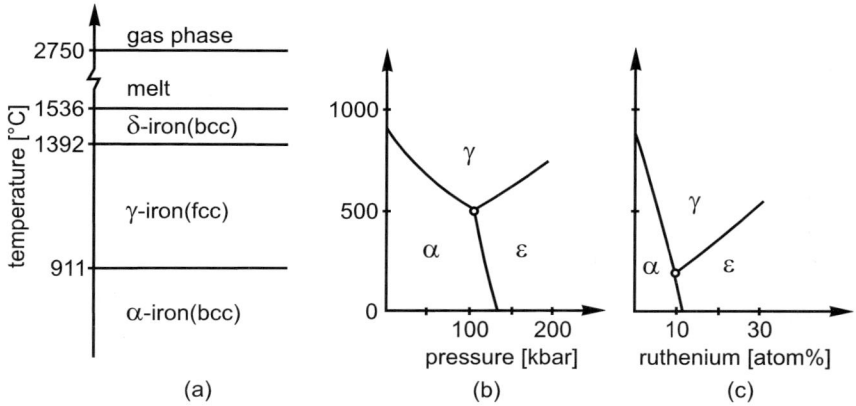

Fig. A.1.1 Constitution of iron: (a) Temperature-dependent ordering at standard pressure, (b) influence of pressure and temperature (from F.P. Bundy), (c) influence of internal pressure (from H. Schumann), this pressure is generated by the larger ruthenium atoms, an iron homologue, i.e. it has a comparable outer electron shell (atomic diameter $Ru = 0.268$, $Fe = 0.248$ nm). The external and internal pressures stabilise the closer-packed iron phases, namely fcc-γ and hcp-ε.

manufacturing processes, e.g. during cooling of a thick-walled casting in a thermally insulating sand mould or during heating of a heavy forging ingot in a gas-fired furnace. In contrast to these near-equilibrium manufacturing steps, atomisation or welding requires rapid temperature changes, which means that these systems are well away from equilibrium conditions. Such non-equilibrium states are intentionally brought about by heat treatments such as hardening; however, they will be discussed in the following chapters. Which equilibrium state is reached depends on the three state parameters and is shown in phase diagrams. They form the basis for the subsequent discussions.

A.1.1 Pure iron

The constitution of pure iron is shown in Figure A.1.1. At atmospheric pressure, the phase diagram is limited to the temperature axis (Figure A.1.1a). Slow cooling of the disordered states in the gaseous and liquid phases results in two solid phases with crystalline ordering that both exhibit a cubic unit cell as the smallest structural element. The face-centred cubic (fcc) crystal lattice has one atom at each of the eight corners and one atom in the centre of each of the six faces. In contrast, the body-centred cubic (bcc) lattice has one atom in the centre of the cube instead of in the faces (Figure A.1.2).

The fcc arrangement of the atoms is known as γ-iron. The bcc structure occurs in two temperature regions and is known as δ- and α-iron. The fcc or bcc ordering of the iron atoms leads to differences in the structure of the crystals. These differences are summarised in Tab. A.1.1. Because $\frac{1}{8}$ of a corner atom and $\frac{1}{2}$ of a face atom belong to one unit cell, the fcc cell contains twice as many iron atoms as the bcc cell. Its edge length, lattice parameter a, is correspondingly larger, as can be seen from the relationship between a and the atomic diameter d in Figure A.1.2. The ratio between the volume of the unit cell a^3 and the number of atoms it contains gives an $\approx 8\,\%$ lower atomic volume for γ-iron compared to α-iron at room temperature. As a result of this closer atomic packing, the thermal expansion of γ-iron is higher so that the transition from α- to γ-iron at 911°C only lowers the volume by $\approx 1\,\%$. The higher packing density of the atoms in γ-iron is also reflected in the higher coordination number (number of neighbouring atoms in contact with an atom). There are gaps (interstices) between the iron atoms that are either of octahedral or tetrahedral shape (Figure A.1.3). γ-Iron has larger interstices and this leads to its greater solubility for small atoms, such as carbon, nitrogen and hydrogen. α-Iron, on the other hand, has a larger number of interstices on account of its less closely packed atoms, and this facilitates diffusion. As can be seen in Figure A.1.2, the cubic crystal has closely packed directions in which the atoms are in contact. Such directions are denoted by Miller indices in angle brackets, whereas planes are enclosed in curly brackets (Figure A.1.4). The atoms are loosely packed in the {100} face of the bcc cube, which is why is it is a common cleavage plane for brittle fracture. In contrast, the atoms are staggered and closely packed in the {111} fcc plane, and this favours ductile shearing.

Increasing the pressure stabilises the denser phase, thus leading to an increase in the melting and boiling points. In the temperature/pressure phase diagram of iron, the phase field of γ-iron grows at the expense of α-iron (Figure A.1.1 b). There is another crystalline iron phase with hexagonal close-packed (hcp) ordering that exists above $\approx 100\,\text{kbar}$. However, this ε-iron is only of technical interest insofar as alloying elements are able to favour the hexagonal lattice (cf. ε-martensite) through internal pressure (Figure A.1.1 c) or by changing the concentration of free electrons. For the sake of completeness, the hcp structure is shown in Figure A.1.2 c. If atomic spheres are in a staggered arrangement within a plane and thus closely packed, they have a hexagonal pattern. If these hexagonal planes are then stacked so that the atoms of one layer lie in the interstices of the layers above and below it and

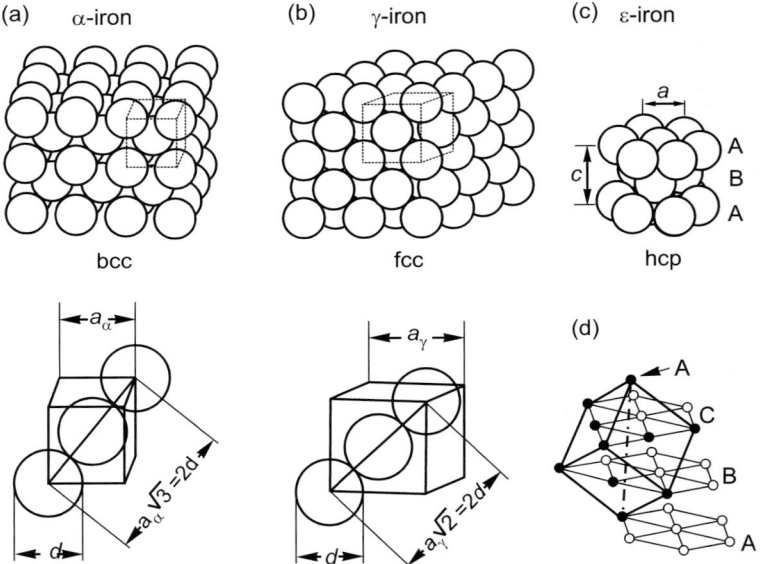

Fig. A.1.2 Crystal structure of iron: (a) α-Iron with a body-centred cubic (bcc) arrangement of the Fe atoms (shown as spheres). In the unit cell (dashed lines), the atoms with diameter d are in contact in the direction of the body diagonals. This gives the edge length a_α (lattice parameter) of the unit cube. (b) γ-Iron has a face-centred cubic (fcc) arrangement and the atoms are in contact within the face diagonals. (c) Unit cell of hexagonal close-packed (hcp) ε-iron, consisting of hexagonal planes of atoms with an ABAB... stacking sequence in a staggered arrangement thus giving lattice parameters a and c. (d) In contrast, an ABCA... stacking sequence produces an fcc crystal (dark mid-points of the atoms).

Table A.1.1 Properties and lattice structure of iron: The data refer to room temperature and standard pressure. The atomic volume of γ-iron was calculated assuming a constant thermal expansion coefficient and using values measured for γ-iron within its actual range of existence at $>911°C$. As a comparison, the thermal expansion coefficient of α-iron increases up to a value of $15.5 \cdot 10^{-6}\,\mathrm{K}^{-1}$ at $911°C$, with a mean value of $14.8 \cdot 10^{-6}\,\mathrm{K}^{-1}$

	α-iron	γ-iron
Atomic number	26	
Rel. atomic mass [g]	55.85	
Density $[\frac{g}{cm^3}]$	7.875	
Modulus of elasticity [GPa]	215.55	
Coefficient of thermal expansion α $[\frac{1}{K}]$	$12 \cdot 10^{-6}$	$23 \cdot 10^{-6}$
Atomic diameter[1] d [nm]	0.2482	
Lattice type	bcc	fcc
Lattice parameter a	$d \cdot (\frac{2}{\sqrt{3}})$	$d \cdot \sqrt{2}$
Atoms per unit cell (UC)	$\frac{8}{8} + 1 = 2$	$\frac{8}{8} + \frac{6}{2} = 4$
Atomic volume $[10^{-3}\,\mathrm{nm}^3]$	$\frac{a^3}{2} = 11.77$	$\frac{a^3}{4} = 10.81$
Coordination number	8	12
Octahedral interstices per UC	$\frac{6}{2} + \frac{12}{4} = 6$	$1 + \frac{12}{4} = 4$
Tetrahedral interstices per UC	$\frac{24}{2} = 12$	8
Size of the interstice[2] [nm]		
octahedral		0.103
- diagonal	(0.156)	
- height	0.038	
tetrahedral	0.072	0.056

[1] smallest distance between mid-points of neighbouring atoms

[2] diameter of a sphere within the interstice

so that the third layer lies above the first (ABAB ...), this gives an hcp arrangement. The other possible stacking sequence, in which the fourth layer lies above the first layer (ABCA ...), gives an fcc arrangement, which is also closely-packed (Figure A.1.2 d).

A.1 Constitution

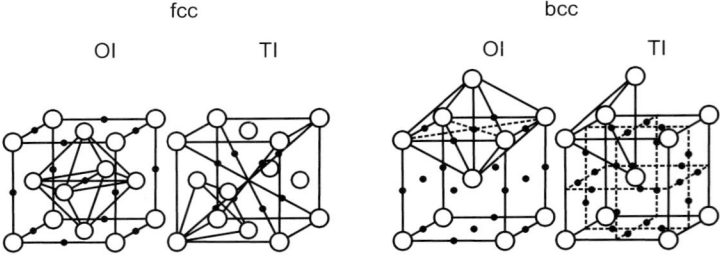

Fig. A.1.3 Lattice interstices in iron: An interstice is surrounded by iron atoms with an octahedral or a tetrahedral arrangement. Their size is given by the diameter of a sphere that just fits inside. For reasons of clarity, only the mid-points of the iron atoms (circles) and interstices (points) are shown (OI = octahedral interstice, TI = tetrahedral interstice).

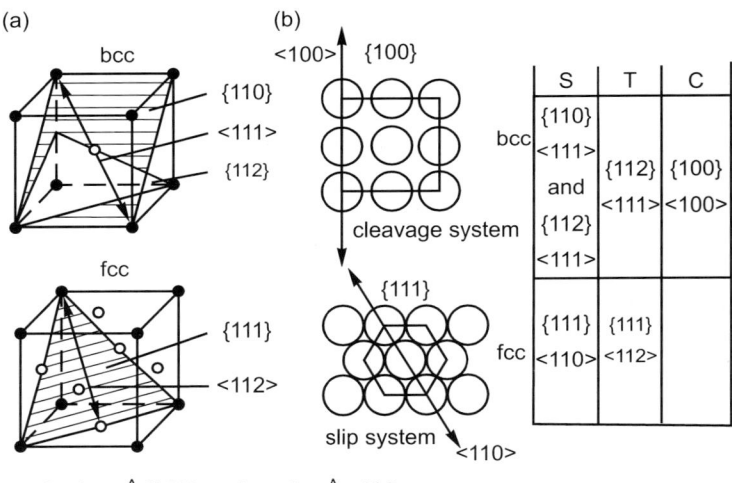

cube face ≙ {100} ; cube edge ≙ <100> ;
face diagonals ≙ <110> ; body diagonals ≙ <111>

Fig. A.1.4 Important planes and directions in an iron crystal: (a) Certain deformation and fracture processes are associated with preferred directions and planes. Their relationship is described by Miller indices (see metallurgical textbooks). Planes are enclosed in curly brackets and directions are enclosed in angular brackets. This groups equivalent planes and directions, e.g. all faces or all edges of a cube. (b) Cleavage C occurs along the less densely occupied cube faces along the edge direction. Slip S occurs along planes in which the atoms are closely packed. Twinning T allows a permanent deformation, particularly at low temperatures where slip caused by the movement of dislocations is hampered.

A.1.2 Iron-carbon

Ferrous materials always contain up to approx. 4%C which is also a tramp element because carbon is used to reduce iron ore. Carbon is dissolved in the molten ferrous material. During solidification, the carbon atoms, which are smaller than iron, fill the interstitial gaps in the iron lattice and thus form a solid solution. As can be seen in Figure A.1.5, these interstices are not large enough to accommodate the C atoms without lattice strain, which thus limits their solubility. On account of the larger octahedral interstices in γ-iron, its maximum C solubility is two orders of magnitude higher than that of α-iron. The insertion of carbon atoms into the interstices of iron produces solid solutions that are commonly termed as follows:

α-solid solution = ferrite (ferrum = iron)
γ-solid solution = austenite (after W.C. Roberts-Austen)
δ-solid solution = δ-ferrite

This squeezing in of the C atoms results in only a slight local distortion in the ordering of the iron atoms; however, the overall ordering remains the same. Therefore, α-iron and ferrite belong to the same phase. New phases are formed if the material contains more carbon than can be dissolved in the solid solution. The excess carbon then precipitates as graphite or as an iron carbide with the composition Fe_3C (cementite), which contains $\approx 6.7\%\,C$. Graphite has a hexagonal layer lattice whereas cementite has an orthorhombic structure. Graphite is regarded as the more stable equilibrium phase in the Fe-C system. However, wherever C atoms start to collect to form a precipitate, Fe atoms have to make room by diffusion so that metastable cementite is formed preferentially if solidification occurs rapidly. Furthermore, graphitic solidification is associated with a volume increase, which means that higher pressures stabilise cementite. This situation may arise, for example, if an already solidified outer shell shrinks around a molten core. As described later on, solidification is influenced in practical applications by the addition of alloying elements: Si and Ni favour graphite formation, whereas Mn and Cr

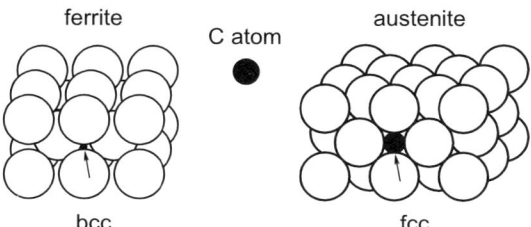

Fig. A.1.5 Solid solution: Comparison of the size of a lattice interstice and a carbon atom in the Fe-C solid solutions of ferrite and austenite.

A.1 Constitution

stabilise cementite. In the solid state and with the corresponding Si content, cementite can be converted into stable graphite by long-term annealing above 900°C. The reaction slows down below this temperature.

The regions of stability of the individual phases are represented by phase fields in the Fe-C phase diagram (Figure A.1.6). The state parameters temperature and carbon concentration are plotted against each other at the standard pressure. Characteristic points are marked with italicised letters (Tab. A.1.2).

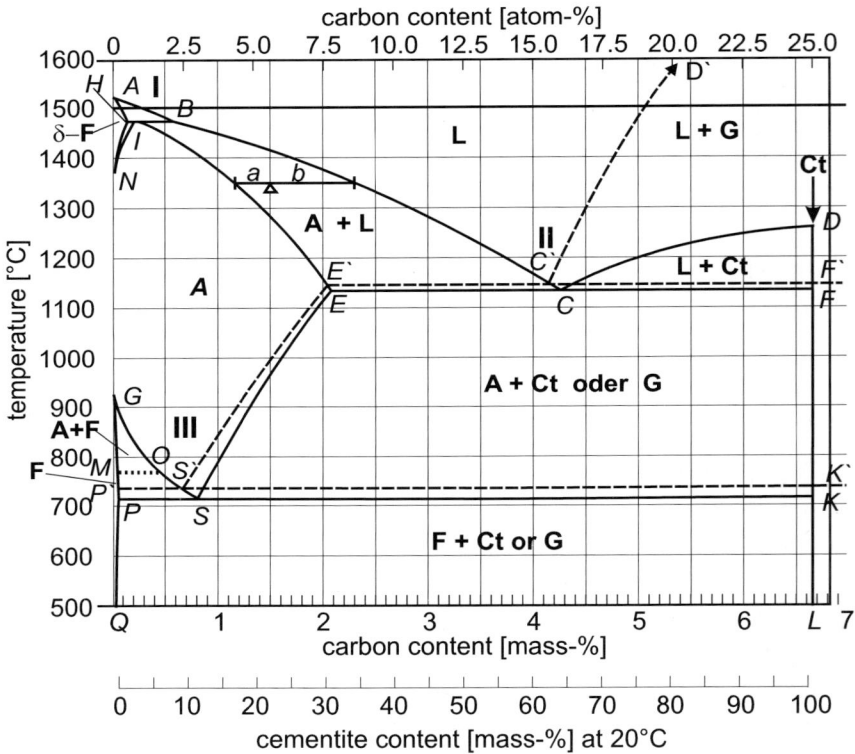

Fig. A.1.6 The iron-carbon phase diagram at standard pressure (see Tab. A.1.2):
— metastable iron-cementite system
- - stable iron-graphite system
··· magnetic transformation
I peritectic, II eutectic, III eutectoid
a,b lever arms of the Lever Rule for 1.5 mass-% C at 1350°C (example)
Phases: melt (liquid, L), δ-ferrite (δ-F), austenite (A), ferrite (F), cementite (Ct), graphite (G)

Table A.1.2 Supplementary data to the Fe-C phase diagram (Figure A.1.6): Positions of the points marked with an italicised letter. A to S: metastable iron-cementite system; dashed letters: stable iron-graphite system (acc. to E. Schürmann and R. Schmid)

		Temperature [°C]	C content [mass%]
A		1536	0.0
B		1493	0.533
C	, C'	1147 , 1153	4.302 , 4.256
D	, D'	1252 , 4000	6.689 , 100
E	, E'	1147 , 1153	2.140 , 2.098
F	, F'	1147 , 1153	6.689 , 100
G		911	0.0
H		1493	0.086
I		1493	0.160
K	, K'	727 , 736	6.689 , 100
M	, O	769 , —	— , —
N		1392	0.0
P	, P'	727 , 736	0.034 , 0.032
Q		≤ 20	≈ 0.0
S	, S'	727 , 736	0.758 , 0.688

A.1.2.1 The iron-cementite system

Let us first consider the metastable iron-cementite system (solid lines). As the carbon content increases, the upper melting temperature (liquidus temperature) decreases along the line connecting ABC, which is why cast iron was produced at a much earlier point in history than mild steel. The lower melting temperature (solidus temperature) is indicated by the line connecting $AHIEF$. Carbon thus causes a solidification interval in which the liquid phase contains solid iron crystals (δ-ferrite above HB and austenite below) to the left of the eutectic (labelled C) or it contains solid cementite crystals to the right between CD and CF. Carbon exerts the same stabilising effect on austenite as on the melt: the line connecting GOS falls, whereas the NI line rises, thus enlarging the phase field. In contrast, the region of bcc stability is destabilised by carbon because even a low C content crops off the phase fields of ferrite and δ-ferrite. The solubility of carbon in δ-ferrite increases to ≈ 0.09 mass-% along NH. In ferrite, it only increases to $\approx 0.03\%$ along QP owing to the lower temperature. However, it suddenly increases (discontinuously) along PS to $\approx 0.76\%$ and is coupled with the transformation to austenite. There is a further continuous increase with increasing temperature along SE. As melting commences at E, the solubility increases discontinuously from ≈ 2.1 to 4.3%

along EC, and then increases continuously in the melt along CD. The bcc- and fcc-ordered iron phases are both present in the triangular fields delineated by NHI and GPS. In the region delineated by $QPSEFL$, there is a soft metal phase (fcc austenite above SK and bcc ferrite below PSK) and a hard cementite phase, which has a hardness of $\approx 850\,\text{HV}$. During slow cooling of an iron melt containing e.g. 5 % carbon, the cementite precipitates in a series of steps as a consequence of the above-described differences in solubility:

primary:	continuously from the melt along DC
eutectic:	discontinuously from C to E, coupled with the transformation of the melt into austenite
secondary:	continuously from austenite along ES
eutectoid:	discontinuously from S to P, coupled with the transformation of austenite into ferrite
ternary:	continuously from ferrite along PQ

The use of melting boundaries (liquidus and solidus), solvus lines (continuous for a phase, discontinuous if coupled with a phase transformation), single-phase solid-solution regions and two-phase solid regions (two iron phases or one iron phase with carbide) make this complicated phase diagram somewhat easier to read. It is essentially composed of three subregions: peritectic (I), eutectic (II) and eutectoid (III). The first applies to the solidification of structural steels and the second to the solidification of cast iron. The third is particularly important in the thermal treatment of steel and cast iron. Carbon steel usually contains $< 1.3\,\%\,\text{C}$, whereas cast iron contains between 2 and 4 % C. The change in state of a particular material is of interest in practical applications. We will illustrate this using the example of the solidification of an iron melt containing 1.5 % C on slow cooling (Figure A.1.6). At the liquidus temperature of about 1415°C, the first austenite crystals containing 0.7 % C start to appear. At 1350°C, the horizontal tie line (conode) shows that the austenite (A) now contains about 1.15 % C and is in equilibrium with the melt (L), which has been enriched to 2.3 % C. This raises the question as to the ratio between the quantities of the two phases: it can be answered by means of the Lever Rule. The conode represents a lever with the fulcrum at 1.5 mass-% C and loaded at each end with masses M of phases A and L. An equilibrium exists when $a \cdot M_A = b \cdot M_L$, where $M_A + M_L = 1$ or 100 %. It then follows that $M_A = \frac{b}{(a+b)} = \frac{0.8}{1.15} \rightarrow 70\,\%$. At a solidus temperature of 1290°C, the last remaining portion of the melt contains 3.0 mass-% C. Between the liquidus and the solidus, the ratio $\frac{a}{b}$ between the lever arms continuously changes in favour of M_A. This example illustrates that carbon, which is evenly distributed in the homogeneous melt, tends to be very unevenly distributed between the two phases within the heterogeneous solidification interval; it is less soluble in the solid phase. Just below the solidus, the differences between 0.7 and 3.0 must be equalised by diffusion to yield homogeneous austenite containing 1.5 % C. There is not enough time for this during technical cooling so that segregation

A.1.2.2 The iron-graphite system

occurs. Below the austenite region, the carbon is distributed heterogeneously again between a solid solution phase and a carbide phase, cementite (Ct). The final room-temperature constitution of the material, which has thus undergone several phase transformations, consists of almost carbon-free ferrite in equilibrium with carbon-rich cementite. The amounts of these phases can be calculated using the Lever Rule: $M_{Ct} = \frac{a}{(a+b)} = \frac{1.5}{6.7} \rightarrow 22.4\,\%$. As this example shows, a phase diagram provides information on the number, ordered state, quantity and composition of the phases in a particular material at a given temperature, and also on phase transformations during slow cooling or heating.

The line joining MO at 769°C indicates the Curie temperature of the ferrite: it is paramagnetic above this line and ferromagnetic below it. This magnetic transformation is based on the orientation of the magnetic moments of the atoms and is independent of their geometric ordering. When the material is heated in a dilatometer, the thermal expansion close to the Curie temperature is slightly retarded by magnetostriction. Cementite also exhibits a weak transformation at its Curie temperature of 210°C. Austenite is paramagnetic.

A.1.2.2 The iron-graphite system

As can be seen in Figure A.1.6, the transition from the metastable (solid lines) to the stable system (dashed lines) only shifts some of the phase fields, and indeed, by surprisingly little if we disregard the high melting point of pure graphite. The different types of cementite precipitation, such as primary, eutectic, and so forth, can be similarly applied to graphite. Nothing changes with respect to the solidification of carbon steels; however, for a cast iron containing 3.5 % C, the high cementite content is reduced to a significantly lower graphite content, e.g. from 52 to 11 volume-% at room temperature (Tab. A.1.3).

Table A.1.3 Phase contents in a cast iron containing 3.5 mass-% C at two temperatures, calculated with the Lever Rule for the stable or metastable Fe-C system. The lever arms a and b were taken from Figure A.1.6. The following densities [g/cm^3] were used to calculate the volume contents: graphite = 2.26; cementite = 7.66; grey cast iron = 7.1; chilled cast iron = 7.6

Temperature [°C]	a [%]	b [%]		Cementite mass [%]	volume [%]	Graphite mass [%]	volume [%]	Remainder
1100	3.5–1.9	6.7–3.5		33.3	30	—	—	austenite
	3.5–1.8	100–3.5		—	—	1.76	5.5	
20	3.5–0	6.7–3.5		52.3	52	—	—	ferrite
	3.5–0	100–96.5		—	—	3.5	11	

A.1 Constitution

A small amount of manganese is added to steel to bind the sulphur, a tramp element. It also stabilises the metastable system. Furthermore, cooling to below 900°C after hot working is usually so fast that undesirable graphite precipitation is suppressed. Therefore, steels can generally be represented by the metastable phase diagram. Depending on the cooling process (wall thickness) and the added alloying elements (Mn/Si ratio), cast iron may undergo metastable or stable solidification, or even both in succession in the case of a rapidly cooling surface region with a core that lags behind. Such a chill-cast component has a wear-resistant outer layer containing a high proportion of hard cementite and a softer core with graphite providing good machinability. However, the sequence of stable solidification and a subsequent metastable progression in the region close to the eutectoid is also common, so that both diagrams apply to cast iron.

A.1.3 Alloyed iron

Alloying elements form a homogeneous melt with iron. In the solid state, interstitial solid solutions are formed by the insertion of carbon, nitrogen and hydrogen atoms into the lattice interstices between the comparatively large iron atoms. Larger alloying atoms replace iron atoms in the lattice and thus form substitutional solid solutions (Figure A.1.7). Their solubility in solid iron greatly decreases if the diameter of the substituted atom differs by more than 15 % from that of iron (e.g. lead, Tab. A.1.4).

Metals with a bcc structure, such as chromium, are more soluble in bcc iron, whereas fcc metals, such as nickel, are more soluble in fcc iron (Tab. A.1.4). The solubility of the alloying elements in iron increases with the temperature because the interatomic spacing grows, which is reflected in the increasing lattice parameter. A precipitate forms if the solubility is limited, e.g. continuously by cooling or discontinuously due to a phase transformation. The alloying atoms are enriched in this precipitate and it has a different ordering. It is thus present as a new phase. The precipitation of elements is rare (graphite, lead, copper) because the excess element usually forms a chemical compound with iron or another element. Non-metallic precipitates, such

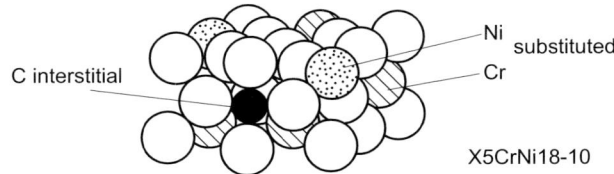

Fig. A.1.7 Iron solid solution: An example of interstition and substitution in a stainless austenitic steel. Enlargement of a lattice interstice by the interstitial C atom. Substitution of Fe by Cr and Ni atoms.

Table A.1.4 Tramp and alloying elements

Element	Atomic diameter [nm]	Lattice structure	Rel. atomic weight [g]	Maximum solubility *) α-iron [%]	γ-iron [%]	Solid solution 4)
Pb	0.350	fcc	207.20	≈ 0	≈ 0	—
Sb	0.322	3)	121.75	10	2.2	S
Sn	0.316	3)	118.69	16	2.0	S
Ti	0.294	hd	47.90	8	0.7	S
Nb	0.294	bcc	92.91	1.2	1.6	S
Al	0.286	fcc	26.98	29	0.6	S
W	0.282	bcc	183.85	35	4.7	S
Mo	0.280	bcc	95.94	31	1.7	S
Mn	0.274	fcc	54.94	3.5	100	S
V	0.270	bcc	50.94	100	1.3	S
Cr	0.258	bcc	52.0	100	12.5	S
Cu	0.256	fcc	63.54	2.1	12	S
Co	0.250	hd	58.93	76.0	100	S
Ni	0.250	fcc	58.71	6	100	S
Fe	0.248	bcc	55.85	—	—	—
As	0.242	3)	74.92	11	2.0	S
Si	0.234	1)	28.09	11	1.7	S
P	0.220	2)	30.97	2.4	0.3	S
S	0.208	3)	32.06	0.02	0.05	S
B	0.178	3)	10.81	0.002	0.005	I/S
C	0.154	5)	12.01	0.03	2.1	I
N	0.148	—	14.01	0.1	2.8	I
O	0.148	—	16.0	0.0008	0.0007	—
H	0.074	—	1.01	0.0003	0.0009	I

1) diamond lattice
2) cubic
3) complex
4) S substitution
   I interstition
5) graphite lattice
*) at elevated temperature

as carbides, nitrides, borides, oxides and sulphides, form predominantly hard and brittle ceramic particles. Intermetallic precipitates of iron and/or metallic alloying elements are usually less hard.

A.1 Constitution

Precipitation starts with nucleation. This requires local assemblies of atoms with the new ordering. The energy of formation changes depending on the volume of the nucleus, but the energy of the new boundary surface to the surrounding solid solution changes with its surface area. Therefore, the nucleus needs a minimum size to resist being redissolved. The number of atoms necessary for this is in the order of 100. The nuclei then grow by diffusion until a phase equilibrium is reached at a given temperature. The smaller interstitial atoms, C, N, H, are able to jump from one lattice interstice to the next. Interstitial jumping is several orders of magnitude faster than the diffusion of larger substitution atoms from one vacancy to the next. This is reflected in the diffusion coefficient D in Figure A.1.8, which represents the jump frequency of the atoms as they move along the concentration gradient. D increases with temperature owing to lattice dilation and the higher concentration of vacancies. The equilibrium is established much faster for compounds of type Fe-X (where X = C, N) than for compounds with one alloying metal of type M-X, in which non-equilibrium precursors are often formed. In chromium- and carbon-containing iron, compounds such as Fe_3C can be enriched with Cr to give $(Fe, Cr)_3C$ without it losing its orthorhombic structure. It eventually forms the hexagonal equilibrium carbide $(Cr, Fe)_7C_3$. The metal fractions of such compounds are often combined and represented

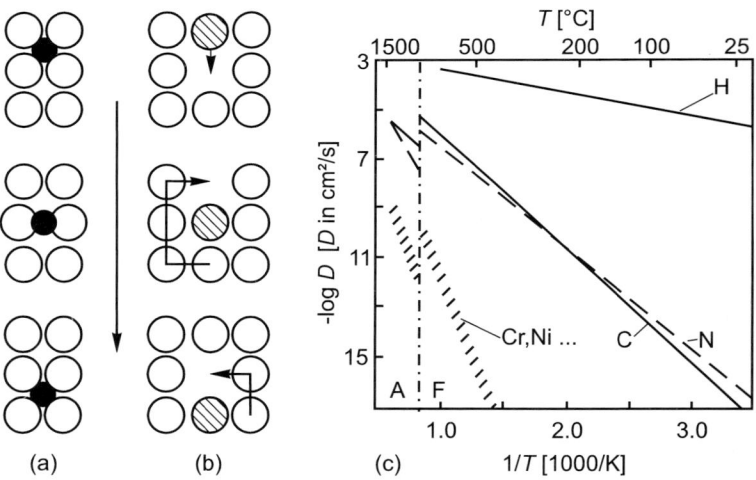

Fig. A.1.8 Diffusion: (a) Interstitial atoms diffuse faster via interstices than (b) substituted atoms via lattice vacancies (schematic). The key factor governing migration is an activity or concentration gradient. (c) The diffusion coefficient D is a measure of the mobility or jump frequency over time. A plot of $\log D$ against $1/T$ gives the activation energy Q of diffusion. In more closely packed austenite A, the diffusion rate is about 100 times slower than in ferrite F.

by M, so that these phases are termed M_3C or M_7C_3. The X fraction can also be present as a solid solution, e.g. in the cubic carbonitride precipitate (V,Fe)(C,N), abbreviated to MX or, more generally, M_aX_b.

The alloying elements can dissolve in the ferrite and austenite iron phases (Figure A.1.7) or form a number of new non-metallic or intermetallic phases. These can be exploited to tailor certain material properties or they may be avoided because they are detrimental. The amounts of added alloying elements can vary from less than 0.1 mass-% to several tens of mass-%. They confer ferrous materials with their enormous range of service and processing properties: high-strength, heat-resistant, ductile at low temperatures, corrosion-resistant, non-magnetisable, castable, cuttable, weldable, cold-workable, etc. The complex constitution of alloyed ferrous materials can be visualised in the following ways: When an alloying element is added as a third component, the two-dimensional phase fields in the binary Fe/C system become three-dimensional phase volumes in a ternary system. Because three-dimensional systems are difficult to draw, they are often represented in two-dimensions as isothermal sections (Figure A.1.9 a). Isoplethal (constant concentration) sections are also commonly used in multicomponent systems: a basic composition is kept constant and only one further element is changed (Figure A.1.9 b). Basic thermodynamic data and experimental results on the constitution of ferrous materials are continuously being added to databases and are used in calculation programs to generate phase diagrams. However, they only apply to a limited extent in practice because they are valid for pure components and extremely slow temperature changes. Nevertheless, phase equilibria are

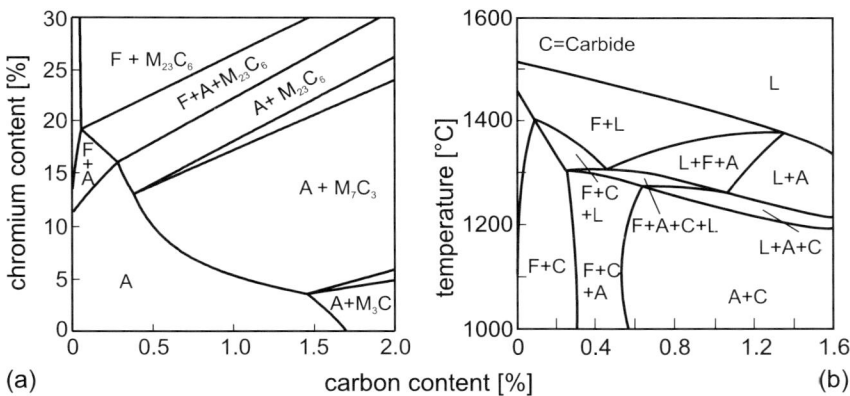

Fig. A.1.9 Examples representing multicomponent systems: (a) Isothermal section through the Fe-Cr-C ternary system at 1050°C, (b) isoplethal section through the multicomponent system Fe - 6 % W - 5 % Mo - 4 % Cr - 2 % V - C that includes the high-speed steel HS 6-5-2 with 0.9 % C (from E. Horn and H. Brandis).

established more quickly as the temperature increases so that the agreement between the diagram and practical applications usually improves.

On slow cooling, all Fe-C materials undergo a phase transformation from austenite to ferrite just above 700°C (Figure A.1.6). This is only slightly affected by low concentrations of alloying additives. However, it is possible to suppress this transformation by adding sufficient quantities of ferrite-stabilising elements, such as Cr, Mo, V, Si, or austenite-stabilising elements, such as Mn, Ni, N, C. In the first case, the phase fields of ferrite and δ-ferrite coalesce and there is no austenite. In the second case, austenite remains stable down to room temperature and there is no ferrite. Such transformation-free ferrous materials are used, e.g. for stainless, non-magnetisable or cryogenic components.

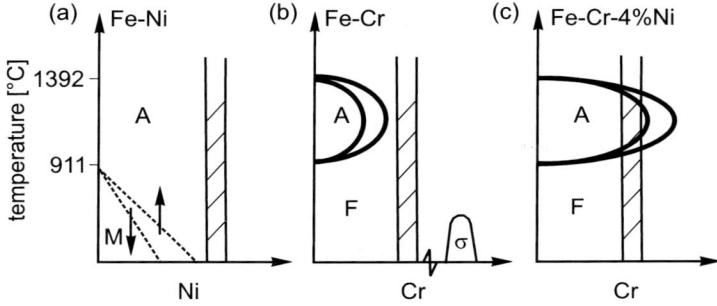

Fig. A.1.10 Transformation-free steels: Examples illustrating the influence of austenite- and ferrite-stabilising alloying elements are given schematically for the Fe-Ni and Fe-Cr phase diagrams. (a) By alloying with $>30\,\%$ Ni, the austenite field can be extended to below room temperature to produce a transformation-free austenitic alloy (hatched bar). On sub-zero cooling, the alloy transforms to martensite (M, Chap. A.2, p. 42). It is not in an equilibrium state (effective phase diagram of Fe-Ni). There is a temperature hysteresis between the forward transformation on cooling (downward arrow) and the backward transformation on reheating (upward arrow). (b) Alloying with $>\mathbf{12\%}$ chromium crops the austenite field and results in a transformation-free ferritic steel. (c) Alloying with a combination of Cr and Ni produces a transformable steel, though.

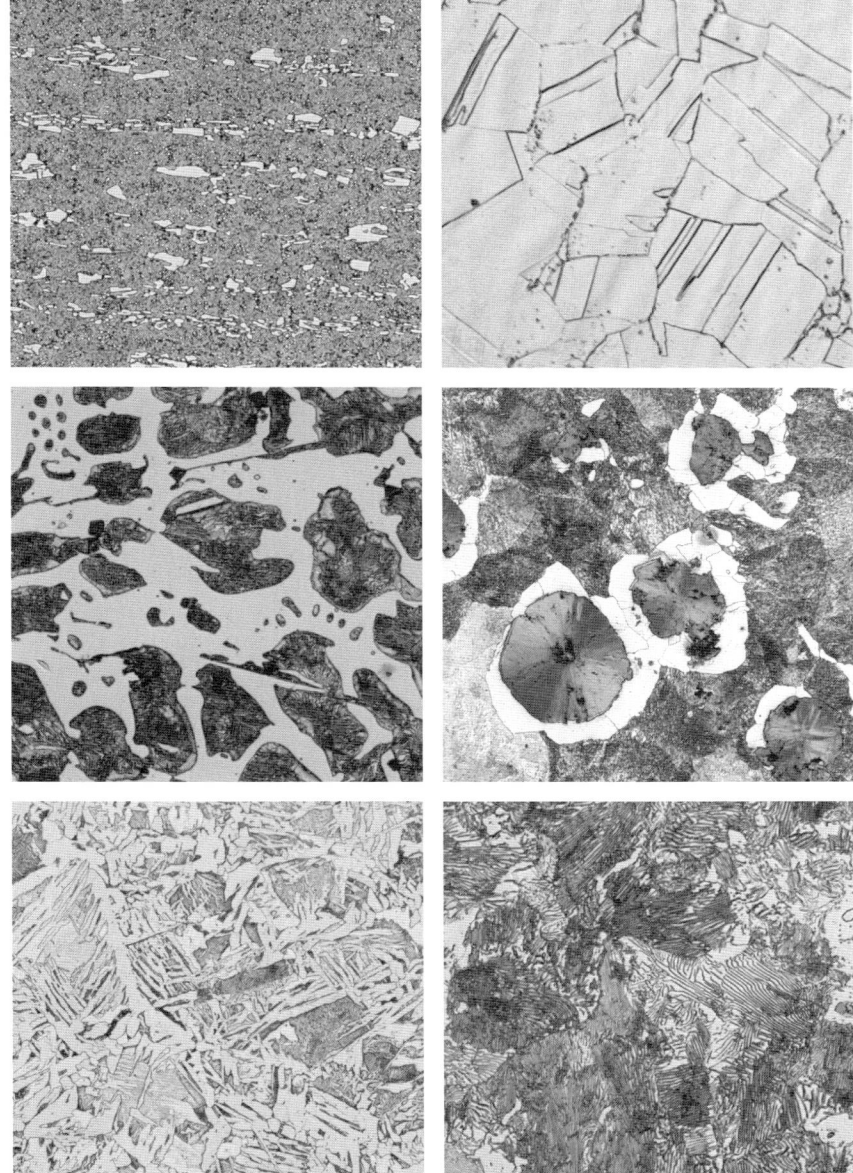

Microstructure of ferrous materials from the SCHMOLZ + BICKENBACH Group

A.2 Microstructure

The constitution of ferrous materials shown in phase diagrams describes the type and quantities of the phases that are in thermodynamic equilibrium; however, it does not provide any information on their appearance and spatial distribution. These morphological aspects are grouped together under the term 'microstructure', which comprises building blocks of the different phases. These building blocks are joined together through their interfaces. They contain disordering and are also affected by anisotropy and segregation. The near-equilibrium microstructure created by slow temperature changes may be completely different from a non-equilibrium microstructure such as that produced by rapid self-cooling of a laser-welded seam or that intentionally obtained by quenching after a heat treatment process.

The **phase building blocks** generally consist of small crystals, known as crystallites or grains, that are joined together via their grain boundaries. Figure A.2.1 shows individual or mixed grains of the ferrite and austenite iron phases in transformation-free stainless steels. In addition to the elongated banded arrangement of the ferrite shown in Figure A.2.1 a, network, dual phase and duplex microstructures consisting of various iron phases are also possible (Figure A.2.2).

If the proportion of one phase is low ($\approx 10\%$), we find grains of this phase (I, dark) as a coherent network or as a three-dimensional shell around the grains of the other phase (II, light-coloured) or dispersed in triple points. The network microstructure has no contacts or grain boundaries between phase II grains. In a dual phase microstructure, there are no phase boundaries between

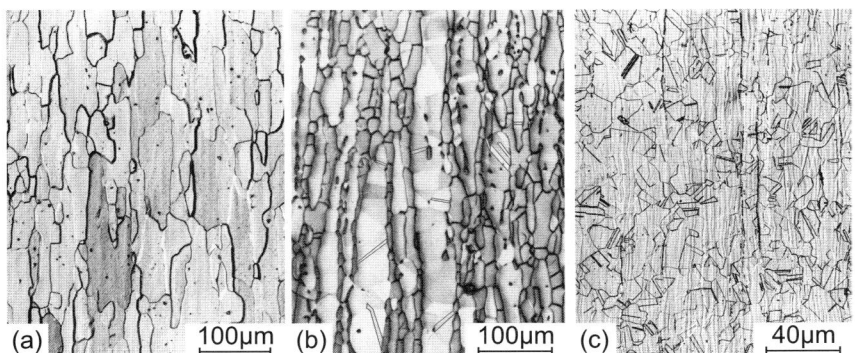

Fig. A.2.1 Ferritic and austenitic microstructure (longitudinal metallographic section of rolled steel): (a) elongated ferrite grains in steel X6CrTi12 dotted with MX precipitates, (b) ferritic/austenitic duplex microstructure in steel X2CrNiMoN22-5-3; austenite bands are recognisable by their twinning boundaries. (c) equiaxial austenite grains with twinning boundaries in steel X5CrNi18-10; there are narrow inclusion bands and banding caused by elongated microsegregations.

A.2 Microstructure

the grains of phase I. A duplex microstructure contains approximately half and half of phases I and II, and there are grain boundaries within each phase and phase boundaries between them.

The matrix is made up of iron phases in which non-metallic or intermetallic phases are precipitated. The morphological aspects are shown in Figure A.2.3 using the example of hard carbidic phases. Spherical grains of

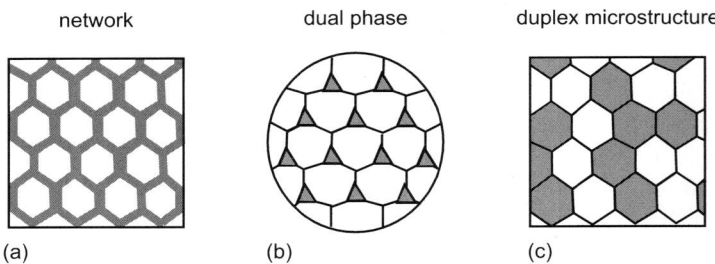

Fig. A.2.2 Arrangement of two iron phases in the matrix (schematic): (a) Network of ferrite grains around pearlite grains or around austenite grains in a stainless steel, (b) dual-phase microstructure comprising martensite grains dispersed in ferrite, (c) ferrite and austenite grains in a duplex steel

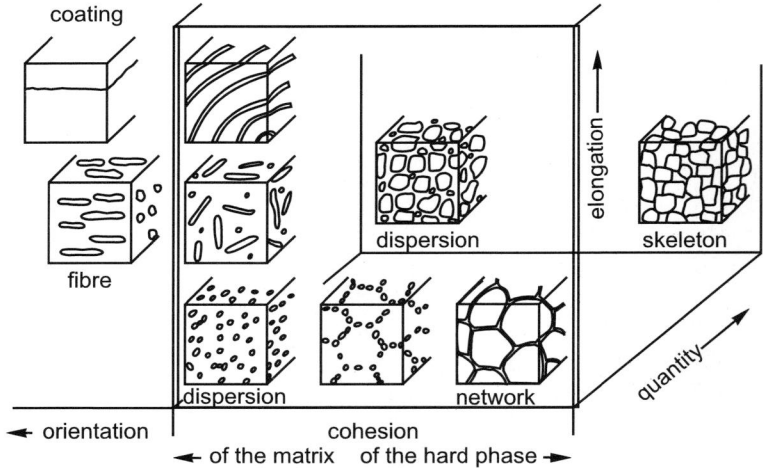

Fig. A.2.3 Arrangement of hard precipitates in an iron matrix(schematic): The microstructure depends on the cohesion (dispersion → network), elongation (sphere → rod → plate) and banded orientation (fibres, see Figure A.2.6) of carbides as well as the quantity of precipitates.

isolated carbides are dispersed within the matrix. This dispersion may be retained even if the amount of carbide increases. If the cohesion between the carbide grains increases, this produces a carbide network or skeleton. The carbide nodules can be elongated to rods or platelets and may also be oriented. It is logical that the carbide morphology affects the properties of the material. It is also clear that the information required to describe the microstructure includes the type, quantity, size, shape, distribution and orientation of the phase building blocks.

The ideal ordering of the atoms in the iron phases is disturbed by zero- to three-dimensional **lattice defects** (Figure A.2.4). Their frequency (concentration) can be intentionally varied by working the metal or by heat treatments

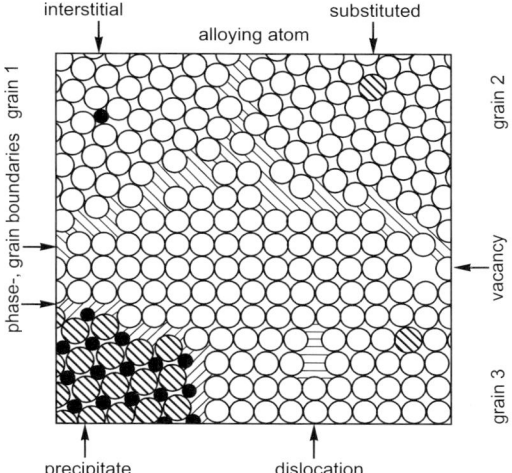

Fig. A.2.4 Lattice defects in iron (schematic): The perfect ordering shown in Figure A.1.2 only extends over tiny regions of the crystal because it is interrupted by lattice defects. These defects are classified according to their dimensionality. **Zero-dimensional or point defects:** individual atoms dissolved in iron (e.g. interstitial carbon atom, substituted niobium atom) or vacancies. **One-dimensional or linear defects:** e.g. an edge dislocation perpendicular to the plane of the paper along the terminated lattice plane (half-plane) in iron. The horizontal hatching indicates the elastic expansion of the lattice in the vicinity of the missing half-plane. These dislocation lines continue to the surface or they combine with screw dislocations to form rings. **Two-dimensional or planar defects:** grain boundary surfaces between the differently oriented grains or phase boundary surfaces between the iron phase and a precipitated phase, such as niobium carbide. In actual fact, this is a disordered (amorphous) zone that is several atoms thick; **Three-dimensional or spatial defects:** fine precipitates of a different phase, e.g. niobium carbide particles with a diameter of $< 10\,\text{nm}$. Coarse precipitates (e.g. $> 1\,\mu\text{m}$) of a different phase are not regarded as lattice defects.

A.2 Microstructure

that produce a hard or soft microstructure. Substituted atoms or vacancies are zero-dimensional (point) defects. If a lattice plane suddenly terminates within a crystallite, a one-dimensional (linear) defect is produced along the half plane and this is known as an edge dislocation. The grains of the iron matrix are orientated randomly, i.e. their lattices meet at an angle to produce a two-dimensional (planar) defect as a disturbed seam, the grain boundary. Precipitates have a different lattice arrangement of the atoms which only rarely matches that of the matrix, i.e. in the case of a coherent interface (Figure A.2.5). If the structure is completely different, this produces an incoherent interface. Furthermore, there are also partially coherent interfaces in which the mismatch between the different crystallites is compensated e.g. by dislocations. The term **anisotropy** refers to the directional dependence of particular properties. An individual crystalline grain is anisotropic because the atoms only touch each other in certain directions (Figure A.1.2, p. 6), and they thus exhibit e.g. a higher modulus of elasticity or a higher thermal expansion in a particular direction. A technical polycrystal averages out the anisotropy of the individual grains and is thus quasi-isotropic. Realignment of the random orientation of the grains in a polycrystalline single-phase into a preferred orientation produces crystalline anisotropy (texture). This is used e.g. in deep-drawing or transformer sheets. On the other hand, if the second phase is oriented in one direction, e.g. sulphides in rolled steel, this is known as microstructural anisotropy (banding) (Figure A.2.6). The oriented elongated carbides shown in Figure A.2.3 are also a type of microstructural anisotropy. This results e.g. in a lower ductility at right-angles to the metal-working direction. **Segregation** is a local variation in the average content of alloying and tramp elements. The mean distance between the concentration maxima or minima is known as the segregation spacing. This may vary by several orders of magnitude, and is thus referred to as macro-, micro- and defect segregation (Figure A.2.7). Metal working may elongate macro- and microsegregations to bands that contribute to the microstructural anisotropy.

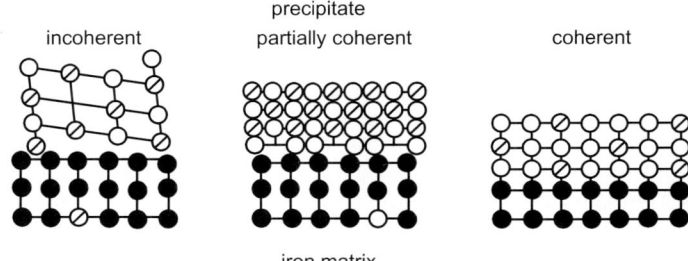

Fig. A.2.5 Coherence of precipitates (schematic): As the mismatch between the crystal lattice of the steel matrix and that of the precipitate increases, the coherence decreases and the energy of the defect increases (see Tab. A.2.1)

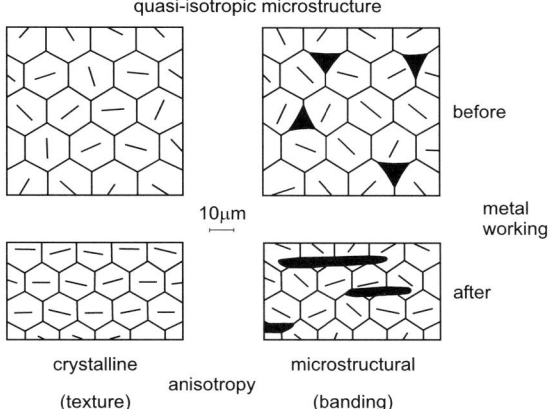

Fig. A.2.6 Anisotropy caused by hot working: In the left-hand diagrams, the random orientation of the steel grains is aligned into a preferred orientation (indicated by the lines that point along e.g. a cube edge, see Figure A.1.4). The right-hand diagrams illustrate the elongation of a second phase (e.g. sulphides). The steel grains have retained their equiaxial shape during recrystallisation. However, they can also be elongated (s. Figure A.2.1 a) without development of a texture.

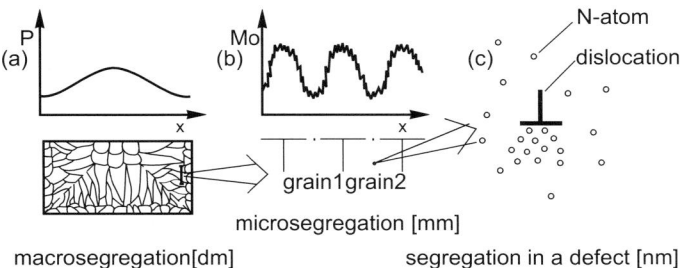

Fig. A.2.7 Segregation: Alloying and tramp elements are not homogeneously distributed in steel. (a) The term macrosegregation refers to variations in the concentration between the surface region and the core within a solidification cross-section (example: the tramp element phosphorus). (b) Microsegregation refers to variations in the vicinity of individual primary grains (example: the alloying element molybdenum). (c) Lattice defects, such as edge dislocations and grain boundaries, attract interstitial atoms (C, N, H) because e.g. they provide more space (hatched areas in Figure A.2.4). The symbol for the edge dislocation shows a half-plane perpendicular to a slip plane. The dimensional unit given in square brackets indicates the order of magnitude of the segregation distances between locations of the greatest variations in concentration within an ingot. Hot working elongates macro- and microsegregations to bands.

A.2 Microstructure

A.2.1 Near-equilibrium microstructure

In parts with a thick cross-section undergoing a moderate heat transfer (air-cooling, gas heating, sand mould), the phase transformations take place under near-equilibrium thermodynamic conditions, as shown in a phase diagram. Here, we focus on Fe-C materials with a low alloy content. We will use Figure A.1.6 to consider steels containing < 1.3 mass% C and then cast irons with 2 to 4 mass% C.

A.2.1.1 Steel

(a) Solidification

Weldable structural steels containing $< 0.09\,\%$ C solidify to δ-ferrite (Figure A.1.6, point H, p. 10) that transforms to austenite on cooling. Up to approx. $0.5\,\%$ C (point B), both austenite and δ-ferrite are formed from the melt. The latter then transforms to austenite as it undergoes a peritectic transformation. The δ-ferrite/austenite transformation reduces the grain size. The grains then grow more slowly in austenite than in δ-ferrite owing to the slower diffusion rate (Figure A.1.8). Grain growth is driven by the lowering of interfacial energy stored in the grain boundaries as the system tries to attain a minimum energy level. On the other hand, a fine-grained microstructure improves the mechanical properties.

Micro- and macrosegregations arise during solidification owing to a departure from equilibrium conditions (Figure A.2.8). Solidification starts with heterogeneous nucleation on the colder wall of the mould. Its growth front is very serrated within the solidification interval. Dendrites with a spacing of λ grow into the melt from the wall of the mould. They contain more iron than the nominal composition so that the remaining melt between them is enriched with other elements. This leads to microsegregation and ultimately to macrosegregation (Figure A.2.8). As λ decreases, the lengths of the diffusion paths required to homogenise the alloy also decrease e.g. during diffusion annealing or heating prior to hot rolling. The relationship between λ and the local cooling rate v is $\lambda \sim \frac{1}{\sqrt{v}}$. Solidification in a thin-walled component is thus favourable to reduce microsegregation by annealing; however, it is already deviating from equilibrium conditions, and will therefore be discussed later in Section A.2.2.

Macrosegregation may be quite distinct in the core, where it is known as centreline segregation. This horizontal macrosegregation is superimposed by vertical segregation within the ingot. The convection currents break off individual arms from the forest of dendrites and the 'cooler current' then carries them downwards where they collect in a heap. Because the dendrite arms have a higher purity than the remaining melt, the concentration of alloying elements and/or carbon is initially lower in the heap. As solidification progresses, these concentrations gradually increase as the heap extends upwards into a central

A.2.1 Near-equilibrium microstructure

zone. Segregation channels may form in a V-shaped arrangement within the central zone. A possible explanation for these local macrosegregations are hot cracks caused by shrinkages in the solidifying central zone that suck in the enriched remaining melt. In the upper part of large ingots, A-shaped segregation bands are also observed. Macrosegregations cannot be eliminated after solidification owing to the long diffusion paths.

Three different regions can usually be identified within an ingot cross-section, based on the shape of the grains: a narrow, fine-grained outer zone, a middle zone with columnar grains and a globular core zone with equiaxial grains (Figure A.2.7 a). The surface zone consists of randomly oriented grains with a dendritic structure. Dendrites with the most favourable orientation have grown against the heat flow as columnar crystals. Each columnar grain consists of one dendrite, which may have developed many primary dendrite arms, particularly at high cooling rates. The above-mentioned detachment of dendrite arms by the convection current during solidification enriches the

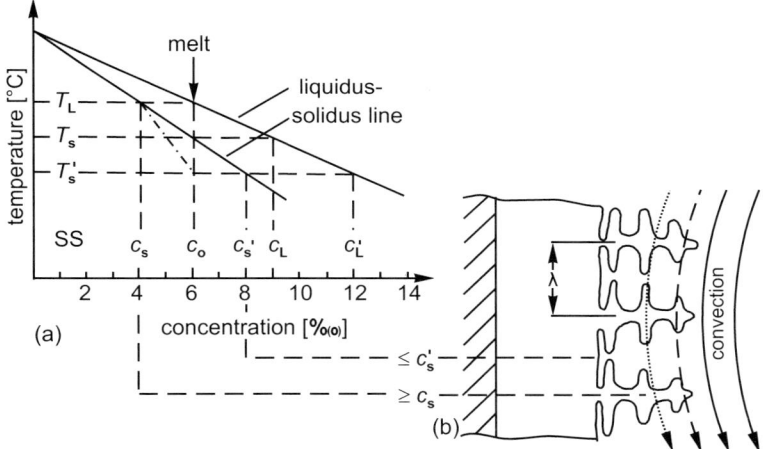

Fig. A.2.8 Genesis of macro- and microsegregation: (a) according to the schematic phase diagram, solidification begins with the precipitation of a solid solution (SS) at the liquidus temperature T_L. The concentration of this solid solution c_s lies below that of the melt composition c_0. By the end of solidification, SS should be enriched to c_0. However, at technical cooling rates, solid-state diffusion is too slow. The solidus then follows the dash-dotted line. The solidus temperature drops from T_S to T'_S thus extending the solidification interval to T_L - T'_S. The degree of segregation $S = c_{max}/c_{min}$ is $1 < S < c'_S/c_S$. (b) Microsegregation is not complete because enriched residual melt between the dendrites is carried away by convection and is then redistributed in the remaining melt. The melt concentration thus gradually increases with increasing volume fraction of SS thus causing macrosegregation towards the core as solidification progresses.

remaining melt with crystallites that grow as cooling progresses. They ultimately slow down columnar solidification and lead to a core zone of randomly oriented grains with a dendritic structure.

Before and during solidification, micro- and macrosegregation may enhance oversaturation of the remaining melt and precipitation of non-metallic inclusions, gases and other phases, e.g. eutectic carbides. Primary inclusions, such as alumina and silicates, already precipitate before the steel itself starts to solidify. They may agglomerate and rise to the surface or be pushed by the solidification front into interdendritic spaces where they collect. Secondary inclusions, such as oxysulphides, are formed during solidification as a result of microsegregation and are trapped by the dendrite arms as they grow. The size of the secondary inclusions increases with decreasing cooling rate and with the degree of macrosegregation. Gases, such as hydrogen or carbon monoxide, are liberated by a sudden change in solubility during solidification. They either rise to the surface or remain in the steel as bubbles and pipes. Together with shrink holes, which are caused by insufficient feed material during casting, they form cavities that may have a negative effect on the properties of castings, in particular. In ledeburitic steels, such as high-speed steels and chromium steels, the maximum carbide size and the mesh size of their networks increases with the cooling time, i.e. with the ingot cross-section.

In summary, rapid solidification leads to decreases in certain microstructural parameters, such as inclusion size, spacings between dendrite arms and the grain size. The tendency to centreline segregation is counteracted during continuous casting by electromagnetic stirring as the melt solidifies. Powder metallurgy completely prevents macrosegregation.

(b) Hot working

Heating the metal before working it reduces microsegregation to a certain extent and precipitates are able to redissolve. During hot working, pores and shrink holes are closed as long as they do not have a layer of scale due to contact with the surrounding air. In contrast to cast steel, rolling and forging steel can also undergo further grain refinement. The activation of slip systems, as shown in Figure A.1.4, leads to a high concentration of dislocations providing energy (Figure A.2.4) that triggers the formation of nuclei with a perfect lattice. Thus, as the system energy is minimised, one deformed grain leads to many recrystallised grains with very few dislocations. This recrystallisation occurs during (dynamic) or directly after (static) each rolling pass or press stroke. At high final hot working temperatures ($>1000°C$), the fine-grained microstructure is lost again owing to spontaneous grain growth. This is why the final rolling temperature of structural steels is usually $<900°C$. Particular textures can be obtained by controlled hot working and recrystallisation (Figure A.2.6).

The advantages of hot working must be weighed against the drawback of producing a banded microstructure due to elongation of non-metallic inclusions and microsegregations. This microstructural anisotropy has an adverse effect on the ductility transverse to the metal-working direction. In hypereutectoid steels with carbon contents to the right of point S in Figure A.1.6, slow near-equilibrium cooling may result in the precipitation of carbide networks on the austenite grain boundaries (see Figure A.2.3), and this causes embrittlement.

(c) Austenite/ferrite transformation

After solidification and hot working, the microstructure of low-alloy steels initially consists of austenite grains that gradually undergo a eutectoid transformation to ferrite and cementite on further cooling. This section of the Fe-C phase diagram (Figure A.1.6, p. 10), which is also important in heat treatment processes, can be adapted schematically for a low alloy content (Figure A.2.9). Besides certain shifts in the phase fields and the corner points, this isoplethal section of Fe-C-E has an additional three-phase field $F+A+M_3C$. The added alloying element E is not specified here; however, it can be inserted into thermodynamic calculations so that this schematic representation becomes an adjusted phase diagram. Important transformation lines and temperatures are denoted A_1 to A_4. The additional character "r" represents "refroidissement" and indicates that the measurements were carried out under slow cooling conditions. Measurements are preferentially carried out under slow heating conditions (additional character c that represents "chauffage"). The transformation begins at Ac_{1b} as the material enters the three-phase field, which ends

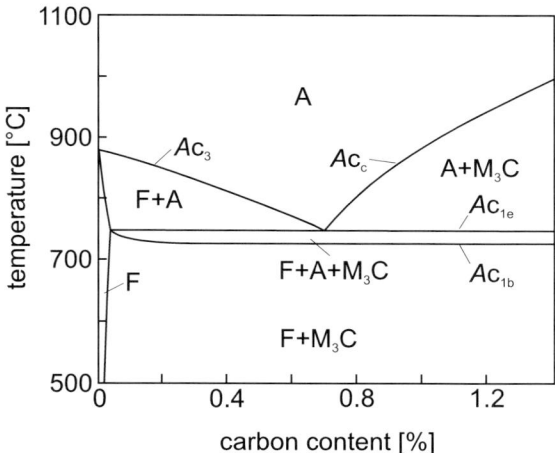

Fig. A.2.9 Austenite transformation: Section of the Fe-C-E phase diagram of a low-alloy steel (schematic, see Figure A.1.6, p.10). The alloying element E is dissolved in ferrite (F), austenite (A) and carbide (M_3C). This gives rise to a three-phase field.

A.2 Microstructure

at Ac_{1e}. In hypoeutectoid steels to the left of the eutectoid point, the transformation is complete at Ac_3, whereas in hypereutectoid steels to the right, the transformation is complete at Ac_c. A_2 represents the magnetic transformation along the MO line (Figure A.1.6), and A_4 indicates the transformation of austenite to δ-ferrite

Let us now return to the cooling of austenite and consider the example of an unalloyed steel containing 0.4 mass% C, for which Ar_{1e} and Ar_{1b} coincide (Figure A.1.6). The transformation starts at Ar_3 with the formation of ferrite grains with a low C content. The remaining austenite is thus enriched with carbon until it reaches the eutectoid composition of 0.76 % C at Ar_1. Ferrite lamellae grow from the grain boundaries within these austenite grains. The excess carbon is precipitated between these lamellae as eutectoid cementite, which also has a lamellar structure. This coupled transformation and precipitation reaction means that carbon has a short diffusion path. Discontinuous precipitation starts at the grain boundaries because the disordering here provides interfacial energy for the nucleation process. The lamellar microstructure is known as pearlite because etched surfaces have a pearly shimmer. Slowly cooled forged materials contain carbide lamellae with a diameter of $\approx 10\,\mu m$. In the present case, the room-temperature microstructure is about half ferrite, half pearlite. Unalloyed steels with less carbon contain more ferrite grains, and those with 0.76 % C contain only grains of pearlite. During cooling of steel C105, secondary cementite is continuously deposited along the grain boundaries until the carbon content of the austenite has dropped to 0.76 %. This is followed by a eutectoid transformation to pearlite (Figure A.2.10).

Fig. A.2.10 Ferritic/pearlitic microstructure (SEM): Produced by slow cooling from the austenite field. (a) Lamellar transformation requires only short diffusion paths for carbon (see arrows), (b) hypoeutectoid steel C45, (c) hypereutectoid steel 100Cr6.

A.2.1.2 Cast iron

Carbon precipitated as cementite gives fracture surfaces a light-coloured metallic appearance (white cast iron), whereas graphite makes them look grey (grey cast iron). Further characteristic microstructural features are the shape of the graphite precipitates (flaky to nodular) and the steel-like matrix consisting of pearlite and ferrite. Because the grey microstructure is obtained by alloying with 1 to 3 %Si, Figure A.2.11 shows an isoplethal section through the Fe-Si-C phase diagram (based on Figure A.1.6) for the example of an alloy containing 2.5 mass% Si. The carbon activity increases with the silicon content so that both contents can be combined to give the following (simplified) carbon equivalent: $CE = \% \, C + \frac{1}{3} \, \% \, Si$.

(a) Solidification

Grey solidification of a cast iron containing 3 % C and 2.5 % Si can be followed on Figure A.2.11. This hypoeutectic composition is $\approx 0.5 \, \% \, C$ to the left of the eutectic at E and is thus at the same distance as $CE \approx 3.8 \, \%$ in the binary system (Figure A.1.6). Solidification commences at the liquidus temperature T_L with the precipitation of austenite, which grows as dendrites into the melt. The remaining melt between the dendrites is enriched with carbon until graphite starts to precipitate at the eutectic temperature T_E. Oxides and sulphides act as nucleation agents for graphite flakes growing in the close-packed direction a of the hexagonal layer lattice, whereas their thickness in the c direction is limited by the growth of austenite in the eutectic cell. Nucleation can be enhanced

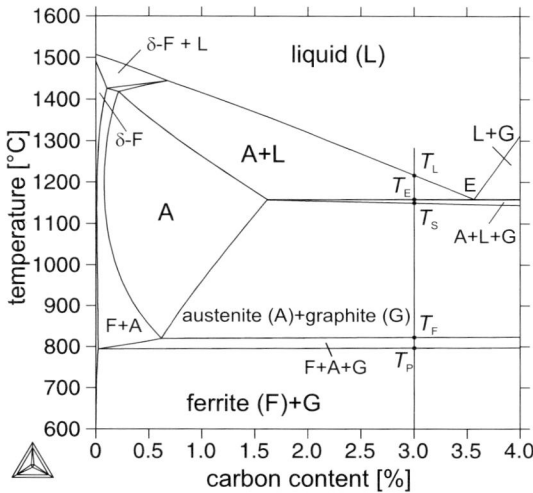

Fig. A.2.11 Phase diagram of Fe - 2.5 % Si - C (see Figure A.1.6, p.10)

by adding oxide-forming and sulphide-forming elements such as Al, Ca, Ce, Ti, Zr, etc. to the ferrosilicon inoculant. Inoculation influences the size and distribution of the flakes and also binds residual sulphur with Mn. Segregation may occur during solidification, similar to steels (see Section. A.2.1.1). There may be enrichment of substitution elements, such as Mn and Si, in separate segregation zones, and these influence the C distribution. Manganese lowers the carbon activity, attracts C atoms and increases their solubility. Si, on the other hand, increases the carbon activity and forces C out of solution. Solidification is complete at the solidus temperature T_S. Although the flakes do not appear to be connected in a metallographic section, they do indeed form a coherent three-dimensional structure. This leads to an optimum damping of vibrations, which is particularly desirable for machine frames; however, it has an adverse effect on the ductility. Therefore, magnesium is added (and/or Ca, Ce) if the Mn and S contents are low and the CE value is high (4.3 to 4.5) to induce the growth of nodular graphite in the near-eutectic melt. In contrast to cast iron containing coherent graphite flakes (grey cast iron), cast iron with spheroidal graphite contains dispersed graphite nodules. They grow as spheres with a radially aligned c axis in a semi-liquid shell that allows growth on all sides. The change from flakes to nodules considerably increases the ductility; however, it also lowers the damping capability. In order to achieve a balance between both property profiles, variants are used in practical applications that have a compact graphite shape. These variants are also produced by treatment of the melt with Mg, Ca, Ce. The objective is to interrupt the three-dimensional coherence between the graphite flakes (ductility) and still keep the largest possible surface area of the graphite precipitates (damping).

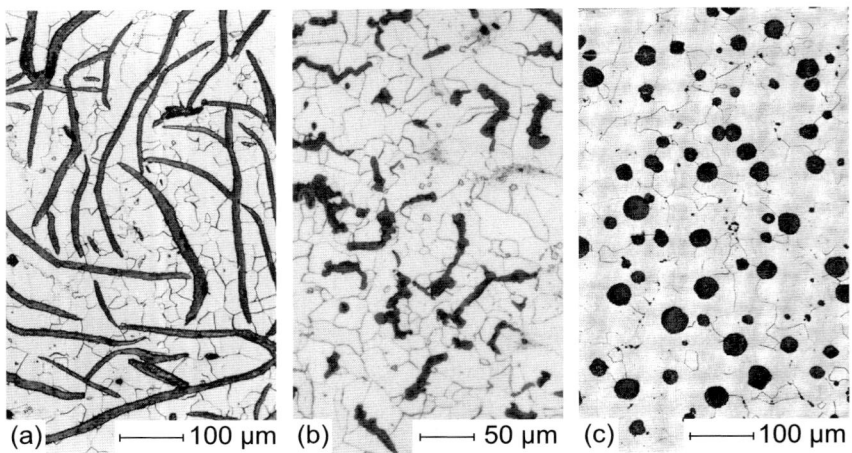

Fig. A.2.12 Typical graphite formation in ferritic cast iron after grey solidification: (a) lamellar, (b) vermicular (compact), (c) nodular.

These materials are also more oxidation-resistant because oxygen penetrates rapidly along continuous graphite flakes at elevated temperatures. Their common name is cast iron with vermicular graphite. Figure A.2.12 shows the three different types of graphite precipitates. The number of precipitates in a metallographic section depends on the solidification rate as well as on the inoculants and alloying additives. Mixed shapes are also observed. The typical range for the surface area of graphite precipitates, measured in mm^2 per mm^3 cast iron, is > 100 in cast iron with graphite flakes, 40 to 70 in cast iron with vermicular graphite and 20 to 30 in cast iron with spheroidal graphite.

The transition from grey to white solidification is facilitated by the following factors: (1) an increasing Mn, Cr, Mo/Si ratio to decrease the carbon activity and promote carbide formation. (2) Faster solidification to supercool the melt, to limit carbon diffusion and to promote precipitation of a phase containing 6.7 % C (Fe_3C) rather than 100 % C (graphite) because the latter also requires volume work as it forms. This supercooling of the melt below the metastable solidus temperature EF (Figure A.1.6) is exploited in practical applications to produce a wear-resistant carbidic surface layer. (3) Pressure to counteract the volume increase associated with graphite precipitation. Kagawa and others have shown that a cast iron with 3.6 % C and 2 % Si has a grey/white transition above 200 MPa and that such pressures can build up in the core of a sufficiently thick solidification cross-section if e.g. the relatively hard, already solidified outer layer of pearlitic spheroidal graphite cast iron shrinks onto the core which is still molten. Pressure may therefore play a role in undesirable chilling in the core of grey cast iron.

White cast iron contains less Si (0.5 to 1.5 %) than grey cast iron and undergoes metastable solidification when quenched against iron chills in the mould (chill casting) or in thin cross-sections. For parts with thicker walls, it is advisable to stabilise white solidification by adding Mn, Cr, Mo (approx. 1 %). The carbon content is usually between 2.5 and 3.5 %, so that primary austenite dendrites grow in this case as well. They are surrounded by a shell of austenite and cementite with a eutectic microstructure (see Figure A.2.3), whereby the latter essentially encloses the austenite. The eutectic portion of the microstructure is known as ledeburite, and is named after A. Ledebur. This continuous hard-phase skeleton makes chilled castings wear-resistant, but also brittle (Figure A.2.13). The addition of $> 8\%$ Cr leads to the precipitation of eutectic M_7C_3 carbides that are harder and less continuous as in M_3C, and thus has a positive effect on the properties. This and other high-alloy cast irons will be discussed in Chapter B.4.3 (p. 259) and B.6.4 (p. 341).

When considering the solidification of cast iron and the resulting microstructure, the difference in the composition of the melt with respect to that at the eutectic is important. This is influenced not only by C and Si, but also by other elements such as P, S and Mn. The CE value previously presented is oversimplified. In practice, the CE value is extended to include at least the influence of P:

A.2 Microstructure

$$CE = \%C + \frac{1}{3}(\%Si + \%P) \qquad (A.2.1)$$

Based on the eutectic content of 4.3 % in the Fe-C system, a value of $(CE-4.3)<0$ indicates a hypoeutectic melt and $(CE-4.3)>0$ indicates a hypereutectic melt. A ratio can be used instead of a differential value. This ratio is known as the degree of saturation, S_c.

$$S_c = \frac{\%C}{4.3 - \frac{1}{3}(\%Si + \%P)} \qquad (A.2.2)$$

$S_c<1$ indicates a hypoeutectic melt, $S_c>1$ a hypereutectic melt.

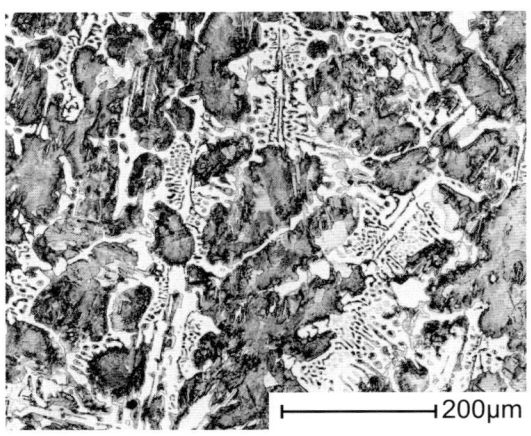

Fig. A.2.13 Typical carbide formation in a hypoeutectic cast iron containing (%) 3.2 C, 0.9 Si, 0.4 Mn after white solidification: Dark areas: primary austenite transformed to pearlite; light areas: eutectic of cementite with pearlite inclusions.

(b) Cementite/graphite transformation

The microstructural dependency on the solidification rate may lead to white solidification in the more rapidly cooling edges or thinner regions in castings made of grey cast iron. This undesirable condition can be brought to a stable equilibrium by subsequent annealing that transforms the cementite crystals to graphite (temper carbon).

However, this heat treatment is much more important for malleable cast iron, which has a composition that is on the verge of white solidification in mostly thin-walled castings and whose cementite can be transformed into graphite by annealing (malleablising) for several hours at 900 to 950°C. This produces compact particles of temper carbon with an irregular outline that are dispersed in a steel-like matrix. With 80 to 150 such carbon clusters per mm^2 in a metallographic section, malleable cast iron is similar to spheroidal graphite cast iron, which usually has approx. 150 graphite nodules per mm^2. Account must be taken of the lower CE value of malleable cast iron (2.5 to 3.5) that is necessary for white solidification. A relatively high Si content (1.5 %) accelerates the tempering process and saves heating costs; however, there is a risk of mixed white/grey solidification. This is suppressed in favour of white solidification by adding 0.01 %Bi for a C content of e.g. 2.4 %. The addition of only a few thousandths of a percent of B or Al enhances the decomposition of carbide.

If annealing is carried out in an oxidising atmosphere, the surface layer is decarburised. In thin-walled castings, the core is affected as well. The temper carbon is consumed and the cast iron becomes steel. The light-coloured, graphite-free zone led to the term "white heart malleable cast iron", in contrast to "black heart malleable cast iron", which retains its dark-coloured clusters of temper carbon on annealing in a neutral atmosphere. White heart malleable cast iron was developed by R.A.F. de Réaumur in 1722 for thin cross-sections. It was a widely used type of cast iron because mild steel could not be produced at that time owing to its higher melting temperature. de Réaumur thus succeeded in producing the first white heart castings with steel properties indirectly via annealing (see Chap.C.2, p. 397). Nowadays, the core of thicker cross-sections is no longer white heart but black heart, and we are interested in e.g. the weldability of decarburised malleable cast iron.

(c) Austenite/ferrite transformation

The graphitic microstructure is produced during solidification or tempering and is thus complete at 900°C. The matrix now consists of austenite. On further cooling to temperature T_F (Figure A.2.11), it starts to transform into ferrite and the resulting insoluble carbon fraction is deposited onto already existing graphite precipitates. This transformation is complete at T_P. The resulting ferritic/graphitic microstructure is soft. To increase the hardness, it is recommended that the cooling rate at T_F is increased to change over to a metastable transformation, as presented for steels in Figure A.2.9, but which is also applicable to the steel-like matrix of cast iron. Ferrite is now in equilibrium with cementite rather than with graphite. The eutectoid austenite transforms into the harder pearlite. This pearlitisation is enhanced by alloying with Cu and small amounts of Sn and Sb because they form a diffusion barrier for carbon as it transforms to graphite. It is logical to establish a consistent nomenclature for the heat treatment of steel and cast iron and to thus correspondingly apply the abbreviations that are commonly used in the metastable system, namely Ac_c, Ac_{1b} and Ac_{1e} (Figure A.2.9) to the stable system (Figure A.2.11). The subscript c in Ac_c is thus no longer the limit of 'cementite', but now represents the maximum solubility of C in austenite. T_P and T_F correspond to Ac_{1b} and Ac_{1e}, respectively.

A.2.2 Non-equilibrium microstructure

As the heating/cooling rate increases, the alloy system moves away from a thermodynamic equilibrium. In addition to rapid heating, e.g. during welding or flame hardening, the influence of rapid cooling on the microstructure becomes particularly important. If the latent heat of transformation cannot be dissipated quickly enough, the imminent transformation of the phase drops below the equilibrium temperature T_{eq}. Although only a few nuclei of the new phase are stable at the onset of transformation during slow cooling, their number increases as the melt supercools. At the same time, their growth slows down and eventually 'freezes'. If a phase is supercooled by certain amounts of ΔT_{sc}, and the times are measured until transformation starts or has progressed half way, this generally gives a C-shaped curve or nose if a log(time) scale is chosen (Figure A.2.14). The importance of the time factor in the microstructural development can be seen in this schematic isothermal time-temperature-transformation diagram (TTT). An equilibrium usually exists prior to rapid cooling e.g. in a melt or at the hardening temperature. At the apex of the C curve, the phases formed from a larger number of nuclei are more finely grained and grow more slowly than at T_{eq}. It also indicates that extreme supercooling can completely suppress the pending transformation.

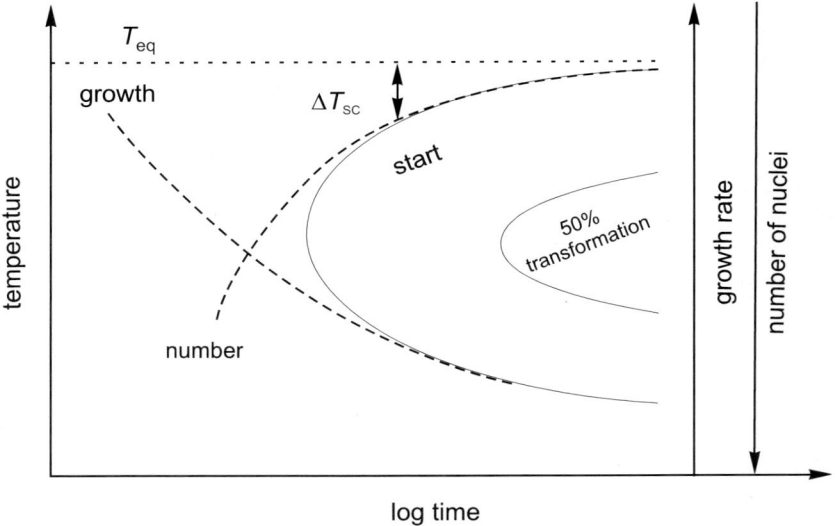

Fig. A.2.14 Isothermal transformation: Although supercooling by ΔT_{sc} below the equilibrium temperature T_{eq} increases the number of nuclei of the new phase, their growth rate decreases. This generally gives a C-shaped curve for the transformation nose (see Figure A.2.23).

A.2.2.1 Shaping

One of the goals of changing over from ingot casting to continuous casting with the thinnest possible cross-section (thin slabs to strips) is to produce hot strips of weldable steel with less metal-working effort. The corresponding refining of the primary grain size by supercooling the melt can be continued even further by subsequent thermomechanical rolling below 900°C and recrystallisation of the supercooled austenite. For grey cast iron, it is not advisable to supercool the melt too far because this could lead to white solidification. However, this is intentionally induced during chill casting by the quenching effect of the steel mould or by laser or arc remelting in order to generate an extremely fine-grained outer layer on a grey cast iron core. The relationship between the solidification cross-section and the grain size can be seen particularly clearly in hypoeutectic alloys because the primary austenite dendrites are surrounded by a network of eutectic carbide whose mesh size d_N can be easily measured in a metallographic section. As Figure A.2.15 shows, d_N decreases by several orders of magnitude from ingot casting to melt atomisation. The carbide network is so fine in the atomized powder that it contracts during hot compaction to small carbide nodules that are much finer than those in a forged ingot. Tool steels produced by powder metallurgy (PM) thus have certain advantages. This can be compared to d_N in the surface layer after electrical discharge machining or welding processes in which localised remelting takes place.

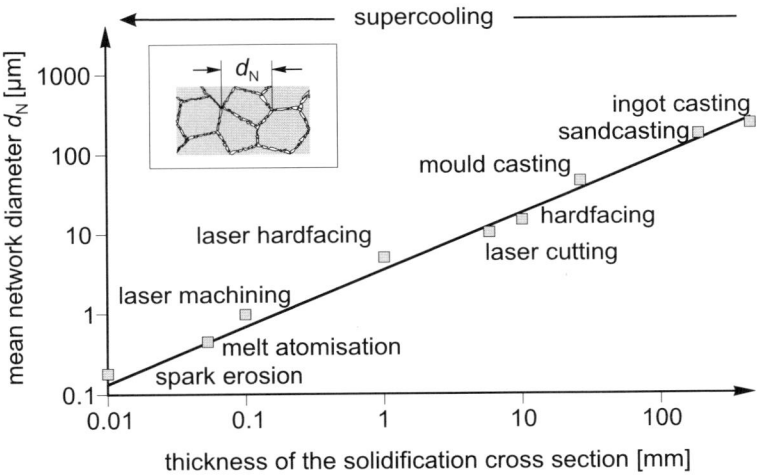

Fig. A.2.15 Primary grain size: The measured mesh diameter d_N of the eutectic network in hypoeutectic hard alloys indicates the size of the microstructural components. A plot of the typical mean mesh diameter against the thickness of the solidification cross-section is given for a variety of manufacturing processes.

A.2 Microstructure

As the hypoeutectic example shows, supercooling not only refines the grains of the iron phase, but also those of the carbide phase. At the same time, the non-metallic inclusions (oxides, sulphides) precipitated out of the melt are also smaller. Refinement also occurs during remelting of an ingot as the molten zone progresses axially through the ingot. The most popular process is electroslag remelting (ESR, Figure A.2.16), in which the faster solidification of the smaller zone volume against a water-cooled copper mould not only reduces the size of the inclusions, but also their content as the slag zone moves forward. The axial growth of the core reduces macrosegregations and shrink holes in the core, which is particularly advantageous in large forging ingots.

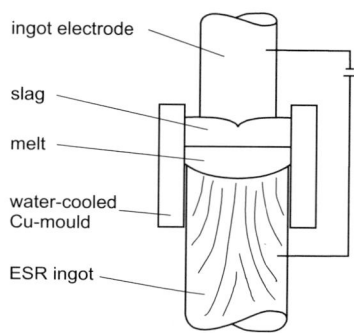

Fig. A.2.16 Electroslag remelting (ESR): Increases the quality of alloyed steels

A.2.2.2 Austenite transformation

After hot working or during heat treatment, austenite is intentionally supercooled to below the GSK line (grey cast iron $S'K'$, Figure A.1.6) in order to suppress the near-equilibrium formation of pearlite (Figure A.2.10) and to generate new microstructures such as bainite (named after E.C. Bain) or martensite (named after A. Martens), both of which have a higher hardness, in particular. As an example, let us consider the low-alloy, hypoeutectoid steel 42CrMo4 with 0.42 mass-% C, which has a near-equilibrium austenitic microstructure at 850°C (Figure A.2.9). It shall now be cooled more rapidly. Natural cooling in air, compressed air, oil or water follows an exponential law that is expressed more simply in practice by the cooling time $t_{8/5}$ from 800 to 500°C or by the quenching parameter $\lambda = t_{8/5}/100$. The harsher the continuous cooling rate of a sample is, that is, as $t_{8/5}$ decreases, the more pronounced is the change in the transformation and the microstructure. This can be seen in the continuous TTT diagram (Figure A.2.17). For slow cooling ($t_{8/5} \approx 70\,000\,\text{s}$, curve I), the transformation is near-equilibrium and corresponds to Figure A.2.9. Up to $t_{8/5} \approx 1000\,\text{s}$, (curve II), the formation of ferrite decreases because it is overtaken by the pearlite reaction that requires shorter diffusion paths for carbon. The pearlite lamellae become thinner for the same reason. The hardness increases. The S shape in some cooling curves indicates the liberation of latent heat of transformation. A further increase in the cooling rate leads to a new microstructural building block: bainite. At $t_{8/5} \approx 200\,\text{s}$, (curve III), it has a volume fraction of 85 %, which decreases with increasing cooling rate in favour of martensite. Below the critical cooling time of $t_{8/5} \approx 3\,\text{s}$, (curve IV), a martensitic microstructure with a hardness of 675 HV is produced from the supercooled austenite between the temperatures M_s and M_f (start and finish). The transformation of supercooled austenite to ferrite or pearlite follows a C curve (Figure A.2.14). The reaction gradually freezes below 550°C and is then taken over by the C curve of the bainite transformation. It meets the martensite transformation curve below it that is not C-shaped

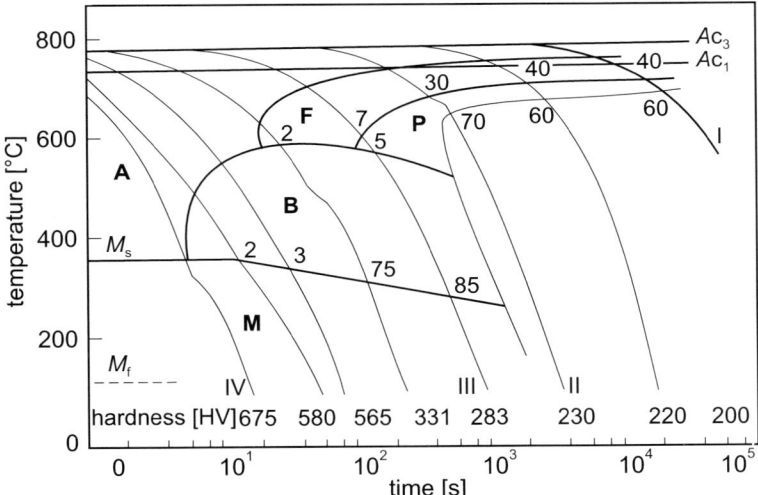

Fig. A.2.17 Time-temperature-transformation (TTT) diagram for continuous cooling, steel 42CrMo4, austenitised 850°C / 10 min (from F. Wever et al.): As the cooling time decreases, the near-equilibrium ferritic / pearlitic microstructure transforms into a non-equilibrium bainitic and martensitic microstructure. The enormous increase in the concentration of lattice defects hardens the material. M_s and M_f refer to the start and finish temperatures, respectively, for martensite formation. A stands for austenite, and F, P, B, M for the regions of ferrite, pearlite, bainite, and martensite formation. The percentages of microstructural constituents are marked along the cooling curves. The diagram is also known as a continuous cooling transformation (CCT) diagram.

but initially starts independently of time at M_s. The displacement of carbon out of bainite and into austenite during transformation means that M_s can decrease with increasing $t_{8/5}$, as shown by the example of hypoeutectoid steel in Figure A.2.17. The austenite is stabilised by the dissolved carbon.

In hypereutectoid steels with a carbon content to the right of the eutectoid, the C curve of ferrite is replaced by a line representing the precipitation of proeutectoid carbide. In cast iron, the content of dissolved carbon in the austenite follows the SE line (white) or $S'E'$ (grey, Figure A.1.6), and, if the Si content is considered (Figure A.2.11) it is less than 1 % at < 900°C. The majority of the total carbon content is already present as cementite or graphite, which does not participate in the pearlite → bainite → martensite transformation. Cast iron thus has a steel-like matrix that behaves similarly to a hypereutectoid steel on supercooling. The proeutectoid precipitation of carbon lowers the content of dissolved carbon in the hypereutectoid austenite, which is thus destabilised so that M_s increases (Figure A.2.18). If the three-phase field of grey cast iron containing Si is rapidly traversed (Figure A.2.11), ferrite formation is suppressed in favour of pearlite with metastable cementite. If the $t_{8/5}$ time is increased, the carbon is deposited as stable graphite

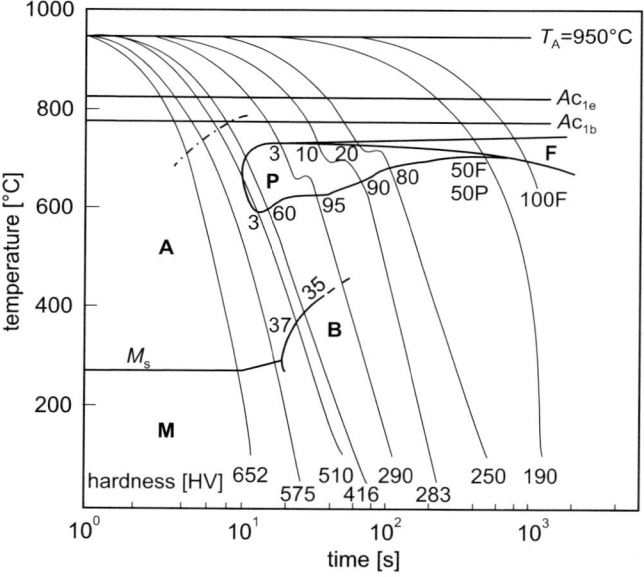

Fig. A.2.18 Continuous TTT diagram of a spheroidal graphite cast iron containing (%) 3.7 C, 2.3 Si, 0.25 Mn, austenitised at 950°C (for abbreviations, see Figure A.2.17). The dash-dotted line indicates proeutectoid precipitation of carbon, which increases M_s (from K. Röhrig, W. Fairhurst, supplemented).

A.2.2 Non-equilibrium microstructure

on already existing eutectic graphite particles and the ferrite fraction grows at the expense of the pearlite fraction.

If the hardness is to be increased, the cooling time must be reduced. A corresponding quenching regime makes the temperature in the component decrease from the core towards the outer surface. The associated decrease in the specific volume leads to thermal stresses. These are superimposed by transformation stresses caused by an increase in volume arising from the transition of fcc austenite into the bcc microstructure (Tab. A.1.1, p. 7 and Figure A.3.7 a, p. 73). Local changes in volume result in a hydrostatic stress that corresponds to the mean of the principal stresses (see Figure A.4.1 p. 80). The difference between them leads to shear stresses that cause plastic flow and thus distortion if they exceed the yield point of a phase. The hydrostatic portion of the internal stresses acts as the state parameter *pressure* on the equilibrium position of a transformation (see Figure A.1.1 b p. 4). If this value is negative, the more closely packed austenite is stabilised. On the other hand, a positive hydrostatic stress facilitates its transformation. Independently of whether they are positive or negative, internal shear stresses enhance lattice shearing that occurs during martensite formation (Figure A.2.19) which they thus facilitate. The application of external stresses during the transformation has shown that flow already occurs for stresses below the yield point of the participating phases. This transformation-induced plasticity (TRIP) is also caused by internal stresses; however, it cannot be measured. Nowadays, it can be calculated by finite element modelling and is used together with the classical, non-transformation-related plasticity to calculate the austenite transformation. Simulation provides an insight into the interplay between the transformation process and the internal stresses of which only the residual stresses and distortion remain after cooling. We will come across the abbreviation TRIP in a different context later on.

The magnetic properties also change during the austenite transformation (see Chap. A.4, p. 117). In laboratory experiments, the use of state-of-the-art superconducting magnets and a field strength of 10 to 15 T significantly accelerated the transformation rate. This is attributed to the significantly higher level of activation of the ferromagnetic transformation microstructure compared to that of paramagnetic austenite. The M_s and M_f temperatures increase, the bainite transformation is shifted to shorter times and the ferrite/pearlite transformation also takes place more quickly so that e.g. the microstructural banding of ferrite and pearlite is less pronounced. We will have to wait and see whether the influence of a magnetic field will become an established technique in heat treatments using continuous processes.

A.2.2.3 Post-quenching morphology

In contrast to near-equilibrium cooling conditions where practically all the carbon is precipitated as cementite in the pearlite (Figure A.2.10), non-equilibrium quenching of a knife blade in brine completely prevents the formation of cementite. The carbon dissolved in the fcc austenite remains dissolved in the bcc martensite even though the lattice interstices are much smaller (Figure A.1.5). This forced dissolution leads to stresses in the iron lattice and it also increases the hardness. The temperature range between pearlite and martensite allows both, forced dissolution and carbide precipitation in bainite.

(a) Martensite

This microstructural building block is formed by a diffusionless displacive transformation of supercooled austenite (Figure A.2.19). As the contents of carbon and alloying elements increase, M_s decreases and the morphology changes from upper to lower martensite.

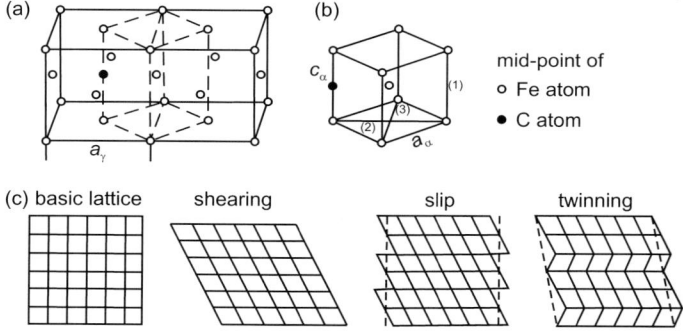

Fig. A.2.19 Austenite-martensite transformation:(a) The atoms in the fcc unit cells of austenite that can be used to make up a bcc martensite cell are joined by dashed lines. (b) In order to produce the corresponding bcc cube, this rectangular parallelepiped must shrink in height (1) to a_α and expand in the face diagonals (2, 3) to $a_\alpha\sqrt{2}$. We can see from Figure A.1.2 that $a_\alpha\sqrt{3} = a_\gamma\sqrt{2}$.
- shrinkage (1): $a_\alpha = (\sqrt{2}/\sqrt{3})\,a_\gamma$; factor of 0.816
combining with $\sqrt{2}$ gives
- expansion (2, 3): $a_\alpha\sqrt{2} = (2/\sqrt{3})\,a_\gamma$; factor of 1.155

A carbon atom dissolved in the fcc octahedral interstice is then found in the bcc octahedral interstice of martensite (Figure A.1.3). Therefore, as the carbon content increases, the unit cell expands (Figure A.1.5) leading to a tetragonal distortion $c_\alpha > a_\alpha$ of the body-centred lattice (bct). (c) The total deformation caused by the transformation manifests itself in a volume increase (Figure A.3.7 a, p. 73) and in lattice shearing, which can lead to permanent deformation by slip of dislocations or – particularly at a low M_s temperature – by twinning.

A.2.2 Non-equilibrium microstructure

Upper martensite is known as *lath (massive) martensite*. It is prevalent in unalloyed and low-alloy steels with less than ≈ 0.4 % C as well as in iron alloys with < 25 % Ni. It grows within the austenite in the form of packets of parallel laths < 1 μm wide and does not leave any retained austenite (Figure A.2.20 a).

Film recordings have revealed that the laths grow suddenly and independently, one next to the other, until the packet reaches another packet or the former austenite grain boundary. The adjustment between the fcc and the bcc lattices takes place via dislocations with a total length of up to 10^{12} cm/cm^3. Above ≈ 0.2 % C the dissolved carbon atoms cause a slight tetragonal distortion (bct) of the body-centred cubic lattice.

Lower martensite is known as *plate or lenticular martensite* because it grows in the form of mutually inclined plates. In low-alloy steels with > 0.4 % C it initially occurs together with lath martensite. Above ≈ 0.8 % C it alone determines the microstructure. This type of martensite is also found in iron alloys containing more than 30 % Ni after sub-zero cooling. The martensite plates subdivide the austenite grain into increasingly smaller regions (Figure A.2.20 b), and stabilise the austenite by means of compressive stresses arising from the volume increase of martensite plates. This transformation is autocatalytic: the plates grow at almost the speed of sound and trigger the formation of nuclei that are capable of growth in the surrounding austenite. The size of the plates formed as growth proceeds decreases and the austenite in the remaining angular areas may be incompletely transformed leaving

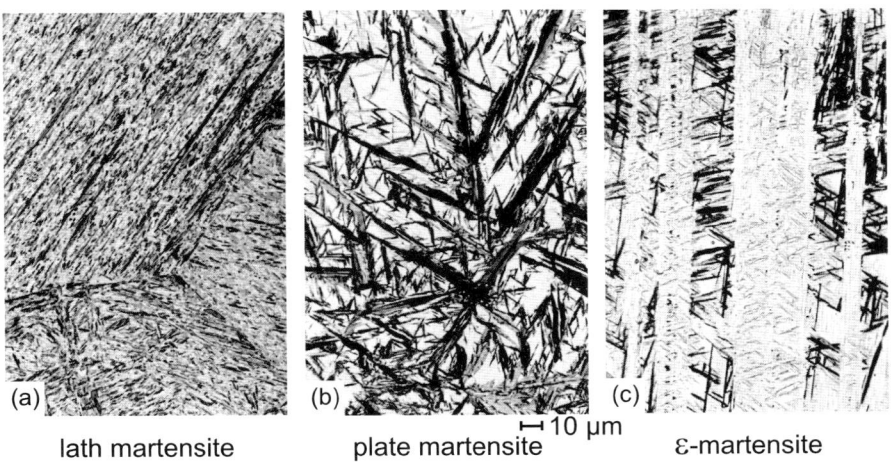

(a) lath martensite (b) plate martensite ⊢10 μm (c) ε-martensite

Fig. A.2.20 **Martensitic microstructure** (LOM): (a) low-alloy steel with 0.17 % C, 1200°C/hot bath, (b) supercarburised surface layer of a case-hardening steel 1100°C/hot bath, (light background = retained austenite), (c) decarburised surface layer of Hadfield manganese steel X120Mn12 (dark areas = α-martensite).

A.2 Microstructure

retained austenite. The lattice adjustment occurs not only by dislocations, but also by twinning as the temperature drops (Figure A.1.4). Their associating streaking effects can be seen in the martensite (Figure A.2.21 a). Hardness and tetragonal distortion increase with the carbon content. In contrast, plate martensite is body-centred cubic and soft in substitutional solid solutions, such as iron-nickel. In high-carbon steels with coarse austenite grains, the impingement of one martensite plate onto another may trigger microcracks of a length corresponding to the plate thickness. Transformation of the fcc to the bct lattice increases the diffusion rate of the carbon. Whereas in carbon steels, holding just above the M_s temperature does not affect the martensite hardness, slower cooling below M_s leads to carbide precipitation, particularly in the initially formed martensite regions and thus to a decrease in the hardness. This process is known as auto-tempering (Figure A.2.22). If it is completely suppressed by sub-zero cooling of thin samples, the martensite can be even more ductile at e.g. -70°C than that aged at room temperature. With respect to residual microstresses in the microstructure, such ever present ageing in this material is regarded as a further cause of microcracking.

In addition to the *bcc/bct* variants known as α-*martensite*, some manganese and chromium-manganese steels exhibit hexagonal close-packed (hcp) ε-*martensite* (Figure A.2.20 c). Alloying elements decrease the stacking fault energy thus favouring the formation of this phase, which is related to ε-iron

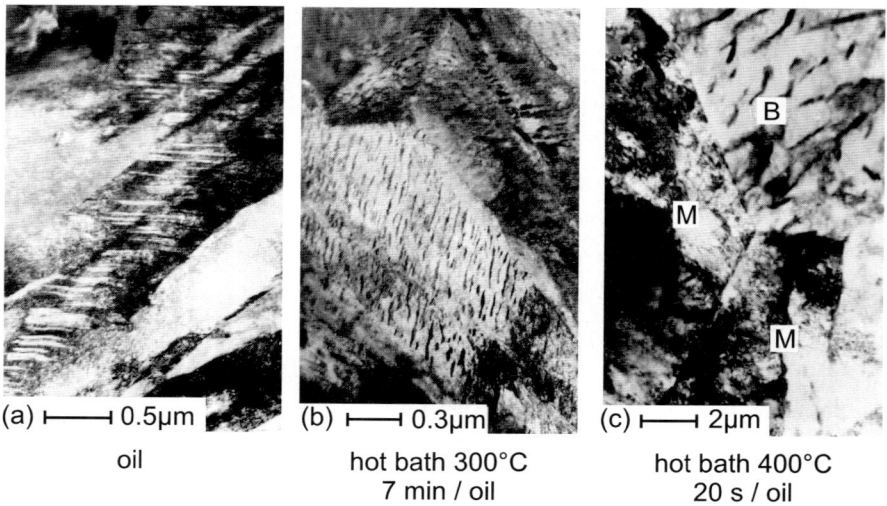

Fig. A.2.21 Martensitic and bainitic microstructure (TEM, 55Cr3, 860°C, 20 min): (a) Martensite plate with twinning bands (light areas), (b) lower bainite with fine, oriented carbide precipitates (dark areas), (c) a grain of upper bainite (B) with coarse cementite precipitates (cf. scale) next to two grains of martensite (M).

A.2.2 Non-equilibrium microstructure

(see Figure A.1.2 c, p. 6). It is not magnetisable and is produced from the austenite with a slight decrease in volume and an increase in the electrical resistance, which is in contrast to α-martensite. Stabilisation with substituted elements softens ε-martensite. On the other hand, its hexagonal structure makes it more brittle than an equally hard α-martensite. ε-Martensite is thus only used in functional materials, such as shape memory alloys, or for strain-hardening of austenitic steels.

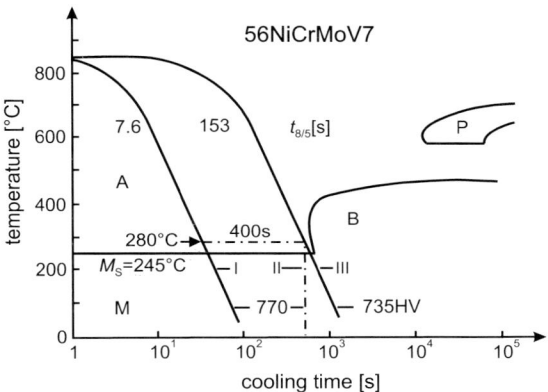

Fig. A.2.22 Auto-tempering: Cooling curve I: after austenitising at 860°C cooling at $t_{8/5} = 7.6$ s produces martensite with a hardness of 770 HV. Cooling curves I/II: a holding time of 400 s just above M_s followed by further cooling II = I (logarithmic scale!) leads to the same hardness. Cooling curve III: cooling at $t_{8/5} = 153$ s leads to a lower hardness. If the cooling rate is increased from 280°C according to curve II, the resulting hardness is higher again. These measurements lead to the conclusion that carbon remains dissolved in the supercooled austenite. In contrast, slow cooling of martensite produces carbide precipitates due to auto-tempering.

(b) Bainite

The morphology of martensite and bainite are similar. In contrast to martensite formation, the bainite transformation is not diffusionless and is associated with carbide precipitation and a lower dislocation density. The formation of bainite (and pearlite) is limited to steels containing carbon. In *upper bainite*, the carbon is enriched on the lath boundaries where it leads to retained austenite or cementite bands, which can grow up to about 2 μm long after isothermal transformation of steel 55Cr3 at 400°C, for example. In contrast, *lower bainite* is formed at 300°C. The size of the oriented precipitated cementite particles in its plates is only about 0.2 μm (Figure A.2.21 b, c). Compared to martensite, less carbon remains in the forced solution and the hardness is lower.

A.2 Microstructure

In summary, Figure A.2.23 illustrates the influence of an increasing degree of supercooling on the microstructure of a low-alloy hypoeutectoid steel as an isothermal TTT diagram. Within the pearlite nose, the interlamellar spacing decreases with supercooling, whereas within the bainite nose the carbide length decreases with supercooling, which is even lower if there is autotempering. Overall, supercooling refines the carbides by two orders of magnitude.

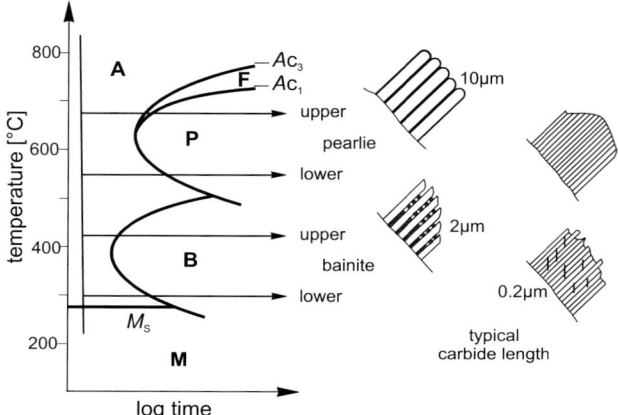

Fig. A.2.23 Isothermal transformation of austenite (schematic):A thin steel sample is austenitised, rapidly cooled to its transformation temperature and then held at this temperature (isothermal treatment). As the holding temperature decreases, the size of the precipitated carbides also decreases until they are no longer present in martensite. The figure is an example of an isothermal TTT diagram, also known as an IT diagram.

(c) Influence of alloying

Nitrogen atoms are a little smaller than carbon atoms (Tab. A.1.4) and are thus more soluble in the interstices of the austenite lattice. Nitrides such as $Fe_{16}N_2$ or $Fe_{2-3}N$ precipitate in ferrite (Tab. A.2.1) with kinetics similar to carbides. Depending on the degree of supercooling, hardening of nitrogen-alloyed steels can thus lead to martensite as well as a bainite- and pearlite-type microstructure. The use of nitrogen to increase the hardness is confronted by the problem of its low solubility in the iron melt. This can be improved by adding large amounts of manganese, chromium, etc., as well as by high pressures and powder metallurgy. On account of these complications, alloying with nitrogen is only used for special applications

A.2.2 Non-equilibrium microstructure

The supercoolability of austenite depends on the alloying levels. The addition of alloying elements can shift the transformation lines in the TTT diagram to longer times. Martensite can then also be formed under milder cooling conditions, i.e. in thicker cross-sections. At the same time, the higher strength of the alloyed austenite leads to a greater resistance to a martensitic transformation. This can only be overcome by a greater degree of supercooling, thus shifting the region of martensite formation between M_s and M_f to lower temperatures. If the M_f temperature drops below room temperature, part of the austenite is retained because it escapes being transformed. At high alloying levels, if M_s drops just below the ambient temperature, an unstable austenitic microstructure is produced. A further increase in the alloying content can lower the M_s to such an extent that it becomes what is known as an austenitic steel.

The effect of alloying elements on non-equilibrium microstructural components is shown by a Schaeffler diagram (Figure A.2.24). It was originally developed for welding materials, and thus corresponds to rapid cooling conditions from a high temperature. The ferrite- and austenite-stabilising alloying

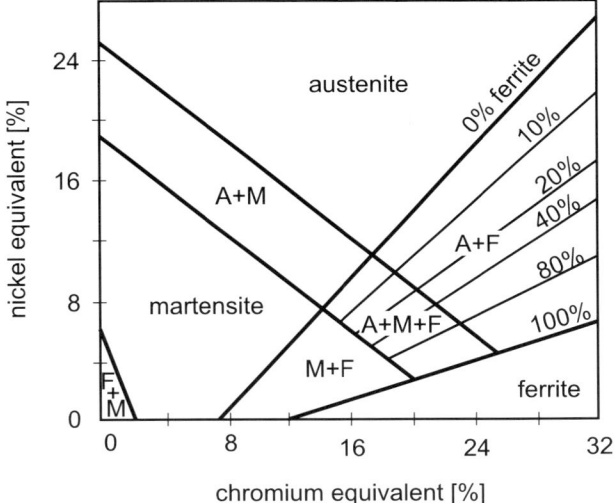

Fig. A.2.24 Schaeffler diagram for rapid cooling from a very high temperature (e.g. weld pool): Nickel and equivalent elements stabilise austenite, chromium equivalents stabilise ferrite (Figure A.1.10, p. 18). A combination of both expands the austenite field. A.L. Schaeffler and W.T. Delong have developed weighting factors for low-carbon steels
— Ni equivalent = % Ni + 30 · % (C + N) + 0.5 · % Mn
— Cr equivalent= % Cr + 1.4 · % Mo + 1.5 · % Si + 0.5 · %Nb + 2 · % Ti
As the C and N contents increase, the weighting factor of 30 decreases

elements are given empirical weighting factors and summarised to an equivalent. Whether martensite, austenite or ferrite predominates in quenched steels depends on the Ni and Cr equivalents. Similar to the use of a phase diagram to predict slowly cooled microstructural building blocks, Schaeffler diagrams are used to predict the microstructure of rapidly cooled materials.

A.2.2.4 Reheating of quenched microstructures

Rapid cooling from an austenite field produces a similar quenched microstructure in steels and cast irons. The non-equilibrium microstructural building blocks, martensite and bainite, can be brought back to near-equilibrium conditions by heating. This process is called tempering. Depending on the constitution of the ferrous material, the tempered microstructure approaches a stable or metastable equilibrium as the temperature increases. Silicon-containing cast iron may be expected to precipitate temper graphite, whereas steels contain only temper carbides.

(a) Steel

As the tempering temperature increases, there is a step-wise degradation of the supersaturated solution of carbon in martensite. In tempering stage zero, carbon segregates into defects, even at temperatures below 100°C (see Figure A.2.7 c). Orthorhombic η-Fe_2C carbide precipitates in carbon-rich steels held at 120°C for several weeks. Above this temperature, hexagonal ε-Fe_2C carbide is formed in all steels in the first tempering stage. In the second tempering stage at up to 300°C, residual austenite completely transforms to bainite, and in the third tempering stage above 300°C, Fe_2C is converted to cementite Fe_3C. The cementite bands spheroidise if the temperature is increased even further. At $\approx$500°C, they just reach a spherical shape with a diameter of about 0.1 µm. At even higher temperatures, they begin to coarsen as a result of Ostwald ripening. Driven by a reduction in the interfacial energy, the smaller carbides are dissolved in favour of the larger ones. The tempering of iron-nitrogen alloys leads to the precipitation of nitrides in a similar way, but in differing stages.

Residual stresses in the matrix are relieved as the tempering temperature increases. Recovery starts above ≈ 400°C, and recrystallisation occurs within the martensite plates above ≈ 600°C. However, an equiaxial ferrite grain is produced only after a further temperature increase. The martensite morphology is thus lost only at temperatures close to Ac_1. Owing to the high dislocation density in martensite, recrystallisation leads to nucleation at many sites. A single original grain of austenite produces several ferrite grains and thus has a refining effect. Close to 300°C, blue brittleness develops as a result of cementite formation and the strain hardening behaviour changes (Figure A.2.25 a). At 500°C, enrichment of phosphorus, tin, arsenic and antimony on the grain

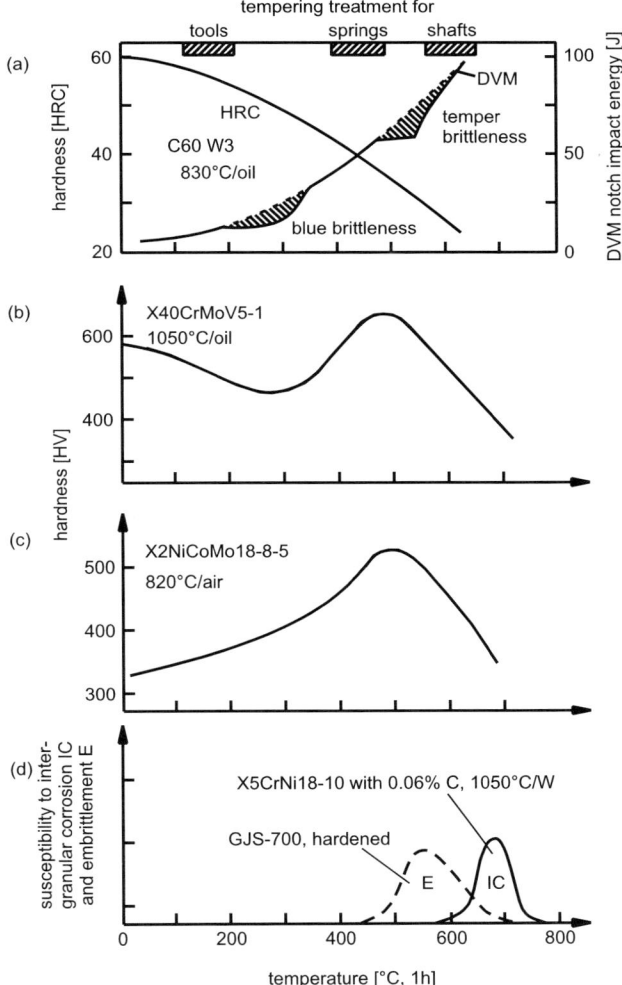

Fig. A.2.25 Reheating of a quenched microstructure: (a) Tempering curve of an unalloyed steel after martensitic hardening. Depending on the tempering temperature, the steel can be used for thin-walled tools, springs or ductile machine components. In this case, temperatures of around 300°C (blue brittleness) and 500°C (tempering embrittlement) must be avoided. (b) As above, however, this applies to a secondary hardening hot-work tool steel. The martensite is hardened by precipitation of MC and M_2C carbides (4th tempering stage). (c) The maraging nickel steel is hardened by precipitation of Ni_3Ti and Fe_2Mo. (d) Susceptibility to intergranular corrosion (IC) due to grain boundary carbides $M_{23}C_6$ or embrittlement (E) due to temper graphite.

boundaries results in temper embrittlement. This grain boundary segregation leads to low-energy intergranular fracture modes.

The addition of alloying elements can shift the temperatures of the tempering processes. Silicon inhibits e.g. carbide growth, because it is not soluble in cementite. The tempering stages and the blue brittleness are thus shifted to higher temperatures. In steels with special carbide-forming alloying elements, such as chromium, molybdenum, niobium, vanadium and tungsten, there is a fourth tempering stage above 450°C that triggers secondary hardening (Figure A.2.25 b). In practically carbon-free high-alloy steels, a series of intermetallic phases or metallic copper are precipitated at temperatures of $\approx$ 500°C. Because their distribution is so fine, the relatively soft martensite starts to undergo precipitation hardening (Figure A.2.25 c).

Quenching produces a dislocation-rich, martensitic solid solution. Reheating the material rapidly heals the dislocations in the matrix. In contrast, the precipitated carbides need longer in alloyed steels to reach a composition that corresponds to an equilibrium. Alloy carbides may be precipitated via several precursors. For example, in a creep-resistant martensitic steel with 12 %Cr, a series of carbide phases are observed in the following order:

$$M_3C \rightarrow M_7C_3 \rightarrow M_{23}C_6 \quad (A.2.3)$$

In this series, the chromium content in the carbides increases at the expense of the iron content. The time required for this process is determined by the diffusion rate of chromium. M_7C_3 can grow (in situ) from a M_3C particle or a new grain can grow by nucleation. In this case, M_3C dissolves in favour of M_7C_3.

During reheating of solution-annealed and quenched materials with a non-transforming austenitic and ferritic microstructure (see the Schaeffler diagram), any changes that occur depend on the content of rapidly diffusing interstitial elements such as carbon and nitrogen, which are precipitated as carbides and nitrides. Thereby austenite may become susceptible to corrosion (Figure A.2.25 d) or it can be destabilised to such a degree that it partially transforms to ferrite or martensite. Intermetallic phases are also precipitated. Similar to alloyed martensitic steels, precipitation sequences can also take place here as equilibrium conditions are approached. Not only the composition of the carbides and nitrides changes, but also that of the intermetallic phases. A plot of the onset time of precipitation of the individual phases against the temperature produces C curves as shown in the TTT diagram (Figure A.2.14). Owing to the lower diffusion rate, the precipitation processes are slower in austenite. In non-transforming steels, there is hardly any difference between the C curves determined during reheating and those determined during cooling.

The distribution of the precipitates after reheating depends on the possible nucleation sites. In agreement with its definition, the new phase must have a different atomic ordering. A phase boundary thus develops around the first nucleus (Figure A.2.4), which requires energy of formation. Some of this can

A.2.2 Non-equilibrium microstructure

be withdrawn from lattice defects. A grain boundary can directly provide the required interfacial energy and is thus a preferential precipitation site. On the other hand, a high dislocation density favours nucleation within the grain. Moreover, the energy consumed during development of the phase boundary is also an important factor. If the mismatch between the matrix and the nucleus is low and the coherence is thus large (Figure A.2.5), nucleation within the grain requires very little energy from lattice defects.

Many precipitates in reheated steels are incoherent. In coherent precipitates, matching is lost with increasing size. Coherence becomes rare as the equilibrium is approached. This leads to the following possibilities with regard to the distribution of a precipitate: incoherent particles are preferentially precipitated as a grain-boundary network, unless a high dislocation density promotes their growth within the grains. This can be induced by a martensitic heat treatment as well as by semi-hot or cold working. Coherent particles can be precipitated more easily as a dispersion within a grain, even without this treatment. Tab. A.2.1 gives details of various precipitates. The shape of a precipitated particle varies between a sphere, a disk and a band, depending on the minimum interfacial and distortion energies. The sphere produces the greatest distortion owing to volumetric differences between the matrix and the precipitate. A disk requires a large interfacial surface area for a given particle volume. As the coherence increases, the interfacial energy decreases.

(b) Cast iron

The decomposition of martensite during tempering of Si-containing grey cast iron initially starts with carbide precipitation, similar to steels. Increasing the reheating time and the Si content lowers the tempering temperature for the start of the transformation of metastable cementite into finely dispersed graphite (450 to 600°C). This has an adverse effect on the strength and ductility (Figure A.2.25 d). At a higher tempering temperature (700 to 760°C), the fine graphite particles vanish and are redeposited on the coarse eutectic particles. During this annealing, the graphite is in equilibrium with ferrite, which contains only a little carbon along the solubility limit QP' (Figure A.1.6 and Figure A.2.11). This is in contrast to annealing at 900 to 950°C, where the austenite dissolves almost 1 % C along $S'E'$.

On reheating pearlitic grey cast iron, the volume increase (expansion) associated with the cementite/graphite transformation may adversely affect the dimensional stability in high-temperature applications. Alloying with Cr $\leq$ 1 % stabilises the cementite in the pearlite and the tempered microstructure. This also applies to white cast iron, although its cementite has a lower tendency to decompose owing to its lower Si content. Its matrix exhibits similar tempering processes to steel when reheated.

Table A.2.1 Precipitates obtained on reheating a quenched microstructure: Where X = C, N and M = metal(s), Fe-X and M-X are precipitates with interstitial elements (carbides, nitrides), M-M are intermetallic precipitates. The diffusion rate of X is several orders of magnitude higher than that of M. It is also higher in ferrite and martensite than in austenite (Figure A.1.8). Approximate holding temperatures can be derived from this that have a perceptible precipitation effect (precipitation hardening, embrittlement, corrosive attack) for a holding time of ≈ 1 h. The match between the crystal lattices of the matrix and the precipitate are roughly classified into incoherent (-), partially coherent (+) and coherent (++). The transitions between them are gradual. For example, the lattice parameter of cubic precipitates may be higher than those of the matrix by a factor of one– (γ'-phase), two– (α"-nitride), three– ($M_{23}C_6$) or four times (G phase). In contrast, the atomic ordering in a hexagonal precipitate such as ε-carbide is similar in only one direction (see also Figure A.2.5 and 'Atlas of Precipitates in Steels').

Precipitate			Matrix	
			bcc	fcc
Fe-X			$\geq 150°C$	$\geq 250°C$
Fe_2C	hex	ε-carbide	+	
Fe_3C	orh	cementite	+	–
$Fe_{16}N_2$	bct	α"-nitride	++	
Fe_4N	fcc	γ-nitride	+	++
M-X			$\geq 450°C$	$\geq 600°C$
MC	fcc	V, Nb, Ti	+	–
M_2C	hex	Mo, V	+	
M_7C_3	hex	Cr	–	–
$M_{23}C_6$	fcc	Cr	–	+
M_6C	fcc	Mo, W	–	–
MN	fcc	Cr	+	
MN	hex	Al	–	–
M_2N	hex	Cr	–	
M-M			$\geq 450°C$	$\geq 700°C$
NiAl	bcc	B_2-phase	++	
Ni_3Al	fcc	γ'-phase		++
Ni_3Ti	hex	η-phase	–	–
$Fe_7(Mo,W)_6$	rh	μ-phase	–	
$Fe_2(Mo,W)$	hex	Laves-phase	–	+
FeCr	tetr	σ-phase	–	–
$Fe_{36}Cr_{12}Mo_{10}$	bcc	χ-phase	–	–
$(Fe, Ni)_{16}Ti_6Si_7$	bcc	G–phase	+	

A.2.3 Morphology of cementite and graphite

In Chapt. A.1, p. 12, a carbon-enriched cast-iron melt was used to show that cementite is precipitated in several steps from primary to ternary as the temperature drops. This has a decisive effect on the size and morphology of the carbides. Primary cementite crystals that grow freely in the melt may be as long as 100 to 1000 µm owing to the high temperature and rapid diffusion in the liquid state within a sand mould. The carbides of the ledeburite eutectic also grow out of a melt; however, their length is only 10 to 100 µm because of the lower temperature. This also applies to primary and eutectic graphite. Owing to the slower diffusion in the solid state, the precipitation of carbon after solidification leads to thinner cementite or graphite morphologies, although they may have a considerable length. In cast iron, secondary precipitates are generally deposited on existing primary and eutectic particles, which does not require nucleation. In contrast, after low-alloy steels have solidified, they pass through a homogeneous austenite field (Figure A.1.6, p. 10 and A.2.9, p. 29). Because there is a lack of self-nucleation, the secondary cementite in hypereutectoid steels is precipitated along the austenite grain boundaries whose disorder facilitates nucleation. This grain-boundary cementite is only tenths of a micron thick; however, it is elongated and may enclose the grains as a continuous coat, i.e. it forms an embrittling carbide network (Figure A.2.3 and Figure A.2.10 c). The eutectoid cementite lamellae in pearlite reach a typical diameter of 10 µm (Figure A.2.23) and the small amount of ternary cementite or graphite also chooses the grain boundary if there are no self-nuclei available. An important objective for hot working and heat treatment is to reduce the maximum carbide size in low-alloy steels to less than 1 µm. These 'fine' cementite precipitates differ from the 'coarse' cementite or graphite particles precipitated from the melt in cast iron, which are larger than 10µm. Similar to the situation in steel, fine carbides are precipitated from a hypereutectoid austenite in white or grey cast iron with compact graphite as a result of a full heat treatment with a martensitic or bainitic transformation of the matrix. On the basis of the microstructural conditions, it can be expected that the differences between the properties of steel and cast iron depend on the coarse particles in the latter because the austenite phase is comparable in both. For a given carbon content, the volume content of the coarse graphite particles in the grey cast iron is distinctly lower than that of the coarse cementite particles in white cast iron (Tab. A.1.3, p. 13). This differentiation between coarse precipitates from the melt and fine precipitates from a solid solution is, at best, only a useful simplification for conventionally produced ferrous materials. Special processes, such as melt spraying (atomization), can be used to refine the primary and eutectic precipitates in white cast iron and ledeburitic cold-work tool steels to such a degree that their size is of the same order of magnitude as that of fine precipitates (Figure A.2.15). This differentiation between coarse and fine secondary precipitates applies only by analogy to high-alloy ferrous materials.

Solution-annealing of a housing made of stainless austenitic cast steel (SCHMOLZ + BICKENBACH Guss GmbH & Co. KG, Krefeld, Germany)

A.3 Heat treatment

Most ferrous materials are not subjected to a separate heat treatment. Their microstructure results directly from solidification and/or controlled hot working and cooling. Semi-finished steel products (strip, section, pipe, wire) undergo a combination of hot working and heat treatment during thermomechanical processing, which is also used e.g. for drop-forged parts. The desired microstructure of cast iron is often obtained by tailoring the melt composition to the solidification cross-section, i.e. the cooling rate in the mould. Another option is to remove the workpiece from the sand mould while it is still hot e.g. to promote growth of pearlite.

However, a separate heat treatment is frequently worthwhile because the microstructure produced during shaping (e.g. by casting, working, welding, sintering, etc.), may not have optimum manufacturing properties (e.g. machinability) or service properties (e.g. fatigue strength). The heat treatment process is therefore integrated into the manufacturing process so that it is independent of shaping and so as to allow the microstructure to be adjusted to the operational requirements of the finished component. For example, tensile and compressive stresses act over the entire cross-section, whereas bending and torsional stresses as well as chemical and tribological loads act primarily on the surface layer of a workpiece. Correspondingly, either the entire workpiece is given a uniform heat treatment or only the surface layer is treated. We differentiate between: (a) annealing processes that eliminate lattice defects and segregations or which bring the material back towards the equilibrium state. (b) Hardening processes that take the material away from the equilibrium state, produce lattice defects and thus increase the hardness. The fact that steel can be soft-annealed to facilitate processing before it is hardened for the intended application is a major advantage. Annealing usually affects the entire workpiece whereas hardening can be applied to the whole material or only to the surface layer.

Cast iron contains eutectic graphite or cementite precipitates in a steel-like matrix. In principle, this matrix can be subjected to heat treatments similar to those used for steel. However, there are limitations caused by e.g. the instability of cementite, a eutectic-related susceptibility to cracking if the material is subjected to a thermal shock and the lack of oxidation resistance of grey cast iron. We will now describe heat treatment processes for steel and then discuss special cases that apply to cast iron, starting with annealing processes ordered according to increasing treatment temperature. The main emphasis is on low-alloy ferrous materials. Additional information on high-alloy materials is given in the sections dealing with the individual groups of materials. The definitions of terms used in the heat treatment of ferrous materials are given in EN 10052. DIN 17022 provides information on some hardening processes.

A.3 Heat treatment

A.3.1 Annealing processes

A.3.1.1 Baking

Objective: To prevent hydrogen embrittlement. Hydrogen originates e.g. during manufacturing (welding, pickling, electroplating) or as a result of corrosion (Chapt. A.4, p. 114 and Chapt. B.2, p. 178). The hydrogen content refers to the amount of H atoms dissolved in the steel or to the amount of molecular H_2 measured by hot extraction or melt extraction:
1 ppm H = 1 µg H / 1 g Fe = 10^{-4} % H = 1.11 cm^3 H_2 / 100 g Fe.
The first method gives an approximate value of the amount of diffusable hydrogen; the second gives the total content.

Method: Components are baked directly after a surface treatment while they are still warm, e.g. after chrome-plating at 65°C, and are heated for a few hours at 200 to 250°C. Owing to the strong attractive forces between H atoms and lattice defects, the room-temperature diffusion velocity in ferrite is a hundred times faster than in martensite. At 200°C, the H atoms leave their sinks and diffuse to the surface.

Applications: Tools and high-strength fastening elements with a low-tempered martensitic microstructure. A forging die burst one week after hard-chromium plating and prior to commissioning. Cadmium-plated pneumatic nails in a suspended ceiling started to fracture months after being installed. Electrogalvanised, highly prestressed screws failed after only a short service life. Baking can limit such damage. A modified form of this heat treatment was frequently used in the past to avoid flaking on semi-finished products. These penny-shaped incipient cracks are caused by hydrogen absorption during melting. Since the introduction of steel degassing, this type of cracking induced by recombined hydrogen (H_2) has become rare (see HIC, Chapt. A.4, p. 114). However, it has been observed in a sharply delimited core segregation of continuously cast high-alloy steel.

A.3.1.2 Stress-relief annealing

Objective: To relieve internal stresses and to avoid distortion originating from the dissipation or redistribution of residual stresses during subsequent heat treatment or during machining. Asymmetrically distributed residual stresses caused by previous cold working, straightening or welding operations are particularly unfavourable because they may induce changes in shape when the material is heated.

Method: Lowering of the yield point by heating, but without significant alteration of the microstructure or strength. The residual elastic strain is transformed into plastic deformation, thus forestalling distortion. If the machining allowance is insufficient, the part must be restraightened and subjected to further stress-relief annealing. The treatment temperatures should be as high as possible, e.g.

- soft-annealed steels: just below Ac_{1b}
- ferritic/pearlitic steels: 600 – 650°C
- quenched and tempered (QT) steels: $\approx$ 30°C below the tempering temperature
- non-transforming steels: it is not possible to make any generalisations because the precipitates in these materials make them prone to embrittlement and corrosion. Temperatures of $\approx$ 580°C are frequently used to avoid 475°C–embrittlement, σ–phase embrittlement and regions that promote intergranular corrosion (see Figure B.6.7, p. 316 and Figure B.6.4, p. 314)
- grey cast iron: the temperature used for annealing depends on the cementite stability,
  high Si $\rightarrow$ < 500°C, low Si $\rightarrow$ < 600°C, added Cr $\rightarrow$ < 650°C.

Applications: (a) Thin components and tools. For example, a drive shaft made of cold-straightened bar steel already starts to distort during rough turning or during heating to the hardening temperature if the residual stresses caused by straightening are not eliminated beforehand by stress-relief annealing. (b) Fatigue-loaded components with unfavourable residual tensile stresses that increase the mean stress. (c) Components that are prone to stress-corrosion cracking.

A.3.1.3 Soft annealing of steel

Objective: To obtain the low hardness required for subsequent cold-working or cutting of steel. This is carried out on transformable steels containing carbon and therefore harder microstructural constituents such as pearlite or bainite (martensite) after hot-working or welding.

Method: Hardness can be decreased by means of four metallurgical steps.

I The hardness of the ferritic solid solution is reduced by withdrawing dissolved elements from the matrix into the carbides. Chromium, molybdenum, vanadium, etc. dissolve in the iron carbide or form their own carbides. They also crop off the austenite field so that the Ac_1–temperature increases. This allows higher annealing temperatures and thus accelerates the process.

dissolved in	Ac_1-temperature decreased	increased
matrix	Ni	Si
carbide	Mn	Cr, Mo, V

CrMo-alloyed QT or tool steels can thus be more successfully soft-annealed than those alloyed with Ni.

II Spheroidising of elongated or plate-shaped carbides to nodules. This is driven by a reduction in the interfacial energy. A decrease in the boundary surface concentration means that the dislocations have longer paths to the next obstacle, and this lowers the hardness. The thicker carbide lamellae in upper pearlite are more difficult to spheroidise than the thinner ones in lower pearlite. The carbide bands of upper bainite behave similarly. Spheroidising is accelerated by cold- or semi-hot working (just below Ac_1). This can be circumvented by thermomechanical working of the austenite because spheroidal carbides precipitate directly during slow cooling through the pearlite stage. Soft-annealing of a prehardened material also produces a spheroidal carbide dispersion (Figure A.3.1).

III Coarsening of spheroidal carbides by Ostwald ripening over time t (c = constant).

$$r^3 - r_o^3 = ct \qquad (A.3.1)$$

If the mean initial radius r_o is small compared to the annealed radius r, then $r \approx (ct)^{1/3}$. It is thus logical that long annealing times become uneconomical. The explanations given under II for the driving force and decreasing hardness also apply here. The upper technical limit for the carbide size is governed by the delayed dissolution of coarse carbides during subsequent hardening. Consequently, this limit is not sharp. For an average carbide diameter, it lies between 0.4 and 1 µm.

IV Reduction in the dislocation density in ferrite during recovery and recrystallisation. Lattice defects caused by cold-working or which are present in higher numbers in bainite and martensite are healed and the hardness drops.

Implementation: Low-carbon and nickel-alloyed steels are annealed for several hours just below Ac_{1b} followed by slow cooling to 500°C to avoid distortion. Carbon-rich steels can be more effectively annealed just above Ac_{1e} because the small carbides dissolve in the austenite. During slow cooling to 600°C at a rate of e.g. 10°C/h, the dissolved carbon precipitates on the undissolved carbides. For a typical initial microstructure, the resulting carbide coarsening is thus more pronounced than that produced by Ostwald ripening alone. This can be seen by comparing Figure A.3.1 a and b. Repeated temperature cycling about Ac_1 is known as cycle annealing.

Applications: Low-carbon steels, e.g. case-hardenable as well as QT steels are soft-annealed to improve their cold workability. Many hobbed or cold-extruded shapes are only possible with very soft steels. This state is too soft for cutting because it produces a continuous chip. The treatment process is included in the designation of the material, e.g. 34Cr4 A (annealed) or 34Cr4 AC (spheroidised annealed). However, carbon-rich steels, such as rolling bearing and tool steels, are soft-annealed to improve their machinability. This allows faster cutting speeds and increases the service life of the tools. In addition, a transient

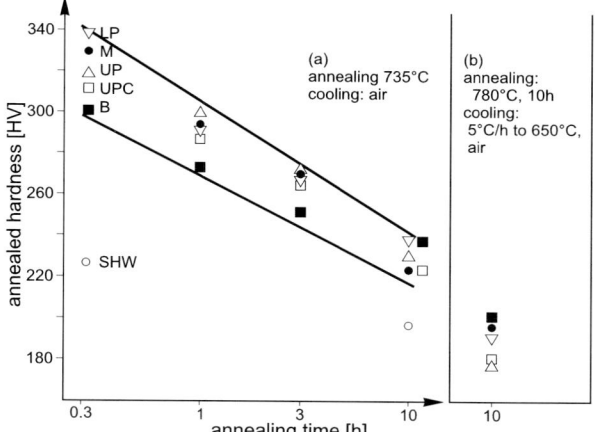

initial condition			soft-annealed, 10 h	
microstructure	treatment		^{1}AC [%]	^2d [µm]
M (martensite)	1100°C, 30 min/oil	(a)	100	0.48
		(b)	100	0.38
B (upper bainite)	1100°C, 30 min/	(a)	60	0.30
	400°C, 80 min/air	(b)	100	0.40
LP (lower pearlite)	1100°C, 30 min/	(a)	100	0.38
	550°C, 60 min/air	(b)	100	0.41
UP (upper pearlite)	1100°C, 30 min/	(a)	35	0.35
	700°C, 15 min/air	(b)	100	0.50
UPC (cold-worked)	UP + 50 % cold upsetting			
SHW (semi-hot worked)	880°C, 15 min / hot upsetting 750°C, 50 % / 735°C	(a)	100	0.76

^{1} Proportional area of spheroidal cementite in the metallographic section, remaining carbides are lamellar or rod-shaped.
^{2} Mean diameter of the spheroidal carbides

Fig. A.3.1 Soft annealing: Influence of the initial microstructure on the annealed hardness of steel 100Cr6 with Ac_{1b}, Ac_{1e}, $Ac_c \approx 740, 775, 945$°C. A coarse-grained austenite and a complete dissolution of the carbides facilitate the subsequent microstructural evaluation. This is why an austenitising temperature of 1100°C was chosen (except for the SHW sample) to produce the initial microstructure. It is close to the hot-working temperature. (a) Annealing at 735°C in the ferrite / carbide field just below Ac_{1b}, (b) annealing at 780°C in the austenite / carbide field just above Ac_{1e}.

soft-anneal below Ac_{1b} can also be used as an intermediate treatment during cold-working to decrease the effects of strain-hardening.

A.3.1.4 Soft annealing of cast iron

Steel is soft-annealed to spheroidise pearlite and to decrease strain hardening. Both are not relevant to grey cast iron. According to the broader sense of the definition given in EN 10052, soft-annealing means a heat treatment to reduce the hardness, which means that this term can be used as follows:

(a) Ferritising annealing

Objective: Reduction of the pearlite fraction in grey cast iron.
Method: Annealing just below Ac_{1b} (T_P in Figure A.2.11, p. 31) at 700 to 760°C, which transforms the unstable cementite in the pearlite into graphite thus improving the machinability. The hardness decreases, but for a different reason than for soft-annealing of steels.
Applications: For grey cast iron and malleable cast iron. Sometimes also used as controlled cooling through the given temperature range after previous annealing to transform the white eutectic microstructure.

(b) Carbide annealing

Objective: Dissolution and redeposition of carbides in austenitic cast iron with spheroidal graphite.
Method: Annealing between 950 and 1040°C.
Applications: To lower the hardness of castings with a simultaneous increase in the elongation at fracture and thus also in the tensile strength.

Similar to steel, the hardness of chromium-alloyed white cast iron can be decreased by spheroidisation and coarsening of carbides without the carbide transforming into graphite as in case (a).

A.3.1.5 Normalising

Objective: To obtain a uniform and fine grain size in ferritic/pearlitic microstructures. An inhomogeneous grain size can be expected in castings, rolled steel and forgings due to differing metal working or cooling conditions. Coarsely grained zones may occur in the heated but not worked regions of partially forged (e.g. headed) parts or in the heat-affected zone (HAZ) of welded joints.
Method: Heating to a temperature slightly above Ac_3 (above Ac_{1e} in hypereutectoid steels and cast iron) and cooling in static air induces $\alpha/\gamma/\alpha-$transformation in which the number of grains is significantly increased due to nucleation and growth of new grains.

Applications: Unalloyed and low-alloy, i.e. readily transformable steels, that do not have a tendency to harden in air. The designation is e.g. C35 N. In addition to the standardised definition, controlled cooling regimes are also possible. Cooling is delayed for parts with a higher alloy content and slender dimensions to avoid bainitic or martensitic regions. Forced-air cooling can be used to force near-eutectoid steels to transform to the lower pearlite stage to obtain a pearlitic microstructure with fine lamellae. This is also achieved during patenting of wires and strip by means of an isothermal transformation as the material passes through a hot bath. According to Ac_c in Figure A.2.11, the austenite in grey cast iron contains up to 0.8 % C in solution at a normalising temperature of 870 to 900°C. On cooling to Ac_{1e}, some of the carbon precipitates as graphite and the eutectoid remainder subsequently forms pearlite. This eliminates ferritic regions that had formed due to slower cooling in the mould, which thus increases the strength.

A.3.1.6 Temper annealing of cast iron

Objective: Transformation of cementite into graphite.
Method: Annealing at a temperature between 900 and 950°C. This is often called first step annealing, while ferritising annealing (see A.3.1.4 a) is known as second step annealing.
Applications: (a) Transformation of undesirable white eutectic regions in thin cross-sections or at the edges of grey iron castings. (b) Transformation of malleable cast iron, a white solidified eutectic, into compact temper graphite.

A.3.1.7 Solution annealing

Objective: Dissolution of precipitates to obtain a solid solution microstructure.
Method: Heating to a temperature above the precipitation nose, followed by holding for one or more hours, e.g.
 - austenitic steels: 1000 to 1100°C/water
 - maraging nickel steels: 820 to 840°C/air

Applications: This treatment is used for non-transforming steels and cast iron or precipitation-hardenable steels.

A.3.1.8 Homogenising

Objective: To decrease microsegregation and microstructural anisotropy and to improve the transverse toughness of steel.
Method: Annealing at 1200 to 1300°C for 10 to 20 hours to homogenise the distribution of alloying elements. Owing to the high degree of surface decarburisation and grain coarsening, this treatment is used for ingots and slabs, although the segregation spacing λ after prestraining is smaller and the effect is thus greater.

A.3 Heat treatment

Applications: In general, there is always a certain amount of homogenisation when materials are held at a high rolling or initial forging temperature. Homogenising is used, in particular, for the following:
- dissolution of eutectic carbides that may have formed due to segregation in cold-work and roller-bearing steels. Such carbides have an adverse effect on the toughness and the fatigue strength.
- improvement of the transverse toughness in hot-work tool steels by decreasing the amount of segregation and reducing the length of sulphide particles (Figure A.3.2).

It is important that a low final temperature is used during subsequent hot working to suppress embrittling carbide precipitates on the austenite grain boundaries.

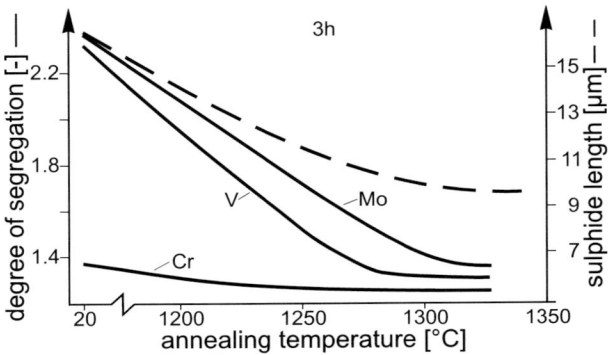

Fig. A.3.2 Reduction of microsegregations by homogenising: The hot-work tool steel X40CrMoV5-1 is used to demonstrate the influence of increasing annealing temperature on the degree of segregation S, measured with a microprobe (see Figure A.2.8, p. 27). The differences in concentration decrease and the banded sulphides are shortened by spheroidising.

A.3.2 Hardening and related processes

A.3.2.1 Hardening

Objective: To produce a considerable increase in hardness by inducing the growth of martensite to a varying depth of the cross-section.

Method: Austenitising and cooling at a rate that produces the desired hardness without excessive distortion and without cracking. If the material is heated slowly in a furnace, it is austenitised at a hardening temperature just above Ac_3 (Ac_{1e} for hypereutectoid steels and cast iron). If the material is heated rapidly, the hardening temperature increases in accordance with the TTA diagram (Figure A.3.3). Materials should be cooled as fast as possible for hardening, but as slowly as possible to avoid distortion and cracking. Depending on the hardenability (see below), water, oil, compressed gas or static air are used — even alternately — as the quenching medium. Step hardening in a hot bath starts with a high cooling rate to avoid pearlite. The temperature is then allowed to equalise over the cross-section. Subsequent air cooling results in a slow transformation to bainite and/or martensite. This process can be used for carbon-rich or high-alloy steels with a low M_s temperature to avoid hardening cracks (Figure A.3.4). The influence of size is important, as shown by the example of a hypoeutectoid steel (Figure A.2.17, p. 39). As the hardening cross-section increases, the $t_{8/5}$ time increases in the surface layer and in the core, and the difference between the two also increases (Figure A.3.5 a). Therefore, a given steel quenched in oil, for example, undergoes full martensitic hardening only up to a certain size. As the thickness of components made of this steel increases, first bainite and then pearlite are formed in the core. The softer core zone increases in size until finally, if the part is very thick, even the surface layer is not hardened. In this case, a more highly alloyed steel must be used with better hardenability, which is represented by the hardness penetration (HP) and depends on the amount of alloying elements dissolved in the

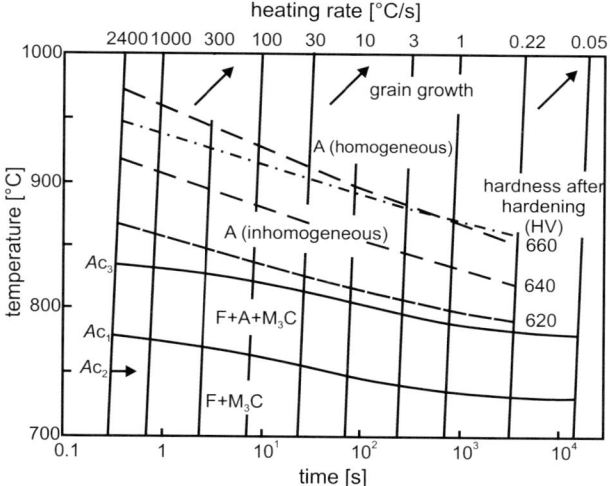

Fig. A.3.3 Continuous time-temperature-austenitising (TTA) diagram, steel 42CrMo4 QT (from J. Orlich, A. Rose, P. Wiest): As the heating rate increases, the transformation is shifted to higher temperatures. Although the carbides have just dissolved in the inhomogenous austenite, their substitutional atoms are not yet uniformly distributed. The dash-dotted line marks the transition to homogeneous austenite.

A.3 Heat treatment

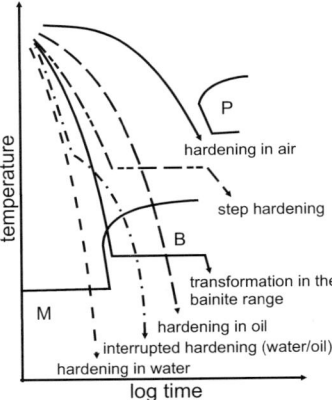

Fig. A.3.4 Types of cooling during hardening: The diagram shows continuous as well as isothermal transformation curves and is thus oversimplified.

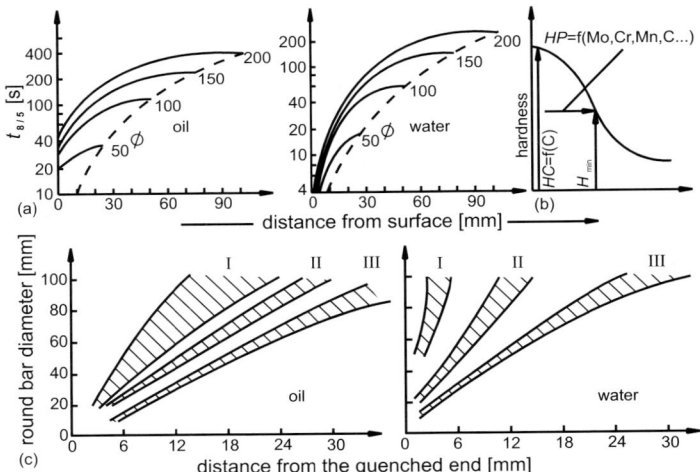

Fig. A.3.5 Cooling time and hardness penetration:(a) The $t_{8/5}$-time increases towards the core of a round bar (from Röchling, Handbuch der Baustähle). (b) This may result in a decrease of the as-quenched hardness due to fractions of bainite and even pearlite. We discern the hardening capacity HC at the surface and the hardness penetration HP defined by the depth below the surface at which a required minimum hardness $H_{\min}$ is met. (c) The same $t_{8/5}$-time and as-quenched hardness of a melt, (i) measured from the water-quenched end face of a Jominy specimen, and (ii) derived for a round bar after hardening in water or oil, lie within the scatter bands for I, II, III = surface, 3/4 radius, core (SAE J406c).

austenite that retard the formation of bainite and pearlite. At the surface, a sufficiently high cooling rate reveals the hardening capacity (HC), which depends mainly on the amount of carbon dissolved in a fully martensitic microstructure. In practice, a specific minimum hardness ($H_{\min}$) is required at a certain depth below the surface. The respective HP is achieved by adjusting the alloy content, i.e. by selecting an appropriate steel (Figure A.3.5 b).

Instead of using complicated microstructural assessments to evaluate the degree of hardening, a much simpler hardness test can be used. The degree of hardening R_H is the ratio between the hardness attained at a distance x from the surface H_x and the maximum attainable hardness $H_{\max}$

$$R_H = H_x/H_{\max} \leq 1 \qquad (A.3.2)$$

In QT steels, the maximum hardness in HRC can be roughly calculated using $H_{\max} = 20 + 60\sqrt{\%C}$ or it can be obtained from the Jominy test (according to EN ISO 642) as $H_{\max} = J_o$. Because the local cooling rate in a component depends on the distance y from the quenched end of the Jominy test piece, the degree of hardening can also be estimated from J_y/J_o.

For example, if we wish to know the maximum possible depth for a particular R_H value or a pure martensitic microstructure in a component, we need to know the local cooling time $t_{8/5}$. This determines whether pearlite is avoided and a hardened microstructure, such as bainite and martensite, is achieved. Figure A.3.5 a shows $t_{8/5}$ curves for different diameters and quenching media.

The $t_{8/5}$ time that is required to avoid the formation of pearlite and bainite in a given steel is shown in a TTT diagram (see Figure A.2.17, p. 39). A comparison of this value with the $t_{8/5}$ values over the radius of the circular cross-section gives the depth of hardening for a full martensitic transformation. However, this estimation is inaccurate because the component and the TTT samples were not produced from the same melt and the type of quenching (water, oil, air) is only loosely defined.

The melt dependency can be taken into account by means of the hardness along a Jominy specimen of the particular melt (Figure A.3.5 c). The measured relationship between the cooling rate and the hardness can then be calculated over the component cross-section. In the meantime, the available data is so extensive that the hardenability can be calculated from the melt analysis and the hardness curve within the workpieces can be derived.

The hardening temperature for hypereutectoid steels and cast iron usually lies at $Ac_{1e} + 50°C$ to cover segregation-related variations in the alloying elements and to complete the austenitising process within a reasonable time (≈ 1 h). Thus along Ac_c (see Figure A.2.9, p. 29 or A.A.2.11 p. 31) the content of dissolved C is 0.7 to 0.8 %, which increases even further as the temperature rises. This leads to an increasing amount of retained austenite in the hardened martensitic microstructure.

Applications: The high carbon content in the surface layer of case-hardenable steels, in surface-hardenable steels, roller-bearing steels, cold-work tool steels and cast iron is utilised in hardening. Therefore, the subsequent tempering temperature is low (usually < 250°C) to retain the high hardness. The designation is e. g. 100Cr6 Q (quenched).

In the other applications, the desired martensitic microstructure is not actually used for 'hardening'. Instead, its high concentration of lattice defects facilitates nucleation and it thus produces a fine dispersion of precipitates on reheating (see Chapt. A.2, p. 48 and Section A.3.2.3).

A.3.2.2 Tempering

Objective: To improve the toughness and dimensional stability of hardened workpieces. Owing to the reciprocal relationship between the changes in ductility and hardness caused by tempering, a compromise must be made for each application case, *e. g.*: tools → *low* tempering temperature → *hard*; components → *high* tempering temperature → *ductile*.

Method: The tempering temperature required for a certain hardness can be read off the tempering curve. This does not take account of the influence of the melt and the degree of hardening. More accurate data can be obtained by stepwise tempering of a Jominy test piece from a single melt. The degree of hardening in the surface layer is obtained from the hardness measured on the hardened component (Figure A.3.6 a). This is used in Figure A.3.6 b to find the suitable tempering temperature.

The core hardness depends on the degree of hardening in the core. If the degree of hardening in the surface layer of a component cannot be measured

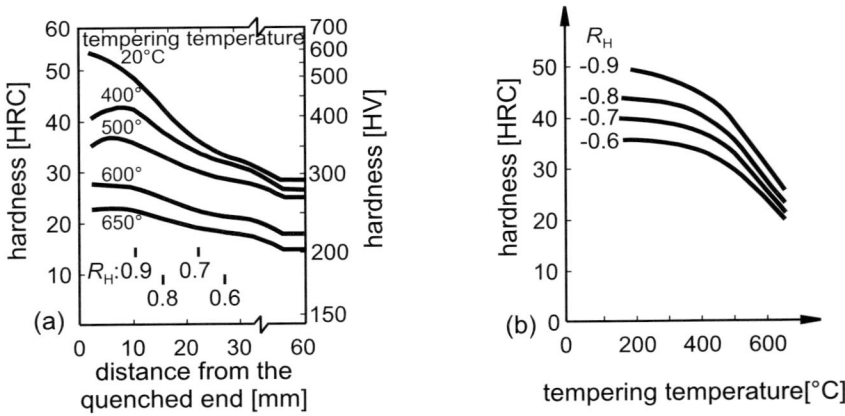

Fig. A.3.6 Influence of the degree of hardening R_H on tempering: Steel 42CrMo4, tempering time 2 h, (a) change in the hardness on stepwise tempering of a Jominy test piece quenched from 850°C (from F. Wever et al.). (b) Tempering diagram obtained from (a) for differing degrees of hardening.

or estimated, it is advisable to use a lower tempering temperature than that given by the tempering curve. A comparison of the measured tempering hardness with the range of nominal values and Figure A.3.6 b indicates whether further tempering is necessary and at which temperature. This ensures that the hardness does not drop below the nominal value, which would necessitate another hardening treatment.

Because the tempering process is diffusion-controlled, not only the temperature must be considered, but also the treatment time. Both can be combined into a single tempering parameter P_T.

$$P_T = T(C + \log t) \qquad (A.3.3)$$

In unalloyed to medium-alloy steels, $C \approx 20$, (for T in K and t in h). A plot of the hardness against P_T is known as a master tempering curve. A complete tempering diagram would have to show the hardness and toughness as a function of R_H and P_T, otherwise regions of embrittlement cannot be identified and thus avoided. However, such detailed information is frequently not available.

Applications: All hardened steels and cast iron, e.g. 100Cr6 T (tempered).

A.3.2.3 QT treatment

Objective: To obtain a good ductility for a given strength.
Method: Hardening followed by tempering in the highest possible temperature range (QT). The treatments discussed under Sections A.3.2.1 and A.3.2.2 are combined into one treatment that improves the quality of the steel by grain refinement and transformation of plate-shaped (pearlite) and banded (bainite) carbides into a dispersion of spheroidal carbides. For example, cementite plates with a diameter of e.g. 10 µm are transformed into nodules with a diameter of 0.1 to 0.5 µm. The necessary tempering temperatures lie between 580°C and 680°C for unalloyed and low-alloy steels. The region of temper embrittlement should be avoided if possible (see Figure A.2.25 a, p. 49). The effects of this treatment are particularly noticeable in the amount of work required in the notched bar impact-bending test (Table A.3.1). However, the yield point increases as well because the soft, carbide-free ferrite grains in the untreated ferritic/pearlitic microstructure disappear.

Just like the band-shaped tempering carbides in martensite, the carbides of lower bainite also spheroidise. If the depth of hardness is insufficient, the resulting upper bainite and pearlite undergo this carbide spheroidisation only at a higher tempering or soft annealing temperature. Therefore the degree of a full QT effect depends on the depth of hardening. As the strength increases, it is all the more important to avoid coarse banded or lamellar carbides in the hardened microstructure. This particularly applies to tempering temperatures < 500°C (e.g. in spring steel) where even fine bands of temper carbides undergo hardly any spheroidisation.

Table A.3.1 Influence of quenching and tempering: Quenching and tempering of specimens increases the proof stress ($R_{p0.2}$) and doubles the toughness (K_U) compared to a normalised ferritic/pearlitic microstructure with practically the same tensile strength and hardness.

material data for steel C60	normalised 850°C/Luft	quenched and tempered 850°C/Öl + 650°C 2 h
tensile strength R_m [MPa]	820	810
proof stress $R_{p0.2}$ [MPa]	480	560
ratio $R_{p0.2}/R_m$	0.59	0.69
elongation A [%]	19	19
reduction of area Z [%]	51	63
notch impact energy K_U [DVM, J]	41	89
hardness [HB]	241	239

Applications: QT steels, creep-resistant structural steels, spring steels and some martensitic stainless steels. A common designation for these materials is e. g. 42CrMo4 QT. Hypereutectoid steels and white cast iron are not usually QT treated. An exception is preliminary hardening and quenching of steels, such as 100Cr6, to limit hardening distortion or to increase the hardening depth for surface hardening. Of the grey cast irons, only alloyed spheroidal graphite cast iron is suitable because graphitisation during tempering is retarded e. g. by molybdenum.

A.3.2.4 Transformation in the bainite range

The objectives of this heat treatment depend on the carbon content of the material. Applications are discussed in subsequent Chapters.
(a) Microalloyed hot strip with $\approx 0.1\,\%$C is cooled so quickly after hot rolling that pearlite does not form. After coiling, the material slowly cools in the coil and passes through the bainite range. The carbides, which are finer in bainite than in pearlite, have a favourable effect on the strength and ductility. (b) If the final rolling temperature of hot strip with $\approx 0.2\,\%$C lies below Ac_3, isothermal holding in the intercritical ferrite/austenite range between Ac_3 and Ac_1 enriches the C in austenite. During further cooling in the bainite range of materials containing $\approx 1.5\,\%$Si, some of the carbon migrates out of the bainite and into the austenite until it is enriched up to $1.5\,\%$C and remains as retained austenite. Its transformation-induced plasticity (see Chapt. A.2.2.4, p. 41) is exploited during cold working or even in a vehicle crash to increase the degree of plastic deformation and the deformation energy (see Chapt. B.2, p. 170, TRIP steel). (c) In a near-eutectoid steel, e.g. 71Si7, transformation in the lower bainite range improves the ductility at a high hardness owing to finely dispersed retained austenite regions that

have a TRIP effect (see Chapt. B.2, p. 200). (d) In a hypereutectoid steel, e.g. 100Cr6, transformation in the lower bainite range reduces distortion and the material is less prone to hardening cracks than for martensitic hardening to a comparable hardness because the transformation in the workpieces takes place with smaller temperature gradients (see Chapt. B.2, p. 204). (e) If carbon-rich steels are alloyed with silicon, austenitised above Ac_c and isothermally transformed in the lower bainite range, the retained austenite (RA) can be enriched by up to 2 % C, and fractions of > 10 vol.-% RA can lead to a surprisingly high TRIP-assisted elongation at fracture (A) in the tensile test (Tab. A.3.2). The hardness is almost 700 HV and the compressive strength > 3000 MPa. At $T_B = 200°C$, the bainite plates grow to a thickness of only 20 to 40 nm. The long transformation time is shortened by alloying with Al and Co. (f) Cast iron with spheroidal graphite contains about 2.5 % Si, which suppresses the precipitation of carbides from bainitic ferrite during isothermal transformation in the bainite range. At the same time, the austenite is enriched to 1.5 - 2.5 % C so that 20 to 40 % remain as retained austenite. Its transformation-induced plasticity under loading produces an even more pronounced TRIP effect than in case (e). If the isothermal transformation temperature of cast iron is reduced from 400 to 250°C, the required holding time increases, the yield strength increases, and the amount of retained austenite and ductility decrease. If the holding time at e.g. 350°C is insufficient, a brittle martensite is produced on cooling to room temperature. If the holding time is too long, carbide precipitates lead to embrittlement (see Chapt. B.2, p. 208). This material is known as austempered ductile iron (ADI). This treatment is also used for thin-walled cast iron with lamellar graphite (austempered grey iron, AGI).

Table A.3.2 Properties of a hardened steel ($\approx$ 80MnCoSiAlCrMo8-6-6-1-1) after transformation in the bainite range by holding at T_B (from H.K.D.H. Bhadeshia)

T_B [°C]	RA [%]	$R_{p0.2}$ [MPa]	R_m [MPa]	A [%]
300	21	1400	1930	9.4
200	17	1410	2260	7.6

A.3.3 Surface layer treatment/Coating

Objective: Intentional tailoring of microstructural differences between the surface layer and the core of workpieces to provide benefits with respect to
  - production, e.g. by (a) reducing distortion and cracking susceptiblity, (b) saving alloying and energy costs, (c) facilitating integration into the manufacturing chain, (d) saving time.

A.3 Heat treatment

- utilisation, by (a) improving the tribological properties (wear resistance) due to a greater surface hardness, (b) improving the chemical properties (corrosion resistance) by creating a higher proportion of solid solution or inward diffusion of chromium or aluminium, (c) increasing the fatigue strength by means of residual compressive stresses in the surface layer.

Method: The surface layer is remelted or heat-treated in the solid state or modified without heating:
  - remelting of the surface layer and remelt alloying with a laser or electric arc to refine the solidified structure or to modify the chemical composition
  - heat treatments that only heat the surface layer (e.g. flame or induction hardening, see p. 217)
  - thermochemical treatments in which the material is fully heated and the chemical composition of the surface layer is changed by diffusion (e.g. case-hardening, see p. 234 and nitriding, see p. 224).
  - physical treatment of the surface layer in which the chemical composition of the surface layer is modified without external heating (e.g. ion implantation).
  - mechanical treatment of the surface layer in which the surface layer is work-hardened without external heating to induce residual compressive stresses (e.g. shot peening, see p. 199 and surface rolling).

In addition to treatments of the surface layer, there are a variety of coating processes, with and without heating of the workpiece (substrate), which have similar objectives. These are divided into:
  - coating processes that have a thermal effect on the substrate (bake-hardening $\approx$ 170°C, see p. 131, hot-dip galvanising $\approx$ 450°C, see p. 128, 137, PVD hard coating $\approx$ 500°C, see p. 286, enamelling $\approx$ 800°C, see p. 128, 137, CVD hard coating $\approx$ 1000°C, see p. 283)
  - coating processes that do not require significant heating (chemical and electrochemical deposition, see p. 137, 175). The temperature increase generally remains low, even with a powder spray-coating system. In thermal spraying, the powder grains are heated or melted e.g. by a flame or a plasma, and are then accelerated onto the surface of the workpiece. In the cold-spraying process, the extremely high impact velocity of the powder grains consolidates the sprayed layer.

Applications: There is not enough space here to describe the wide variety of processes and their applications. They are thus discussed in conjunction with the corresponding materials, e.g. carburising of case-hardening steels, shot-peening of spring steels and surface remelting of cast iron.

A.3.4 Side-effects

The attendant effects of a heat treatment can be classified as follows:
  - thermal side-effects (distortion, cracking, residual stress)
  - thermochemical side-effects (decarburisation, scale growth, salt corrosion).

A.3.4.1 Thermal side-effects

Distortion and cracking are caused by changes in volume due to thermal expansion and/or a microstructural transformation (Figure A.3.7).

With increasing temperature gradients, they lead to internal stresses in the workpiece, particularly on quenching. The relief of these stresses by local yielding leads to distortion. As a result, there is usually some residual stress after a heat treatment. If the internal stresses exceed the rupture limit, the material cracks or fractures. Distortion is made up of two components: size changes $\varepsilon = \Delta l/l$ resulting from a temperature gradient and/or a phase transformation and shape changes that are associated with asymmetric plastic flow and thus alter curvatures, angles and symmetry.

Size changes can be as much as a tenth of a percent. This is particularly significant for hardening of carbon-rich or carburised steels that have only a slight machining allowance, if any, to save costs. The changes depend on the ratio between the temperature gradient and the hardenability and must therefore be considered on a case-by-case basis. This is further complicated by the fact that hardening-induced size changes depend on the workpiece design, microstructural anisotropy and time. The latter also affects the dimensional stability. Tempering processes at room temperature or longer periods at slightly

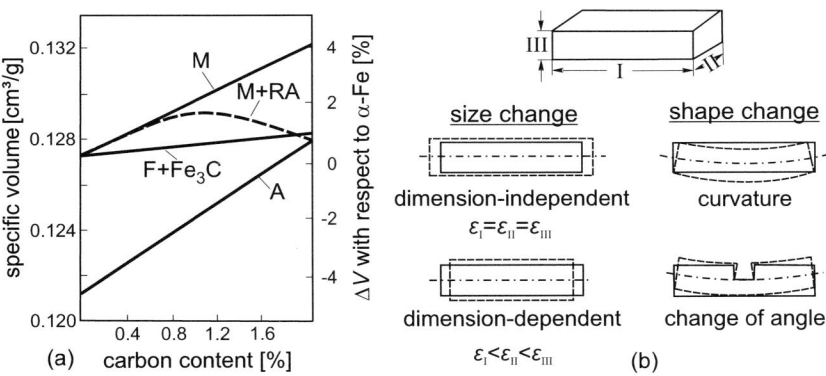

Fig. A.3.7 Distortion on hardening: (a) According to B. Lement, carbon dissolved in martensite M or austenite A (RA = retained austenite) increases the volume to a greater degree than that deposited as cementite from ferrite F. Martensite formation is thus associated with a positive size change. (b) On slow cooling of an isotropic air-hardening steel, the size change is independent of the dimensions. On rapid cooling (water quench), there are additional thermal stresses. They start with tensile stresses acting on the surface that promote a spherical shape as long as the part is not too slender: long axes shrink, short ones grow. The size change becomes dependent on the dimensions. A change in shape can be seen from the loss of symmetry (solid line before hardening, dashed after).

elevated operating temperatures can induce shrinkage of the martensite or growth due to decomposition of retained austenite.

Slender workpieces, in particular, are affected by changes in shape. The resulting deviations in shape may be considerably greater than those caused by size changes and may even be as much as several percent. A change in shape may be caused by relief of asymmetrical residual stresses in the material, one-sided carburisation or decarburisation, uneven heating or cooling, creep due to the dead load or because the material itself was taken excentrically from a rolled or forged bar that contained segregations.

Through-hardening usually leaves residual tensile stresses in the surface layer because martensite starts to form here while the hot core is still able to flow. The subsequent transformation in the core, which is associated with an increase in volume, produces tensile stresses in the hard surface layer. High residual tensile stresses in the surface layer make it susceptible to cracking during further processing or when the workpiece is in service. A low hardening depth (shell hardening) favours pearlite formation in the core over martensite formation in the surface layer thus leading to residual compressive stresses in the surface layer. Such stresses are usually generated in the surface layer by surface hardening and case-hardening as well as by nitriding due to absorbed nitrogen. They lower the mean stress for bending and torsional vibrations and thus increase the fatigue strength.

For hardening of hypereutectoid steels and cast iron, the high amount of dissolved carbon gives rise not only to a large volumetric change but also to a high hardness. The resulting high internal stresses thus act on a brittle matrix. The cracking susceptibility increases with increasing temperature gradients in the workpiece (hardening in air, oil, or water; increasing cross-section). Cast iron undergoes additional embrittlement due to the comparatively coarse eutectic precipitates. In white cast iron, the carbides form a brittle scaffold (Figure A.2.13). In grey cast iron, the graphite acts as a notch for the internal tensile stresses whereby the notch acuity increases from spheroidal to vermicular/temper to flake graphite. Hardening-induced stress cracking can be avoided by an isothermal transformation (see ADI, p. 208) or by surface hardening.

A.3.4.2 Thermochemical side-effects

The ingress of oxygen or oxidising gases during heat treatment of steels can lead to oxidation of iron as well as carbon. The resulting layer of scale and decarburisation (determined according to EN ISO 3887) usually occur together and influence each other.

At temperatures of up to 570°C, scale consists of two layers: magnetite Fe_3O_4 (next to the substrate) and haematite Fe_2O_3 (outer layer). The reaction rate is slow. Above 570°C a third layer, wustite FeO, grows next to the substrate. The metal deficit in wustite allows rapid diffusion of iron via cation vacancies, resulting in rapid growth of this layer, which is in contact with the metal surface and which must therefore be taken into account in a heat treatment strategy. At a given temperature, the layer thickness s increases with time t as follows

$$s = c \cdot t^q \; ; \; 1/3 < q < 1 \tag{A.3.4}$$

The scale growth rate decreases ($q \to 1/3$) for mass transport through the thickening scale layer. If the oxidising gas is able to pass through pores and cracks in the scale and reach the steel surface, the growth rate is faster, often linear ($q \to 1$).

Above $\approx 700°C$ the thermodynamic equilibrium favours the formation of CO over FeO. Nevertheless, iron is still oxidised because the transport of carbon to the surface is not fast enough. In addition, decarburisation via loss of the CO formed on the steel surface is also slowed down by the layer of scale. Pores and cracks in the scale thus accelerate decarburisation.

Atmospheres containing water vapour, e. g. in gas-heated furnaces, increase the scale porosity, facilitate mass transport and promote the oxidation of iron and carbon. The iron is oxidised out of the steel surface by H_2O. The resulting hydrogen recombines with oxygen on the pore walls in the scale layer to regenerate H_2O. The ever active H_2/H_2O mixture thus induces rapid transport of oxygen within the pores. The water-gas shift reaction facilitates a rapid transfer of carbon e. g. in carburisation during case-hardening. For the decarburisation case, it proceeds in the other direction.

Alloying elements influence oxidation. Silicon reacts with iron to give fayalite, an iron silicate, and chromium produces a spinel. The oxidation of Cr and Si have precedence over the reaction with iron. Therefore, fayalite and/or spinel regions are formed on the surface of the metal. They gradually grow into a coherent layer as the alloy content increases. This oxide skin is denser and less permeable. It slows down mass transfer between the steel surface and wustite.

Chromium and silicon affect decarburisation differently. Chromium decreases the activity and the diffusion velocity of carbon. It slows down decarburisation if the replenishment of carbon becomes the rate-determining step. Although silicon also lowers the mobility of carbon, the activity increase is greater so that decarburisation increases with the silicon content (Figure A.3.8)).

Scale adversely affects the roughness and dimensional accuracy. It is also generally unwanted in practice. Decarburisation, particularly in more highly carburised steels, produces an undesirable soft skin after hardening. Internal and external oxidation as well as decarburisation can be avoided by carrying out heat treatments under an inert gas or in a vacuum. The affected regions can also be removed by reworking the machining allowance. Grey cast iron is annealed in an inert gas to prevent oxidation of the coherent graphite flakes. In contrast, the low-carbon microstructure of white heart malleable cast iron is produced by oxidative annealing.

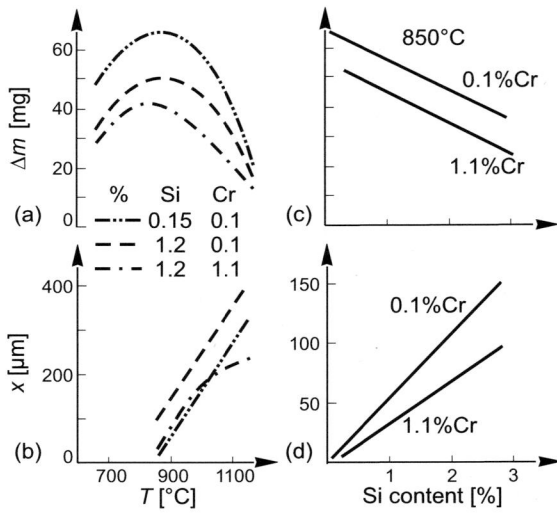

Fig. A.3.8 Growth of scale Δm and decarburisation depth x during heat treatment (from R. König): Steels with 0.55 % C, 0.9 % Mn and alloyed with Cr + Si were annealed for 30 min in damp nitrogen (25 g H_2O/kg N_2). (a) Scale growth and decarburisation influence each other. Iron oxidation has precedence at 700°C. The displaced carbon is enriched to as much as 1 % under the steel surface. Oxidation of carbon has precedence at 1150°C. The replenishment rate of carbon from the core is high. The CO/CO_2 gases that are diffusing away through cracks in the scale prevent the furnace atmosphere penetrating directly to the steel surface. The scale growth rate is lower than that at 700°C. The maximum growth rate lies between these two temperatures, just above Ac_3. This is caused by a decrease in the carbon diffusion velocity by two orders of magnitude after the ferrite/austenite transformation. The carbon replenishment rate and blanketing by the CO/CO_2 gases is lower than at 1150°C. (b) In contrast to the scale growth rate, the decarburisation rate continuously increases with the temperature. (c,d) Silicon lowers the scale growth rate and accelerates decarburisation. In contrast, chromium has an inhibiting effect on both rates.

Plastics mould of stainless tool steel used for the production of closures for beverage bottles
(Deutsche Edelstahlwerke GmbH, Witten, Germany)

A.4 Properties

The properties of ferrous materials ultimately depend on their microstructure, which is in turn dependent on the chemical composition and the manufacturing sequence. The required service properties of steel and cast iron as a structural material are determined by the applied loads, which can be of a mechanical, tribological and/or chemical nature. If they are to be used as functional materials, their physical properties are important.

Manufacturing properties, such as suitability for welding, deep drawing and cutting, are discussed in the chapters on the individual groups of materials.

A.4.1 Mechanical properties

A.4.1.1 Loading

(a) Multiaxiality

A mechanical load results from the action of forces on a body. In general, a force F applied to an area A of a component is expressed as the stress. If the force is applied parallel to a reference plane, it produces a shear stress $\tau = F/A$, and if it is applied perpendicularly, it produces a normal stress $\sigma = F/A$. There are four basic types of applied force - tension, compression, bending and torsion - and these may be active within in a component more than once and in combination. The forces acting from different directions are ordered by referencing them to three mutually perpendicular planes and resolving them into normal and shear stresses. If these reference planes are rotated, one position can always be found in which the shear stresses are zero and octahedral planes are spanned at an angle of 45°. In this principal stress system (shown in Figure A.4.1), the equivalent stress σ_{eq} and the hydrostatic stress σ_h are derived as characteristic parameters of the shear stress and normal stress states under multiaxial loading. These are invariant stresses, which means that they act in all directions i.e. with respect to all crystal planes.

(b) Distribution

The stresses in unnotched tension and compression specimens are distributed homogeneously over the cross-section, whereas they are inhomogeneously distributed in bending and torsional specimens and increase towards the surface of the object. Further causes of inhomogeneous stress distribution are notches due to changes in cross-section, e.g. holes, grooves, and steps as well as cracks, which are extremely sharp notches. Figure A.4.2 gives an example of the stresses in a notch. As the notch acuity increases ($\rho \to 0$), the notch becomes a crack and the situation can be represented by fracture mechanics.

A.4 Properties

Its stress intensity factor $K_{\mathrm{I}} = \sigma\sqrt{\pi a}$ describes the intensity of the elastic stress field in front of a crack of length a and is used instead of the stress concentration factor α_{k}. A plastic zone develops in front of the crack tip as K_{I} increases. A plane strain condition develops in the core with increasing thickness (B) of the disk. Notches and cracks not only lead to an inhomogeneous distribution of the stress, they also increase the multiaxiality $\sigma_{\mathrm{h}}/\sigma_{\mathrm{eq}}$ of the applied load.

(c) Energy and chronological progression

The longer the distance over which the forces act, the higher is the stored elastic energy, which is then available for crack propagation should the material fail. This explains why crack arrest is more likely to occur in a water pipe (hard system) than in a gas pipe (soft system) where the loss of pressure is less rapid. Some of the fracture energy is transformed into heat, the dissipation of

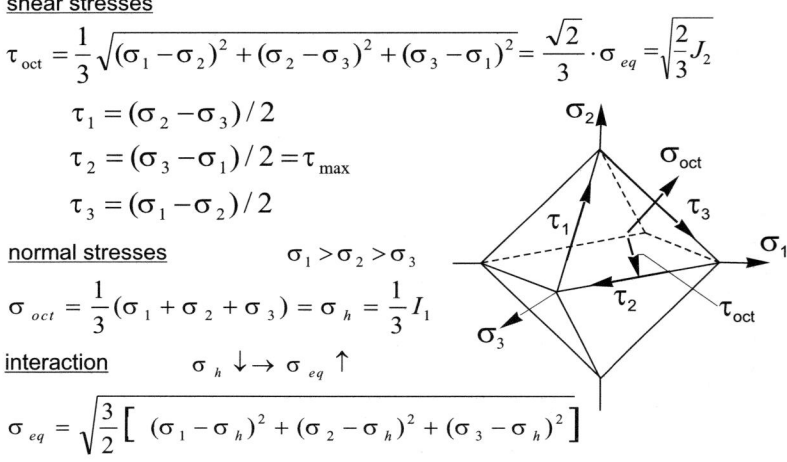

shear stresses

$$\tau_{oct} = \frac{1}{3}\sqrt{(\sigma_1-\sigma_2)^2+(\sigma_2-\sigma_3)^2+(\sigma_3-\sigma_1)^2} = \frac{\sqrt{2}}{3}\cdot\sigma_{eq} = \sqrt{\frac{2}{3}J_2}$$

$$\tau_1 = (\sigma_2-\sigma_3)/2$$

$$\tau_2 = (\sigma_3-\sigma_1)/2 = \tau_{max}$$

$$\tau_3 = (\sigma_1-\sigma_2)/2$$

normal stresses $\quad \sigma_1 > \sigma_2 > \sigma_3$

$$\sigma_{oct} = \frac{1}{3}(\sigma_1+\sigma_2+\sigma_3) = \sigma_h = \frac{1}{3}I_1$$

interaction $\quad \sigma_h \downarrow \to \sigma_{eq} \uparrow$

$$\sigma_{eq} = \sqrt{\frac{3}{2}\left[(\sigma_1-\sigma_h)^2+(\sigma_2-\sigma_h)^2+(\sigma_3-\sigma_h)^2\right]}$$

Fig. A.4.1 Multiaxial stress state: The principle shear stresses τ_1 to τ_3 cause shearing and slip in the crystal lattice without changing the volume (Figure A.2.19, p. 42). They can be combined to an octahedral shear stress τ_{oct} that is proportional to the equivalent stress σ_{eq} and the square root of the second invariant of the stress deviator J_2. Positive principal normal stresses σ_1 to σ_3 change the volume thus dilating the lattice until atomic bonds are broken. These stresses can be combined to an octahedral normal stress σ_{oct} that is proportional to the hydrostatic stress σ_{h} and the first invariant of the stress tensor I_1. σ_{eq} promotes slip and thus plastic deformation, whereas σ_{h} promotes cleavage and thus brittle fracture. Their interaction leads us to expect that ferrous materials become more ductile as the hydrostatic stress decreases. This is exploited in metal working under compression ($\sigma_{\mathrm{h}} < 0$).

which is suppressed as the deformation velocity $\dot{\varepsilon}$ increases. The temperature in the fracture zone is thus raised (e. g. by more than 100°C during breakage of a tough impact bending specimen).

A unidirectional load can be applied slowly (quasistatic), suddenly (dynamic) or cyclically. In the latter case, the stress amplitude σ_a alternates about a mean stress σ_m or follows periodic or completely independent stress cycles. They can be multiaxial, out-of-phase and inhomogeneously distributed. At high frequencies, microplastic deformation may lead to heating as well.

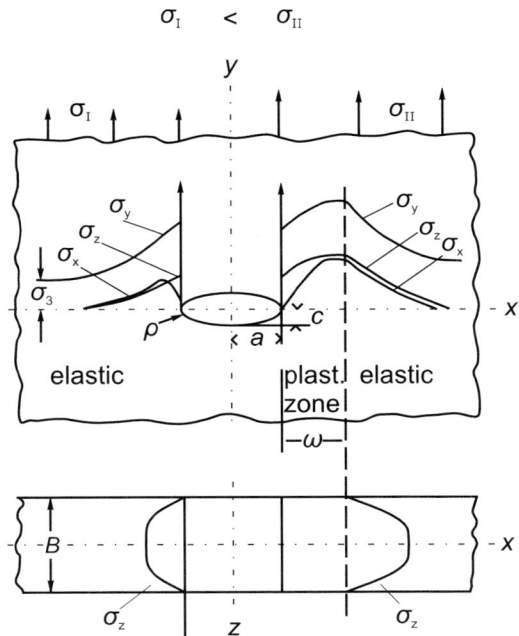

Fig. A.4.2 Notch stresses: At a nominal stress of $\sigma = \sigma_I$ (left half of the diagram), an elastic stress concentration is built up in the notch root of radius $\rho = c^2/a$, and can be described by the stress concentration factor $\alpha_k = \sigma_{y\,max}/\sigma$. The core of the disk is in a plane-strain condition and this changes to a plane-stress condition by the loss of σ_z at the surface. In thin disks ($B \to 0$; $\sigma_z \to 0$) there is a pure plane-stress condition. In this case, $\alpha_k = 1 + 2a/c$. A circular hole ($a = c$) thus gives $\alpha_k = 3$ and $\sigma_{x\,max} = \sigma$. For an oval hole with e. g. $a/c = 3$, the elastic surplus stress is already $\sigma_y = 7\,\sigma$. If the nominal stress is increased to $\sigma = \sigma_{II}$ (right half of the diagram), the yield strength is exceeded locally and a plastic zone of depth ω develops in front of the notch root (semi-plastic state). This may include the entire ligament (fully plastic state) if the disk has a limited size.

A.4 Properties

(d) Delimitation

Calculations of the type, magnitude and distribution of the stresses acting in a body are initially based on the assumption of a macroscopic continuum. The second step considers how the stresses are microscopically resolved within the discontinuum existing in ferrous materials at the atomic level (crystallinity) and at the microstructural level (phases and lattice defects). The latter factor, in particular, highlights an important difference between cast iron and steel, namely, cast iron contains eutectic phases (graphite, carbide) that have grown from the melt to produce coarser particles than in the remaining matrix, which is similar to steel (Chapt. A.2, p. 53). The 'coarse' graphite or carbide particles represent a sudden change in stiffness owing to their larger modulus of elasticity (Young's modulus), and they thus exert an internal notch effect. It therefore appears to be expedient to discuss the mechanical properties of ferrous materials using eutectic-free steels first and then take a look at cast iron, which can be simplified to a steel-like microstructure with internal notches.

The following discussions on deformation, fracture and strength focus on processes under quasistatic and dynamic loading at temperatures $< 200°C$. Additional information is given elsewhere: e.g. cyclic loading is discussed under spring steels and roller bearing steels, creep rupture loading under creep-resistant steels, stress corrosion cracking under wet corrosion.

A.4.1.2 Behaviour of steel

The initial effects of an applied mechanical load are elastic deformations. For stresses, they are associated with lattice expansion and a change in volume. Shear stresses distort the crystal lattice without changing the volume. If the load exceeds the elastically withstandable limit, from the microscopic perspective, the steel can only react in two possible ways: cleavage or slip. Cleavage is associated with breaking overstrained bonds between atoms in the lattice. This takes place on bcc cube planes, which are populated with fewer atoms. Slip is associated with the displacement of atoms with respect to one another by movement of dislocations in well-populated lattice planes (Figures A.1.4, p. 8 and A.2.4, p. 23).

(a) Cleavage or slip?

Even if overloading is slight, cleavage generates crack nuclei that require little energy and propagate at high speed. Such brittle behaviour is dangerous. In contrast, slip due to the movement of dislocations in a large number of slip planes requires a large amount of energy. The component is plastically deformed and does not fracture immediately, but only after it has been subjected to a high degree of overloading and deformation. This ductile behaviour provides safety. Figure A.4.3 illustrates this situation using the example of a ferrite grain subjected to a uniformly increasing tensile stress. The question arises as to which conditions are necessary to initiate slip in the neighbouring grain and thus plastic deformation in order to avoid cleavage. The answer lies in the loading conditions and in the type of material.

Loading conditions that prevent dislocations from moving by lowering their thermal activation promote cleavage. Such conditions include low temperatures or a high deformation velocity. A high degree of multiaxiality σ_h/σ_{eq}, e. g. in front of notches and cracks, expands the lattice until it cleaves and simultaneously decreases shear and the tendency to slip. Thus for a given material condition, both brittle and ductile failure can be found simply by changing the method of testing (Figure A.4.4). The tendency to a cleavage fracture increases as σ_h/σ_{eq} and $\dot{\varepsilon}$ increase and as the temperature decreases.

(b) Transition temperature

The transition from the lower shelf of toughness or ductility to the upper shelf can be easily seen by plotting them against the testing temperature.

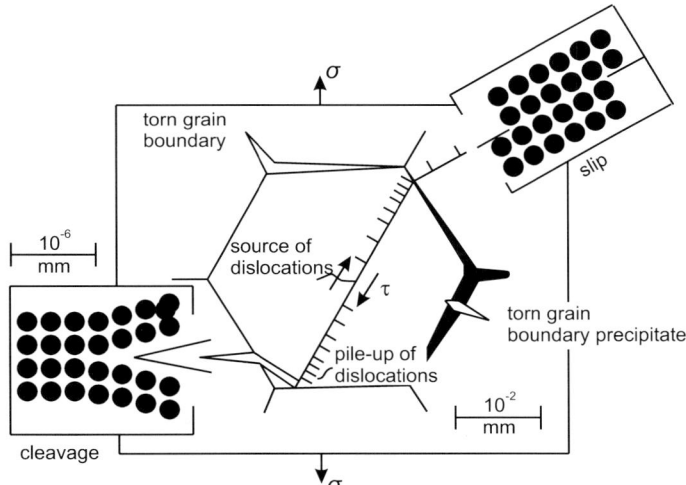

Fig. A.4.3 Cleavage or slip (schematic): As the nominal stress σ increases, a source in a favourably oriented grain (greatest shear stress τ at an angle of 45° to σ is oriented in the direction of slip, see Figure A.1.4, p. 8) starts to generate dislocations that pile up in front of the nearest obstacle, here the grain boundary. The stress is concentrated at the head of this pile-up due to inserted half-planes (indicated by the lines drawn perpendicularly to the slip plane). This may lead to an incipient crack by cleavage of atomic bonds (see the enlargement) or to slip in the neighbouring grain. Cracks can also be initiated by the fracture of precipitates or by the failure of boundary surfaces.

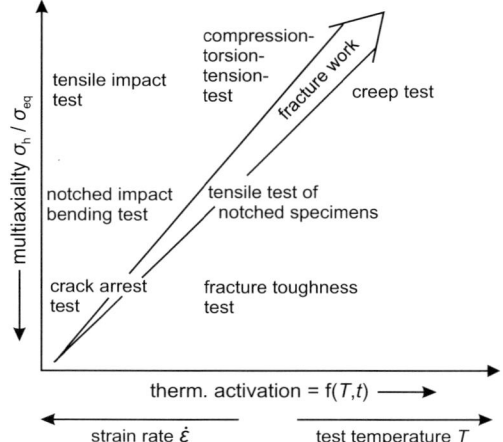

Fig. A.4.4 Influence of test conditions on the toughness (fracture work): In this schematic classification of test methods, the degree of multiaxiality increases from compression to tension or bending tests with unnotched, notched or precracked specimens. In impact tests or those with fast-propagating cracks, the time t for thermal activation is short, whereas in creep tests it is long.

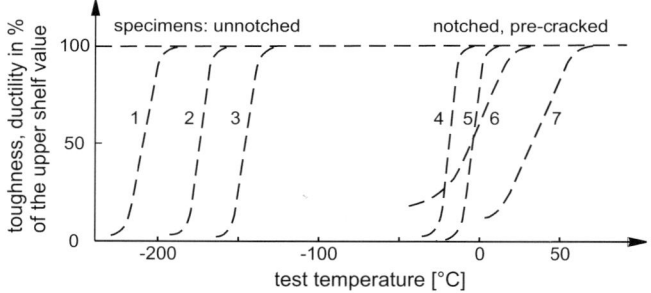

Fig. A.4.5 Transition between the upper/lower shelf regions: A plot of the ductility or toughness relative to the upper shelf against the test temperature for a steel containing $\sim 0.2\,\%\,C$ shows the following approximate transition curves for (1) rotation until fracture in the torsion test, (2, 3) elongation at fracture in the tensile test under moderately fast or impact loading, (4, 5) impact bending energy for specimens with round or ISO-V notches, respectively, (6, 7) fracture toughness under fast loading and impact loading, respectively. In the notched-bar impact bending test, the transition lies in the climatic temperature range so that it can be used to determine the transition temperature T_T of notched components.

As the notch acuity and deformation velocity increase, the transition temperature T_T is shifted upwards and thus into the climatic temperature range (Figure A.4.5).

Between the iron atoms in the crystal lattice there are directional *atomic bonds* (primarily involving the d electrons) and non-directional (primarily involving the s electrons). The former tend to lead to brittle failure if overloading occurs, whereas the latter enhance ductile metallic behaviour. Non-directional bonding allows electrons to move freely. These free electrons not only govern the mechanically ductile character of metals, but they also bestow good electrical and thermal conductivity. Alloying metals such as Ni and Co, which are to the right of iron in the periodic table, increase the concentration of free electrons in the solid solution and lower the T_T. Therefore, e. g. martensitic case-hardening steels with 3.5 % Ni or stainless austenitic steels with 10 % Ni are particularly tough. Elements such as Mn and Cr, which are to the left of iron, have the opposite effect on the solid solution. In contrast to C, which has two p electrons, N has a third, unpaired p electron and this increases the concentration of free electrons in the solid solution. This is exploited e. g. in stainless steels.

The *microstructure* also influences the position of T_T. This starts with the ferrite grain size . The shorter the pile-up length of dislocations in front of an obstacle is, the faster the retroactive effect of the stress concentration in the pile-up slows down the source producing the dislocations (Figure A.4.3). This means that as the grain size decreases, the local stress at the head of the pile-up also decreases for a given external stress. Fine-grained steels use this effect to lower the T_T.

The fracture of *brittle phases* plays an important role in initiating cleavage. Theoretical studies have shown that the resulting crack nuclei generally trigger the cleavage of ferrite. Figure A.4.3 illustrates the initiation of cleavage by the fracture of a grain boundary carbide. The crack nucleus becomes increasingly shorter as the thickness of the hard phase decreases. This reduces the probability of its propagation within the ferrite. Increasing proportions of pearlite as well as an increasing carbide plate thickness raise the transition temperature. Silicon is dissolved in the ferrite and increases T_T. A higher strength also generally leads to a higher T_T. Exceptions are hardening by grain refinement and solid solution strengthening by nickel.

(c) Upper shelf

In the upper shelf region, slip generates internal voids. They are produced by fracture or detachment of precipitates or non-metallic inclusions or by the interaction between dislocations in the iron lattice. If there is further plastic deformation, the void diameter grows exponentially with the degree of multiaxiality σ_h/σ_v. The remaining metal bridges are gradually sheared off and individual voids coalesce to produce a crack that propagates and ultimately initiates shear fracture. The amount of local fracture strain required for this

86 A.4 Properties

depends on the size and quantity of the hard phases and on the deformability of the metal bridges. Figure A.4.6 illustrates a shear fracture sequence. It also shows the difference between loads applied longitudinally and perpendicularly to the rolling direction of the steel (s. Figure A.4.11, p. 93)

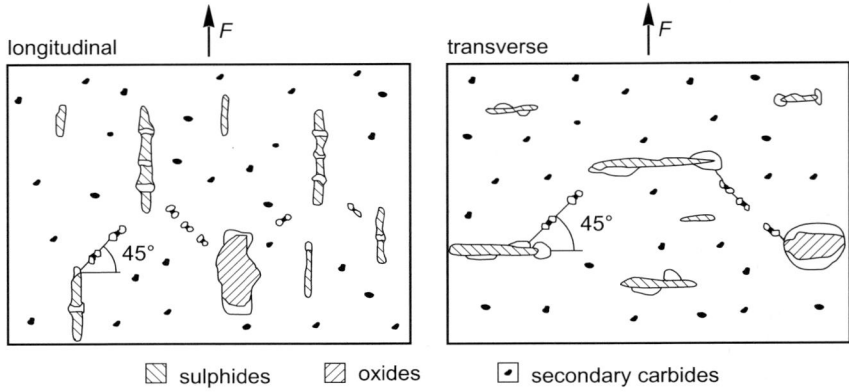

Fig. A.4.6 Shear fracture: In a tensile test of a high-strength steel, matrix detachment and the formation of voids starts at coarse, non-metallic inclusions or carbides. The latter are formed during deoxidation, e. g. Ti(C,N), or by segregation, e. g. M_7C_3 in steel X40CrMoV5-1. In longitudinal specimens, the sulphides tear in half and then in half again. As deformation continues, voids are formed around the finer secondary carbides at an angle of 45° in the direction of slip. These voids coalesce to produce internal incipient cracks that grow steadily until the material fractures. In transverse specimens with the same degree of strain, longer elliptical voids form perpendicularly to the force F. These voids exert a larger notch effect (Figure A.4.2), which reduces necking.

(d) Fracture path

Cleavage and shear fracture pass through the grains, i.e. they are transgranular. However, if the grain boundaries represent the path of least resistance, fracturing occurs along an intergranular route (Figure A.4.7). This can be facilitated by cleavage of atomic bonds or it can be associated with slip and void formation, which is, however, limited to the flanks of the grain boundaries. Intercrystalline fracture thus generally requires little energy.

This microscopic fracture path must be differentiated from the macroscopic one. If the fracture surface is perpendicular to the maximum tensile stress, it is a normal stress fracture. However, if the fracture surface lies at 45° to the normal stress, i. e. in the direction of shear stress, it is a shear stress fracture (Figure A.4.8).

A.4.1 Mechanical properties

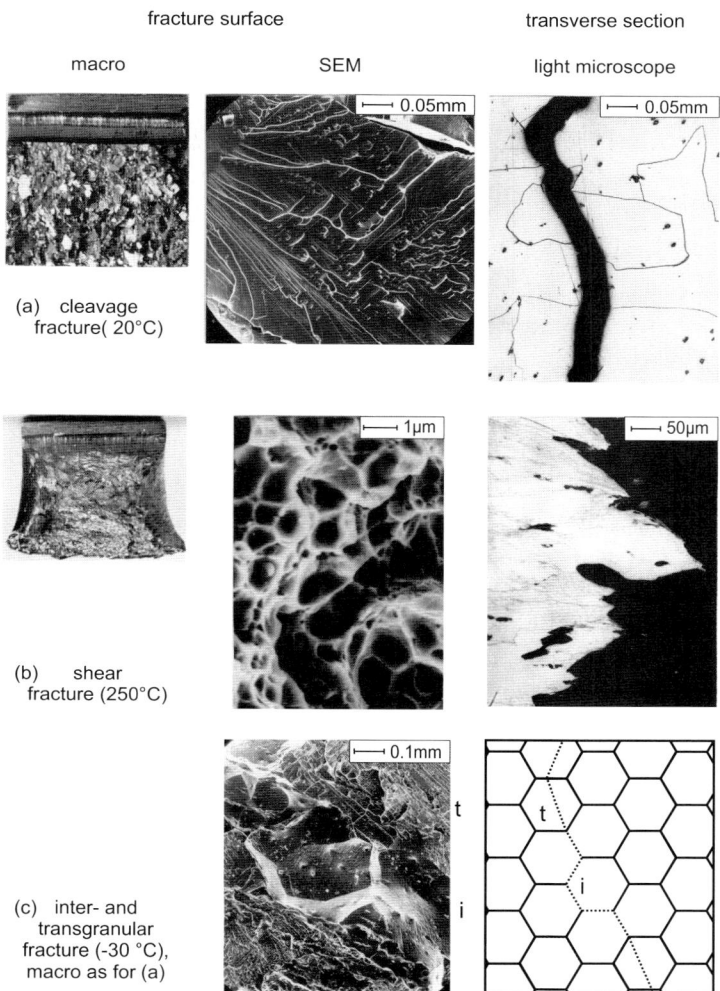

Fig. A.4.7 Crack path:: The specimen shown here is a ferritic chromium steel X6CrTi12 containing coarse grains with ≈ 1 mm diameter subjected to a notched impact test ($T_T \approx 140$°C). (a) The cleavage fracture below T_T is transgranular and there is practically no deformation along the cube plane (Figure A.1.4, p. 8), which has a different orientation in every grain. (b) The shear fracture above T_T is accompanied by lateral deformation that can be seen on the macroscopic scale. The microscopic image reveals honeycombs or dimples that were generated by shearing off of the bridges between voids. (c) At low temperatures, partial intergranular fractures (i) are formed that have a brittleness similar to partial transgranular cleavage (t).

A.4 Properties

Frequently, a normal stress fracture develops under conditions of plane strain in the core, whereas the surface zone under plane stress undergoes a shear stress fracture thus generating a shear lip (mixed-mode fracture).

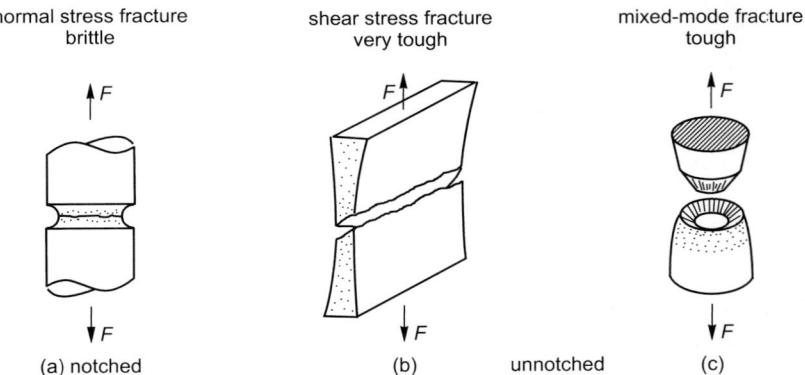

Fig. A.4.8 Toughness of forced ruptures: For a given steel and material cross-section, the deformed volume (dotted) and thus the fracture work required in a tensile test depend on the geometry of the specimen. (a) A crack appears in the notch root that pushes only a small plastic zone through the specimen. Plastic deformation is limited to the flanks of the fracture. (b) A thin flat specimen is essentially in a plane stress state. The lower multiaxiality favours plastic deformation of a large volume of material. (c) A plane strain state develops in the core of a round specimen. A normal stress fracture occurs in the core owing to the higher degree of multiaxiality. This changes to a plane stress state towards the outer surface and produces a shear lip characteristic of a shear stress fracture.

(e) Fracture sequence

Fracturing occurs in three consecutive stages:
- cracking (if not already pre-cracked during manufacturing)
- stable crack propagation (slow, stops when the load is removed)
- unstable crack propagation (fast, difficult to stop in soft systems)

If all three occur directly one after the other as a result of a single load application, this is known as an instant fracture or forced rupture. If, on the other hand, a crack propagates steadily over a long period, it is known as a delayed fracture. This occurs when mechanical loading is assisted chemically or thermally or is applied cyclically.

A.4.1 Mechanical properties

(f) Instant fracture

Breakage of a workpiece by a forced rupture requires energy. This fracture work is a measure of the material's toughness. It depends on the plastically deformed volume. The fracture strain is known as the ductility (Figure A.4.8).

The toughness of a forced rupture essentially depends on the amount of plastic deformation that has taken place until crack propagation becomes unstable. It manifests itself as necking, bending, buckling or twisting of the object's surface. Notches localise the deformation. The deformed volume is reduced and consequently the associated fracture work decreases. Unstable crack propagation requires comparatively little energy. In front of the crack tip there is only a small plastic zone pushed through the material. It is larger in regions with plane stress than in those with plane strain. And this manifests itself as a shear lip on the edge of the fracture. After failure, the toughness of a forced rupture is evaluated in terms of macroscopic factors, i. e. based on the deformation of the outer surfaces and the proportion of shear lips on the fracture surface. Microscopic examination of the fracture surfaces does not allow any clear conclusions to be drawn regarding the total fracture work. Thus, in unnotched components, plastic deformation may occur before a brittle cleavage fracture. In notched components, slip may be limited to the fracture flanks, and shear fracture requires little energy overall. On the macroscopic scale, a brittle fracture resembles a normal stress fracture. On the microscopic scale, it is a cleavage fracture or a low-energy shear fracture, which may also be intergranular. A ductile fracture is characterised macroscopically by shear stress fracture or mixed-mode fracture; microscopically, it is accompanied by slip over a large volume and by dimples.

(g) Delayed fracture

A crack propagates stably under cyclic stress, stress corrosion cracking or creep until the energy released by crack lengthening just equals that being consumed. In this case, the external stress σ, the stress intensity factor K_I and the crack length a reach a critical size

$$K_\mathrm{c} = \sigma_\mathrm{c} \sqrt{\pi a_\mathrm{c}} \cdot Y \qquad (\mathrm{A.4.1})$$

Under plane strain conditions, K_c is known as the plane-strain fracture toughness K_Ic (EN ISO 12737). The geometrical factor Y takes account of the shape of the crack and its layout. The time until an unstable (residual forced) fracture occurs determines the service life of a component. Whilst the key factor for a forced rupture is the amount of plastic deformation experienced by the workpiece until crack propagation becomes unstable, the key factor for a delayed fracture is the service time that has already elapsed. This type of fracture is the most frequent failure mode caused by mechanical loading. This also explains the significance of the fracture toughness as a material property parameter.

(h) Strength properties

The key properties for steel subjected to unidirectional mechanical loading are its resistance to permanent deformation and its resistance to fracture. Both are characterised by a limiting stress. Plastic deformation starts when the yield strength is reached and unstable fracture when the fracture strength is reached. Between these two limiting stresses, the material strengthens itself by forming lattice defects such as dislocations and stacking faults. A discontinuous elastic limit is known as the yield point and a continuous limit as the proof strength (Figure A.4.9). The fracture strength is also known as the fracture stress or the rupture strength, and it includes the compressive,

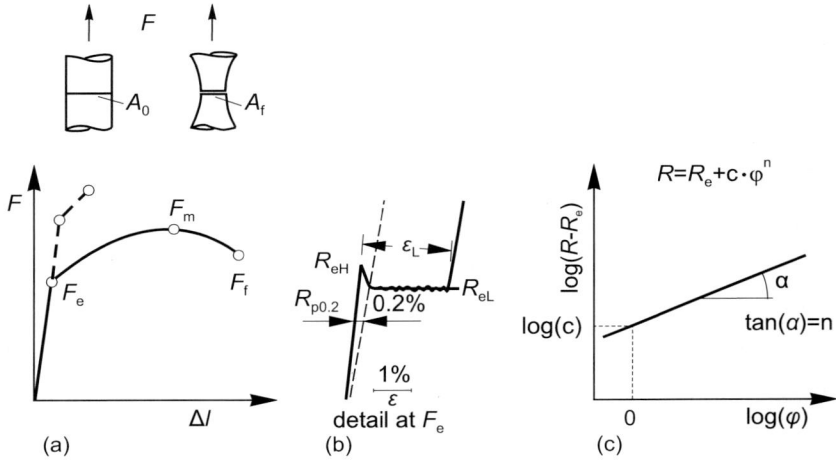

Fig. A.4.9 Characteristic values of the tensile test: (a) Force/extension curves for a soft steel and a hard steel (dashed). Design engineers relate the force F to the initial cross-section A_0 and the extension Δl to the initial length l_0 (technical stress/strain diagram). Thus $\sigma = F/A_0$ and $\varepsilon = \Delta l/l_0$. The fracture work $= \int F\, dl$ corresponds to the area under the curve. The tensile strength $R_m = F_m/A_0$ represents the fracture strength. Although the true fracture stress within the neck $R_f = F_f/A_f$ is a higher value, it is not useful as a design parameter. (b) The yield strength $R_e = F_e/A_0$ is given as the upper or lower yield point R_{eH} or R_{eL} or as the 0.2 % proof strength $R_{p0.2}$. A discontinuous yield point is accompanied by Lüders elongation ε_L. A continuous yield point (dashed curve) is read off at 0.2 % permanent elongation. (c) A production engineer who is interested in cold working needs to know which mechanical force F is required to make the momentary work-hardened cross-section A continue yielding. According to the yield curve, the true flow stress $R = F/A$ increases due to strain hardening with the true strain $\varphi = \ln(l_1/l_0)$ (true stress/strain diagram). R is also known as the deformation resistance k_f and φ as the logarithmic strain. Above uniform elongation at $\varphi \approx n$, a correction is required in the recorded curve to account for multiaxiality in the necking region.

A.4.1 Mechanical properties

torsional, bending, tensile and notched tensile strength. In precracked specimens and components, the fracture strength can be derived from (Eq. A.4.1) as $\sigma_c \sim K_c/\sqrt{\pi a_c}$. The fracture strength thus represents the greatest withstandable force applied to the smallest initial cross-section of a specimen or a component. During metal working, on the other hand, the key factor is the force required to continue deformation in the respective cross-section. This true stress is known as the flow stress or deformation resistance.

For a given microstructure, the strength properties depend on the loading conditions, i. e. on the degree of multiaxiality and on thermal activation. According to Figure A.4.1, σ_h/σ_v equals $-1/3$ in the compression test, 0 in the torsion test and $1/3$ in the tensile test. It reaches even higher values in the notched tensile test as the notch acuity increases. The yield strength and the fracture strength approach each other in this order and plastic deformation decreases. Thus a structural steel does not fracture even after experiencing a high degree of compressive strain, but notched specimens can undergo brittle failure in a tensile test. This tendency increases as the temperature decreases because the resulting increase in the yield strength is greater than that of the fracture strength . They coincide at the transition temperature T_T (Figure A.4.10). Macroscopic deformation approaches zero. An increase in the deformation velocity lowers the time required for thermal activation and increases the transition temperature. Below T_T, the fracture strength can fall below the original yield strength. Hardness tests are carried out slowly and with compressive stresses; $\dot{\varepsilon}$ and σ_h/σ_v are small. During such tests, even hard steels generally remain free of cracking and are independent of T_T. The hardness is thus an easily measurable value of the strength of a particular microstructure. How much of this can be utilised depends on the loading conditions. Below T_T, the steel is still hard but lacks strength (Figure A.4.10).

(i) Taking of specimens

The microstructural banding caused by elongation of microsegregations and non-metallic inclusions (Figure B.1.1, p. 126) leads to differences in the mechanical properties, depending on whether the rolled steel is tested along or across the banding direction. The strength parameters (R_{eH}, $R_{p0.2}$, R_m) are less affected by this than the ductility (elongation at fracture A, reduction of area Z, EN 10002) and the toughness (ISO V notched-bar impact bending energy KV, EN 10045). Figure A.4.11 shows the position of tensile and notched impact bending specimens in round and flat steel. The X, Y, Z directions correspond to L (longitudinal), T (transverse) and S (short transverse) used in the USA.

Roll-formed profiles and open-die forgings (disks, rings, etc.) have a complex banding orientation that has to be taken into account when specifying the position of a test specimen. As shown in Figure B.1.5 (p. 132), the notch impact strength of flat steel may be reduced to only a tenth simply by changing the sample position from XZ to ZX. For tests in the through-thickness

A.4 Properties

direction Z in thin flat steel, supplementary pieces are welded on perpendicularly to the surface so that samples can be taken. For example, load case ZX is obtained by welding a horizontal bending bar onto the rear face of a vertical U section.

As Figure A.2.7 a, (p. 27) shows, the degree of microsegregation increases towards the core of the solidification cross-section until it becomes a macrosegregation. This means that the position of longitudinal and transverse specimens within the cross-section is important, particularly in thicker ingots or forgings. Therefore, like the entire scope of required tests, it has to be agreed upon on

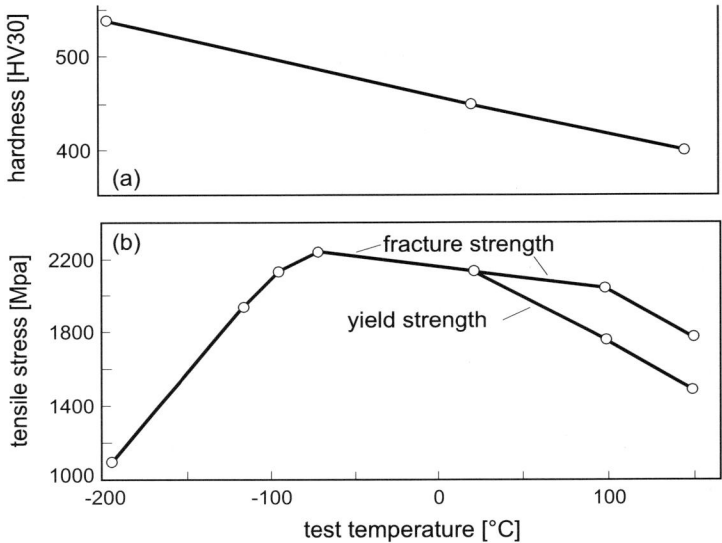

Fig. A.4.10 Hardness and fracture strength: Steel 50CrV4, quenched and tempered at 860°C/oil + 460°C for 2 h to a hardness (EN ISO 6507) of 450 HV. According to EN ISO 18265, this corresponds to a tensile strength of $R_m=1455$ MPa. (a) As the test temperature decreases, the mobility of the dislocations decreases and the hardness increases. (b) In a tensile test of notched specimens with a 4 mm diameter and a 0.5 mm deep circumferential V notch, the applied force is related to the area below the notch. The high degree of multiaxiality at the notch hinders yielding and raises the fracture strength above R_m. In the upper temperature range, the yield strength is indicated by the start of a measurable overall elongation. The local strain in the notch cross-section is higher and increases until the fracture strength is reached. In the middle temperature range, the yield strength and the fracture strength coincide and continue to increase. The plastically deformed volume decreases and is ultimately limited to the notch root. In the lower temperature region, dissipation of the stress concentrated in the notch by plastic flow is limited to such a degree that the surplus stress by α_k (Figure A.4.2) becomes important. The fracture strength drops below R_m, while the hardness continues to increase.

ordering. This also applies to steel castings: although they do not contain any grain orientation due to forging, they may exhibit anisotropy, e. g. along and across columnar crystals (Figure A.2.7 a) and, of course, they may also contain micro- and macrosegregations. If there is insufficient feed of molten material during casting, microvoids may also form in the core. The samples

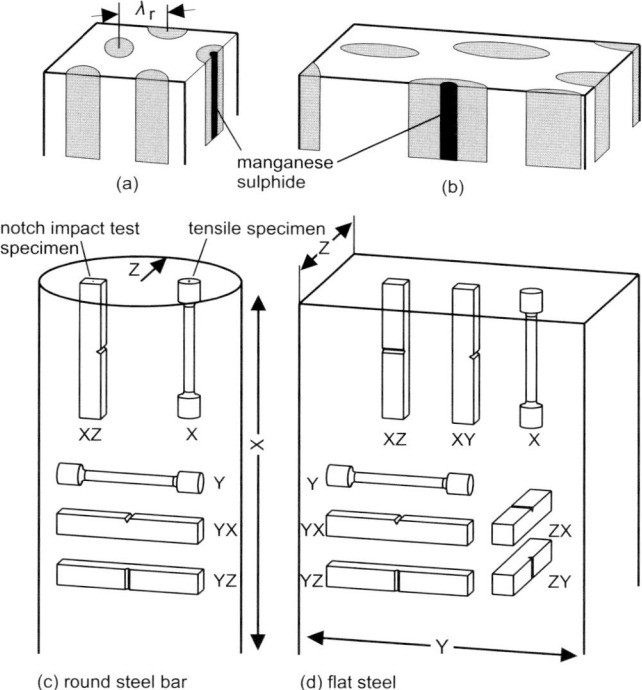

Fig. A.4.11 Position of the test specimen: (a) Volume element taken from a rolled round bar with roughly outlined segregation banding (grey) and sulphides (black) that are elongated in the rolling direction (=longitudinal direction). The interdendrite spacing λ, as shown in Figure A.2.8 (= segregation spacing, see p. 27), is reduced to λ_r by stretch-forming. (b) In flat steel (sheet), segregations are not only elongated as in (a), but widened as well so that they have a tape-like geometry. (c) According to EN ISO 3785, the main direction of stretch-forming (banding) is designated as X and thus a longitudinal specimen is designated as X. Transverse specimens taken radially are designated as Z and those taken tangentially are designated Y. In thicker rods, the off-centre position of the specimen can be specified e. g. at 1/2 radius. In notched or pre-cracked specimens, the direction of crack propagation is given by the second letter. (d) In flat steel, we differentiate between transverse specimens with a Y in the width direction and Z in the through-thickness direction as well as three directions of crack propagation.

A.4 Properties

are either taken from machining allowances of the casting or from cast-on or separately cast test pieces.

(k) Strengthening

Pure iron is soft. It can be strengthened by increasing the number of lattice defects (Figure A.2.4, p. 23). The degree of disruption of a regular lattice structure increases with the concentration of lattice defects. This hampers the movement of dislocations and the yield strength increases. The carbon and nitrogen atoms dissolved in interstices have a much greater effect on solid solution hardening of ferrite than substituted chromium, manganese or nickel atoms. Phosphorus is a very effective ferrite hardener. Cold or semi-hot working strengthens the material by increasing the dislocation density and this leads to mutual hindrance of dislocations. A decrease in the grain size also has a strengthening effect. Precipitation hardening is most pronounced for a dispersion of hard particles with the smallest possible size, but which is just

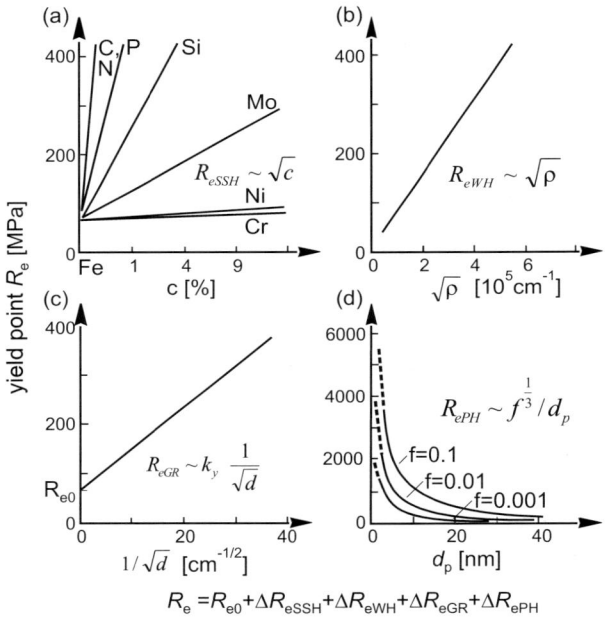

Fig. A.4.12 Strengthening of α-iron by lattice defects (from E. Hornbogen, G. Lütjering): (a) Solid solution hardening (SSH), c = concentration of a dissolved element, (b) work hardening by dislocations (WH), ρ = dislocation density, (c) hardening by grain refinement (GR), d = diameter of ferrite grains, k_y = grain boundary resistance, (d) precipitation hardening (PH), d_p, f = diameter and volume fraction of hard precipitates.

A.4.1 Mechanical properties

short of being cuttable by dislocations (approximate value: 10 nm). Within certain limits, the effects of zero- to three-dimensional lattice defects are additive (Figure A.4.12). These defects can also interact. For example, interstitial atoms congregate in edge dislocations and at grain boundaries. The elastically stored deformation energy in these dislocations and grain boundaries promotes nucleation of precipitates.

Lattice defects are particularly prevalent in non-equilibrium microstructures. This applies to martensite, in particular. Indirectly via austenitisation, carbon atoms are able to dissolve in the bcc lattice well over the actual maximum solubility of 0.03%. Such solid-solution hardening is accompanied by an increase in the dislocation density due to a displacive transformation of austenite. This corresponds to a high level of cold working. These zero- and one-dimensional lattice defects are superimposed by two-dimensional ones, such as packet or lath or plate boundaries.

In addition to increasing the number of lattice defects, the strength can be increased by incorporating another phase that is *coarser* than the fine precipitates. A mixture of soft ferrite grains and harder pearlite grains can produce a microstructure with a medium hardness. In a pearlitic microstructure, the concentration of phase boundaries is extremely high and the free path length s for the movement of a dislocation in the ferrite lamellae is very short. The yield point increases with $s^{-1/2}$ (Figure A.4.13). There is an obvious similarity to hardening by grain refinement in which the free path length extends to the grain boundary and $R_e \approx d^{-1/2}$ (Figure A.4.12 c). However, the relatively large carbide plates in pearlite tend to tear under tensile stress thus lowering the toughness and ductility. As a result, the ferrite in modern multiphase steels is strengthened by less brittle microstructural components, such as grains of bainite, martensite and retained austenite. A fine grain size imparts a low transition temperature to such dual-phase and duplex microstructures (Figure A.2.2, p. 22). The low carbon content improves the weldability (Chapt. B.2, p. 168). The strength in a particular direction can be increased by inducing anisotropy (Figure A.2.6, p. 25). Examples include directional solidification of turbine blades or a rolling texture in deep-drawing steel sheets.

(l) Shifting the transition temperature

Strengthening of ductile steels by solid solution hardening, cold working or precipitation hardening increases the yield strength to a greater degree than the fracture strength. This means that T_T is shifted upwards (Figure A.4.10). The reverse is true of hardening by grain refinement. Particularly in high-strength structural steels, there is always a certain degree of hardening by grain refinement to compensate the increase in T_T due to other strengthening mechanisms. The difference in T_T between longitudinal and transverse specimens increases with the degree of microstructural banding.

The thicker carbide lamellae in pearlite increase T_T (Figure A.4.13). In hard steels, e. g. those used for tools and roller bearings, the low-tempered martensitic microstructure already tends to undergo brittle fracture in unnotched

A.4 Properties

tensile and bending tests as the hardness increases. The higher hardness can only be completely exploited under pure compressive loading. The hardness increase shown in Figure A.4.14 has an effect similar to the temperature decrease shown in Figure A.4.10. The drop of the fracture strength at a higher hardness in Figure A.4.14 depends on the microstructure. Coarse carbides and inclusions, as well as microvoids and sintering pores (here as defects) shift it to the left. If we assume that cracks are initiated at defects and that their length is proportional to the size of the defect, we can obtain further information by applying a qualitative fracture mechanics treatment (Figure A.4.15). This reveals that in hard steels, in particular, a small defect size is necessary in order to take advantage of the high hardness.

(m) Influence of the test method

The tough/brittle transition is a property of ferrous materials that depends not only on the condition of the material but also on its loading conditions. With increasing degree of multiaxiality (e. g. due to notching) and decreasing thermal activation (decreasing temperature and increasing load application rate), the toughness of a given steel moves from the upper shelf region to the lower shelf. The fracture strain is reduced along with the fracture work, although the condition of the material does not change. Even the toughest

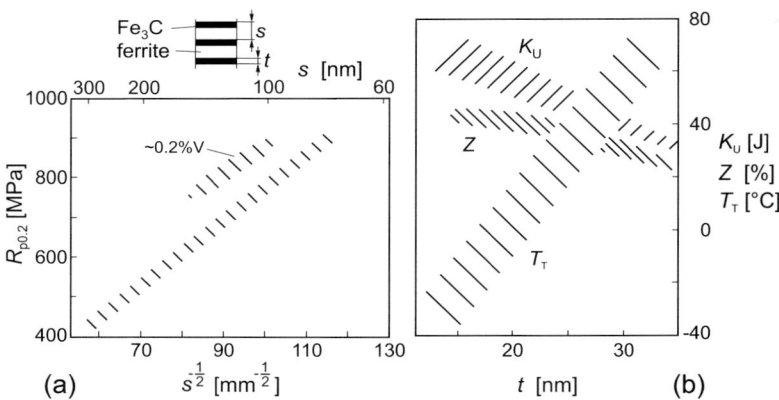

Fig. A.4.13 Mechanical properties of the pearlitic microstructure: The data published by J. Flügge, W. Heller and R. Schweitzer refer to unalloyed and low-alloyed pearlitic steels with 0.55 to 0.99 % C. (a) The proof strength $R_{p0.2}$ increases with decreasing interlamellar spacing s. The proof strength can be increased by microalloying with vanadium. (b) The notched-bar impact bending energy K_U for DVMF impact toughness specimens in the upper shelf region and the reduction of area Z both decrease with increasing carbide plate thickness t, whereas the transition temperature T_T increases.

A.4.1 Mechanical properties

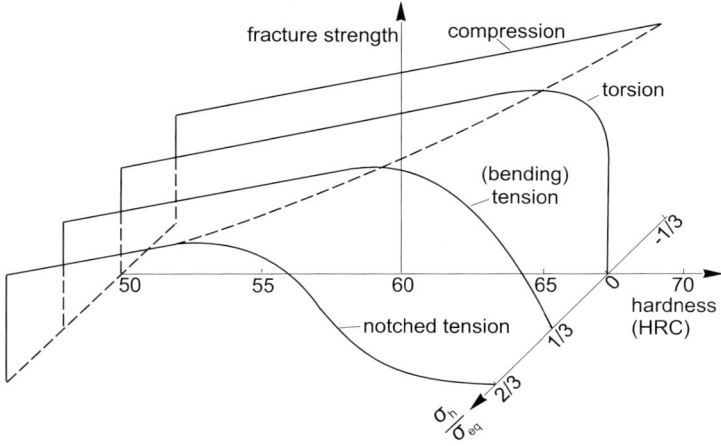

Fig. A.4.14 Fracture strength of hard steels (schematic): The drop in the fracture strength shifts to a lower hardness as the degree of multiaxiality σ_h/σ_{eq} increases.

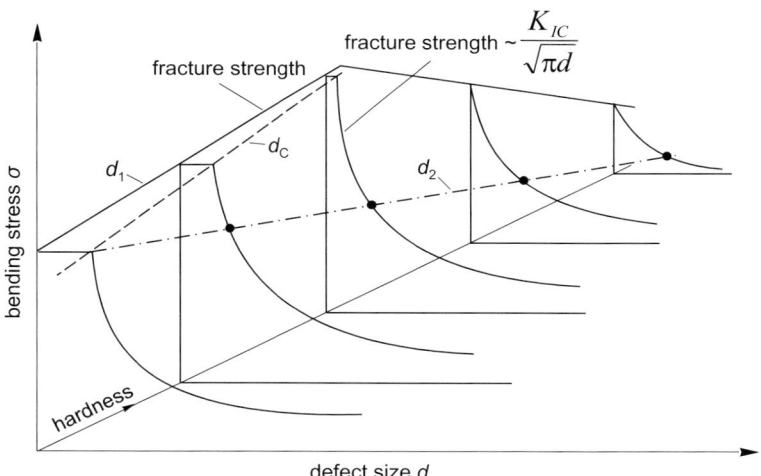

Fig. A.4.15 Influence of microstructural defects in hard steels (schematic): As long as the defect size d remains below a critical value ($d_1 < d_c$), the fracture strength measured in the bending test increases with the hardness. Critical defects ($d_2 > d_c$) cause incipient cracks that lead to a hyperbolic drop in the fracture strength (Eq. A.4.1, p. 89).

structural steel can undergo brittle failure if the external conditions induce a transition from the upper shelf region to the lower shelf (Figures A.4.4 and A.4.5). The strength is also affected by a transition to the lower shelf (Figures A.4.10 and A.4.14). This effect of the loading conditions is superposed by the influence of the material's condition (hardness, cleanness, microstructural banding, etc.) on the tough/brittle transition.

The condition of the material is usually characterised by testing unnotched tensile bars at room temperature because this allows the resistance of the material to deformation and fracture (strength) to be followed in the upper shelf region. However, this does not apply to very hard steels (Figure A.4.14). The fracture work determined in the tensile test (area under the force/elongation curve) is important e.g. for the toughness of unnotched, crash-resistant vehicle components. The notched-bar impact test can be used to raise the transition temperature - which is in the cryogenic range for tensile testing of a structural steel - into the climatic temperature range (Figure A.4.5). The notch acuity is reduced for harder materials to such a degree that unnotched impact bending specimens are used for hot-work tool steels in order to shift the ductile/brittle transition into the desired temperature range. These two test methods thus provide a great deal of information on the material in the upper shelf and transition regions.

Instant fractures are produced during testing, whereas the failure of an actual component is generally caused by a delayed fracture. This has led to the testing of precracked specimens and to the determination of the fracture toughness. Only the work expended until an existing crack starts unstable propagation is of importance for this type of toughness. The deformed volume remains small and is limited to the plastic zone in front of the crack tip and to the flanks of the crack. In contrast, the comparatively large volume of a tensile specimen undergoes considerable deformation prior to crack formation in the core of the necking region. The toughness depends mainly on the fracture work expended before crack initiation. In the notched-bar impact test, the deformed volume is localised beneath the notch and is thus decreased, indeed all the more the closer it is to the lower shelf. At the same time, the amount of fracture work expended before crack initiation decreases and the conditions approach those of the precracked specimen. The toughness measured by the various test methods not only differs considerably with respect to the amount of fracture work, but it also reacts differently to changes in the material's microstructure. For example, a low content of non-metallic inclusions can increase the fracture work in the tensile test but may not necessarily change the fracture toughness because there is only a low probability of finding coarse non-metallic inclusions in the small volume being tested.

The material property data usually includes a contribution from the loading conditions, even if it is only the difference in failure between a tensile specimen with a round cross-section as opposed to one with a flat cross-section (Figure A.4.8). The application of these technical data to actual components requires suitable models and is usually more successful for strength properties than for the toughness. This is why materials testing is supplemented by component testing. The tensile test of a threaded bolt is affected not only by the notch effect of the thread, but also by the residual compressive stress in the cold-roll-formed root of the thread. The bursting test of a pressurised pipe is affected by the multiaxiality of the stress state and by the weld seam.

A.4.1.3 Behaviour of grey cast iron

(a) Notch effect of graphite

A ferritic cast iron contains about 11 volume% graphite (Tab. A.1.3, p. 13) with a lamellar, vermicular or spherical shape. It is distributed in a matrix which corresponds to that of a low-carbon cast steel, albeit with a higher Si content. Under uniaxial compressive stress, solid-solution strengthening by Si (Figure A.4.12 a) increases the yield strength; however, under further compression it initially exhibits a ductile behaviour similar to that of cast steel. The situation is different for uniaxial tensile stress because practically no stress is transferred by the layer lattice of graphite. The graphite precipitates act almost like voids, but hinder their necking. We can see from Figure A.4.2 that a cylindrical borehole through a disk in a plane stress state causes a local elastic surplus stress of $\alpha_k = 3$, which can easily increase to more than 10 if it has a flat oval shape. For a spherical pore, $\alpha_k \approx 2$. Flattening to a lenticular lamella transverse to the tensile stress increases α_k by more than one order of magnitude, which gives rise to conditions similar to incipient cracking. The elastic surplus stress around a spherical pore is completely relieved within a distance of two to three times the pore diameter. However, for a lamella, this distance is smaller than its diameter. A qualitative conclusion is sufficient here: graphite precipitates exert an internal notch effect that increases drastically from a nodule to a lamella. The global uniaxial stress state is multiaxial near the notch, and this significantly increases the overall multiaxiality σ_h/σ_{eq}. As a result, the transition temperature T_T for brittle fracture increases in the matrix (Figure A.4.5) and the fracture work (toughness) decreases (Figure A.4.4). This is aggravated by the fact that T_T is increased even more by Si and that a casting does not have the grain refinement due to recrystallisation that is usual in a rolled steel.

The stress concentration factor α_k depends on the shape of the graphite precipitate. According to Equation A.4.1, for example, its size becomes important as soon as a sharp incipient crack has developed at a crack-like lamella. In the tensile test, a small plastic zone develops locally only in front of the crack tip, without global yielding taking place. The fracture strength σ_c decreases with increasing precipitate or crack size a_c. Because not all the lamellae are isolated, they can also be coherent, the fracture strength decreases even further. Typical for cast bar-shaped specimens of ferritic grey cast iron with a hardness of e.g. 156 HB is a tensile strength of only $R_m = 152$ MPa. Information on the yield point does not apply. The elongation at fracture $A < 1\%$(Tab. A.4.1).

A.4 Properties

Table A.4.1 Mechanical properties of ferritic cast iron. Approximate values give a rough idea of the different effects of spheroidal or flake graphite.

Parameter		spheroidal	flake
Hardness	[HB]	160	120
$R_{p0.2}$	[MPa]	315	—
R_m	[MPa]	440	130
A	[%]	18	0.6
KV[1)]	[J]	15	3
K_{Ic}	[MPa$\sqrt{m}$]	70[2)]	35
Young's modulus	[GPa]	170	90
Rel. damping[3)]		6.4	53

[1)] ISO-V impact toughness
[2)] derived via the J integral: K_{Jc}
[3)] low carbon steel = 1

A macroscopic brittle fracture is characterised microscopically by brittle cleavage of the matrix (Figure A.4.7 a). The compressive strength of 572 MPa is almost four times the tensile strength.

In spheroidal graphite cast iron, $\alpha_k \approx 2$, independently of the nodule diameter; however, for a given graphite volume, the nodule spacing increases with its diameter, i.e. the matrix bridges between the precipitates become thicker. In the tensile test, the local surplus stress is relieved by plastic yielding of the ferritic matrix. After the yield point has been exceeded, further deformation and strengthening occurs in which the matrix detaches from the graphite nodules and the spherical pores coalesce, essentially by shear fracture (Figure A.4.7 b) in a way similar to non-metallic inclusions in steel (Figure A.4.6). This ultimately leads to macroscopic ductile fracture. Typical property data for ferritic cast iron with spheroidal graphite are given in Tab. A.4.1. Although the hardness of this material is comparable to that of cast iron with flake graphite, its tensile strength is approximately three times higher. Furthermore, it has an additional safety margin due to ductile strengthening between $R_{p0.2}$ and R_m, which is similar to that of steel.

(b) Fracture work

The influence of the test conditions on the fracture work (toughness) shown in Figure A.4.4 is superimposed by the internal notch effect of graphite in cast iron. The fracture toughness of ferritic cast iron with spheroidal graphite is higher than that with flake graphite by a factor of only 2 because the sharp precrack is more important than the internal notch effect. With respect to the notch impact energy, the difference is already about a factor of 5, and the area under the force/extension curve from a tensile test gives a factor of > 50 for the fracture work. The influence of vermicular graphite lies between the two because, on the one hand, its stress concentration factor exceeds that of spheroidal graphite, and on the other hand, the continuous network of flake graphite is disrupted.

(c) Stiffness and damping

In cast iron with flake graphite, even a slight external tensile stress leads to local plastic yielding owing to the higher internal stress concentration factor so that Hooke's Law is not obeyed. This means that the determination of Young's modulus E depends on the test conditions. Thus, one of the variables measured is the secant at $1/4$ of the tensile strength. This problem rarely exists for cast iron with spheroidal graphite (Tab. A.4.1). Microplastic deformation at the edges of the flakes also contributes to damping mechanical vibrations. This is enhanced by a soft ferritic microstructure, even if it is only present as a shell around the flakes. However, the chief contribution to damping comes from the movement of dislocations within the graphite flakes, and this increases with the flake size. Internal friction rapidly reduces the vibration amplitude. The effect of spheroidal graphite is much lower, although it is still significant compared to that of steel (Tab. A.4.1).

(d) Strengthening

The matrix bridges become more slender as the diameter of the graphite nodules decreases, the constraint by multiaxiality is enhanced and the yield point increases slightly. In contrast, enlargement of nodules slightly increases the fracture toughness on account of the longer mean free path length in the matrix. Smaller graphite flakes promote crack branching, which may have a beneficial effect on the fracture toughness. However, the decisive contribution to increasing the strength comes from hardening of the matrix, which was assumed to consist of soft ferrite up to this point. As in hypoeutectoid steels, the tensile strength increases with the pearlite content and the fineness of its striped microstructure. In cast iron with flake graphite that has a thin cross-section and a fully pearlitic matrix, the tensile strength can increase to almost three times that of the ferritic cast iron given in Tab. A.4.1 without a loss of fracture toughness. According to W. Glaß, samples with a diameter of 30 mm and cast using an optimised process exhibited $R_{p0.2} \approx 360$ MPa, $A \approx 2.3\%$ and $E \approx 150$ GPa. In pearlitic cast iron with spheroidal graphite it is possible to more or less double the tensile strength compared to that given in Tab. A.4.1; however, the transition temperature increases, and the elongation at fracture and the notch impact energy decrease.

The hardening associated with the pearlitic fraction can be increased even further by a heat treatment. In cast iron with flake graphite, surface hardening is commonly used to obtain a hard martensitic microstructure. It is also used for cast iron with spheroidal graphite; however, this group of materials is generally transformed in the bainite range on account of their favourable combination of strength and toughness.

(e) Sampling

If cast iron with spheroidal graphite is to be subjected to tensile tests and notched-bar impact-bending tests, the specimens can be taken from separately cast pieces or from the casting itself. In the latter case, either a series part is destroyed or cast-on test pieces are used. As long as the microstructure is comparable, the specimens have approximately the same properties as the casting. However, there may be deviations caused by differing solidification conditions or reactions with the moulding material. If the melt is too hypereutetic (high degree of saturation S_c, see Equation.A.2.2, p. 34), flotation of primary graphite leads to microstructural differences. If S_c is too low, volume reduction may result in microstructural differences in the form of pores and shrink holes. The position of the test specimen in or on the casting must be agreed upon. Draft EN ISO 3785 can be used to define specimens. Possibilities for separately cast test pieces that are specified in EN 1563 are rods ⌀ 25 mm (Lynchburg specimen) or vertical rectangular specimens with a cast-on feeder (Y specimen) with an appropriate thickness (Chap. C.1, p. 383).

A.4.1.4 Behaviour of white cast iron

White cast iron may contain up to a third of its volume as eutectic M_3C with a hardness of 900 to 1100 HV (Figure A.4.16).This M_3C forms a network which encloses the pearlite that has grown from the original austenite cells. As a consequence of the greater stiffness, the carbide experiences an overproportional load when an external tensile stress is applied. If we take account of its notched shape and high brittleness, even a low nominal stress can be expected to produce the first incipient cracks in the carbide, which can then easily propagate throughout the continuous carbide skeleton (Figure A.2.13, p. 34). This is why the tensile strength is not stated, only a macrohardness of 500 to 600 HV, which represents a good compressive strength and wear resistance. White cast iron is also suitable for applications at elevated temperatures because both the eutectic carbides and the eutectoid carbides in pearlite forfeit much less hardness than the ferritic matrix when heated.

Increasing the hardness by means of a heat treatment is generally avoided owing to the brittleness of white cast iron. In contrast, by alloying with a few percent of Ni and Cr, it is possible to induce a martensitic transformation and a macrohardness of ≥ 700 HV by slow, low-stress cooling in a sand mould. Such white martensitic cast irons containing M_3C carbide include those marketed as Nihard I and II. In order to produce a white cast iron that survives the stresses applied during subsequent hardening, the coherent M_3C skeleton has to be converted to isolated, non-coherent M_7C_3 carbides. This can be achieved by alloying with $> 8\,\%$ Cr. Its microhardness increases from 1100 to 1550 HV with increasing Cr content. Hardening of a high-alloy white cast iron containing M_7C_3 carbides produces a matrix of martensite and retained austenite and also raises the macrohardness of the cast iron to > 800 HV. These materials

A.4.2 Tribological properties

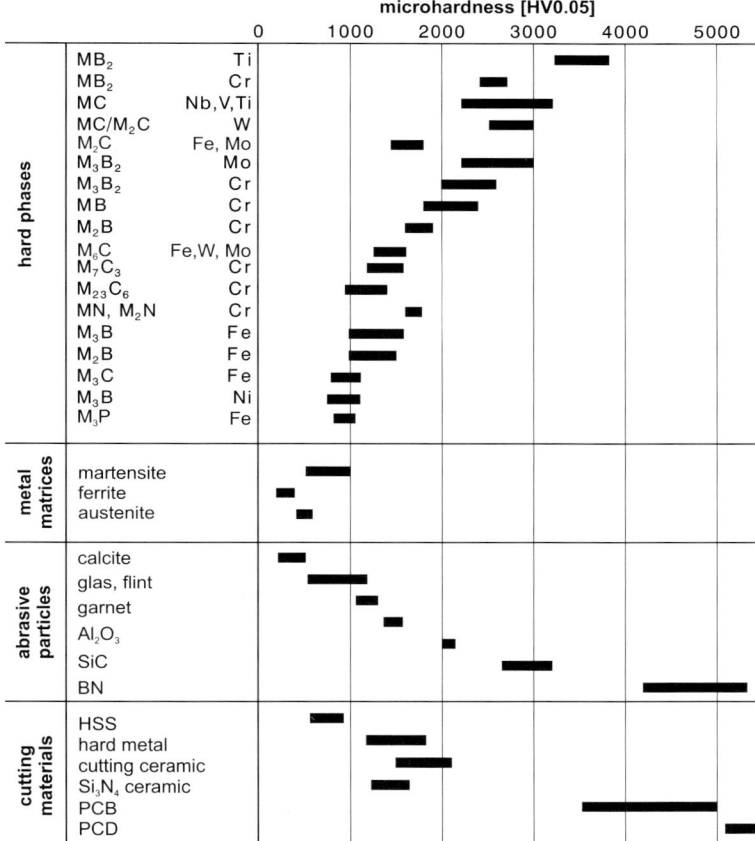

Fig. A.4.16 Microhardness of hard phases, metal matrices, minerals and cutting materials at room temperature. The elements represented by M are listed on the right of the type of hard phase.

are usually gas-quenched to avoid hardening cracks. White cast iron has a Young's modulus of between 180 and 210 GPa, a fracture toughness that is usually less than 20 MPa$\sqrt{m}$ and a damping capacity similar to that of steel.

A.4.2 Tribological properties

Tribology is the science of friction, lubrication and wear. It describes the processes occurring on the surface of a solid object when it is in moving contact with a counterbody. In contrast to Section A.4.1, only near-surface force effects are taken into consideration. They cause friction between the body and the counterbody, and this results in wear. Lubrication reduces friction and thus also lowers wear.

A.4.2.1 Friction

The surface of solid objects always has a certain degree of roughness. The actual total surface area A_{tot} increases with the resolution of the roughness profilometer. Advances in mechanical and optical methods reveal increasingly more detailed information on the roughness. An almost atomic resolution can be achieved with a scanning tunnelling microscope and with the adsorption method. The two-dimensional surface gains a three-dimensional component, expressed as a non-integer (fractal) dimension between 2 and 3. When two solid objects come into contact over a nominal surface area A, they first start to touch each other with the summits of their asperities. These are deformed by the normal force F_N until the real contact area A_r has increased to a load-bearing size. This process depends on the hardness H of the body under consideration and the number n of contact points within A.

$$A_r \sim F_N^m/H \tag{A.4.2}$$

When $n \to 1$, m approaches 2/3 (Hertzian contact). For a low degree of roughness ($n \gg 1$), m approaches 1. In this case, the external force F_N is superposed by an adhesive force, which can be taken into account by e.g. a factor of $((1 - \gamma_{\text{ad}}/H \cdot S)$ in the denominator of Eq. (A.4.2), where, γ_{ad} is the specific adhesion energy and S is the standard deviation of the heights of the asperities. A_r thus increases for strong adhesion between soft bodies. Plastically deformed contact points can undergo microwelding. Application of a tangential force results in static friction (ISO 4378-2). When this force reaches a critical value, adhesion is overcome by shearing off of the interlocking asperities and microwelds. The frictional force F_F is then associated with sliding friction. Over the friction length l, it generates a frictional energy that is available for heating, chemical reactions as well as for the formation and tearing of microwelds. If τ_s represents the mean shear strength of the continuously formed adhesion points, we obtain $F_F = A_r \cdot \tau_s$. Insertion into Eq. (A.4.2) gives the coefficients of friction

$$\mu = F_F/F_N \sim \tau_s/H \tag{A.4.3}$$

High hardness and weak adhesion lower μ. The interfacial energy γ_{ad} is proportional to $\tau_s \cdot l$ and represents the difference between the surface energies γ_A plus γ_B of bodies A and B minus the energy of their common interface γ_{AB}

$$\tau_s \sim (\gamma_A + \gamma_B - \gamma_{AB})/l \tag{A.4.4}$$

A.4.2 Tribological properties

The surface energy decreases in the following order: metal and covalent bonding, ionic bonding, weak bonding (of polymer materials); the interfacial energy increases with increasing dissimilarity between the two bodies. This is the reason for the low coefficient of friction between steel and PTFE (Teflon). F_F can be increased not only by adhesion, but also by abrasion due to a harder counterbody, work-hardened broken bridges or detached particles (Tab. A.4.2). The value of μ greatly increases in a high vacuum, which can of

Table A.4.2 Coefficients of friction: Approximate values of μ_0, μ for static and sliding friction, dry or lubricated (l) under slow testing conditions (average values taken from the literature)

Steel against		μ_0	μ
steel			
finish machined		0.30	0.23
polished		0.15	0.12
	(l)	0.12	0.08
grey cast iron		0.19	0.18
	(l)	0.10	0.06
bronze		0.19	0.18
	(l)		0.07
beechwood		0.60	0.50
graphite			0.10
glass			0.60
PE			0.20
PTFE			0.04

course cause problems in vacuum engineering or aerospace applications. Covering of the boundary surfaces in air with oxygen or water molecules is thus an important prerequisite to lower the coefficients of friction of steel contact pairs below a value of 1. Tribochemical reactions produce deposits that prevent direct contact between the solid bodies. The use of lubricants reinforces this separating effect to such an extent that the partners are only in partial contact (mixed friction). In grey cast iron, the thin graphite layers act as a solid lubricant that is exposed as wear progresses and which is then carried into the contact surface. In plain bearings, for example, a high rotation speed and a corresponding oil flow rate can be used to completely separate the shaft and the bearing shell by means of a coherent lubricating film (fluid friction e.g. for hydrodynamic lubrication).

A.4.2.2 Wear

According to DIN 50320, wear is the progressive loss of material from the surface of a solid body as a result of mechanical actions. This usually refers to friction. It causes deformation and fracturing of the surface layer. The dimensionless wear intensity W increases with contact stress $p = F_N / A$ and decreases with increasing resistance to deformation, expressed by the hardness.

$$W = k \cdot p/H = \Delta m / \rho \, A \, l \tag{A.4.5}$$

Where k is the wear coefficient, Δm is the mass loss and ρ is the density. This includes a hardness increase due to cold working. W^{-1} is the wear resistance. Eq.(A.4.5) provides only a coarse approximation of the complex sliding wear process. There are four different wear mechanisms that cause mass loss (Figure A.4.17).

Abrasion is caused by a scratching load exerted by a harder counterbody (Figure A.4.17 a). The groove is gouged out by the tip of an asperity, a work-hardened abraded particle, a piece of hard phase that has broken off or a mineral grain. The material from the groove can be driven into a bead or chipped away. We therefore differentiate between micro-ploughing and micro-cutting. Micro-fatigue also occurs if there is repeated ploughing. Very hard carbide-rich steels and white cast iron may also undergo a slight amount of micro-cracking, e.g. detachment of eutectic carbides. Micro-cutting, however, is the main wear mechanism for steels and grey cast iron. A high hardness lowers the depth of the groove and thus reduces wear. Martensitic hardening of carbon-rich steels is one way of achieving this. However, its maximum hardness of $\approx 900\,\text{HV0.05}$ is exceeded by many mineral counterbodies (Figure A.4.16).

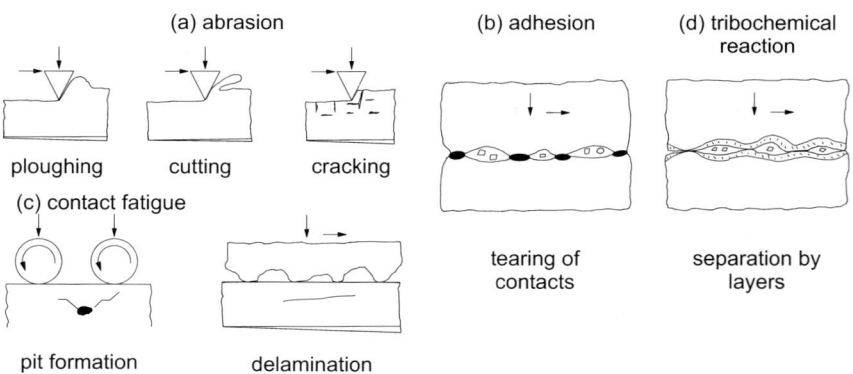

Fig. A.4.17 Wear mechanisms: Abrasion, adhesion, tribochemical reaction and contact fatigue (schematic)

In such cases, the precipitation of a harder phase may help. Carbides of the alloying elements chromium, niobium and vanadium are significantly harder than iron carbides. In order to be effective, they must be harder than the counterbody and of sufficient size, namely approximately the width of the groove, which is kept small by hardening the matrix. However, the hard particles are cut out if they are too small. A homogeneous dispersion of coarse hard phases (Figure A.2.3, p. 22) leads to the smallest matrix groove length and thus to the lowest wear contribution compared to a network arrangement. An increase in the volume fraction has the same effect; however, the wear rate can increase again due to micro-cracking as a consequence of increased brittleness. Another method is to produce a hard coating, e. g. thermochemical deposition of a nitride or boride layer or electroplating of a hard chromium layer or vapour deposition of titanium carbide or nitride layers. There are two vital factors that must be taken into account here: a) the substrate must have a sufficient hardness to support the layer and b) the groove depth $\ll$ coating thickness.

Adhesion refers to the formation and separation of adhesive spots (cold-welding, galling) within the interface (Figure A.4.17 b). Depending on the strength, separation can occur at the interface, in the component or in the counterbody. This means that material can be transferred. Adhesion of ferrous materials is particularly prevalent in metallic counterbodies if there are no separating layers. The hardness exerts a more indirect effect: asperities are less deformed so that the real contact area remains smaller. This means that higher flash temperatures in the contact are conceivable, and these in turn promote separating oxide layers. For the case of sliding wear, e.g. for metal-working tools used in the metal industry, the tendency to cold welding is reduced by the precipitation of hard phases. There is no adhesion between the carbides in the tool steel and the metallic workpiece (Eq. A.4.4). In addition, abrasion of the softer matrix in the tool surface produces a particular topography, which is advantageous for a separating lubricant film. Similar principles apply to multiphase bearing materials. Therefore, some materials with a microstructure similar to that used when abrasion is the main mechanism have proven successful - although for other reasons - for types of wear based on adhesion. Coatings can also help to reduce adhesion. The adhesion tendency of white cast iron is reduced by its high carbide content and that of grey cast iron by self-lubricating graphite.

Contact fatigue is caused by crack initiation and subsequent propagation as a consequence of cyclic tribological loading that leads to spalling of the surface (Figure A.4.17 c). If the surface is subjected to repeated rolling, this can cause fatigue due to Hertzian contact stress with low tangential forces. A high matrix hardness increases the elastically withstandable fraction of the load and thus the service life under rolling contact conditions. However, the harder the matrix, the more effective are the stress concentrations around the hard phases (e. g. non-metallic inclusions and carbides) at initiating cracks. These cracks usually form beneath the surface. They then propagate to the surface to form pit-like break-outs. Compared to ploughing and sliding wear, more finely grained hard phases are desirable for rolling wear (Figure A.4.14). If a periodic point contact is accompanied by high friction-induced tangential forces, as in e. g. sliding of a blunt body, detachment of thin flakes from the strongly deformed surface can be expected in materials with a low hardness. This is known as delamination.

Tribochemical reactions lead to the formation of reaction products due to a chemical reaction between the body and the counterbody as well as the surrounding media under a tribological load. They frequently have a wear-reducing effect by forming separation layers (Figure A.4.17 d). Tribooxidation layers are supported more effectively by a hard substrate, thus decreasing the adhesion associated with sliding wear.

In a four-component tribological system (body, counterbody, interfacial medium, surrounding medium), wear is usually caused by a combination of mechanisms. In practice, a system analysis and visual examination of the wear will indicate which type of mechanism is active (Figure A.4.18). It is obvious that e. g. abrasion in a gravel chute (grooving wear) or in a slurry pump (erosive wear) will produce grooves with very different depths. This is why the microstructural requirements for a wear-resistant steel differ according to the application. The wear resistance is therefore a system property and not a material property. The selection of suitable materials thus requires the most detailed knowledge possible of the wear system as well as experience with similar systems. Some applications are discussed in Chapters B.1 to B.6.

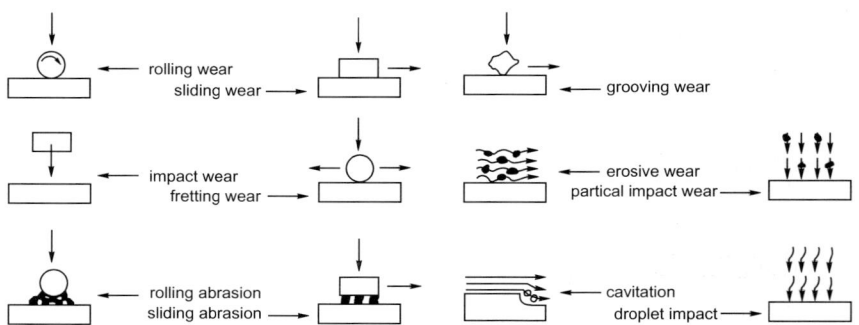

Fig. A.4.18 Types of wear: Classified according to the structure of the system (e. g. two- or three-body wear or solid / fluid, also with particles) and type of tribological load (e. g. sliding, rolling, impact).

A.4.3 Chemical properties

The following section discusses chemical reactions of ferrous materials with the surrounding media. Corrosion is defined in EN ISO 8044 and DIN 50900-2 as the modification of a material due to undesirable chemical or electrochemical attack of the surface. It is based on the oxidation of the ferrous material and simultaneous reduction of an oxidising agent contained in liquid electrolytes (e.g. aqueous solutions, molten salts and metals) or in solid scale. Important types are wet corrosion in aqueous media and high-temperature corrosion in oxygen-containing gases. Corrosion-resistant ferrous materials are discussed in Chapter B.6. Corrosion-protection layers are discussed in Chapter B.1, p. 136 and B.2, p. 175. The following discussions refer to the metal matrix of steels and cast iron.

A.4.3.1 Wet corrosion

The corrosion of an iron electrode is based on two half-reactions:

a) anodic half-reaction (to the left: oxidation)

$$Fe^{++} + 2e^- = Fe \qquad (A.4.6)$$

The electrode accepts electrons e^- (positive pole) and Fe^{++} ions go into solution (cations)

b) cathodic half-reaction (to the right: reduction)
The electrode releases e^- (negative pole) e.g. by

- producing hydrogen from an acid

$$2H^+ + 2\ e^- = H_2 \qquad (A.4.7)$$

- reduction of oxygen in aerated water

$$2H_2O + O_2 + 4\ e^- = 4(OH)^- \qquad (A.4.8)$$

According to (A.4.6), anodic dissolution of iron is only possible if the released electrons are consumed by redox reactions, such as those shown under point b), which do indeed frequently occur. Hydrogen ions or oxygen act as the oxidising agent in these reactions. The prerequisite for the flow of a charge nF is a potential difference ΔU of the overall reaction, which is comprised of the potentials U_C and U_A of the cathodic and anodic half-reactions, respectively.

$$\Delta U = U_A - U_C \qquad (A.4.9)$$

Combining the half-reactions given in (A.4.6) and (A.4.8) gives the overall reaction, which proceeds to the left.

$$2Fe^{++} + 4(OH)^- = 2\ Fe + 2\ H_2O + O_2 \qquad (A.4.10)$$

A.4 Properties

Such equations are generally written with the oxidised form of the metal (ox) and the negative charges (e⁻) on the left-hand side and the reduced form (red) on the right

$$\text{ox} + n\text{e}^- = \text{red} \qquad (A.4.11)$$

(a) Thermodynamics

The product of the charge nF (in Ah) and potential U (in V) is an energy – the free enthalpy (Gibb's energy)

$$\Delta G = -n\ F\ U \qquad (A.4.12)$$

where n represents the number of exchanged electrons per atom or ion and F is the Faraday constant (F = 26.8 A · h · mol^{-1}). The enthalpy change of a chemical reaction can also be written as

$$\Delta G = \Delta G^0 + RT\ \ln(c_{\text{red}}/c_{\text{ox}}) \qquad (A.4.13)$$

where c_{red} and c_{ox} are abbreviations for the concentrations (or activities) of the participating substances. Dividing by $-nF$ gives the temperature- and concentration-dependence of the Nernst potential

$$U = U^0 + \frac{RT}{nF}\ \ln(c_{\text{ox}}/c_{\text{red}}) \qquad (A.4.14)$$

The corrosion range can be determined by considering four particular cases.

$\Delta U > 0$: corrosion is possible

$U = U^0$: all concentrations are unitary (dissolved ions = 1 mol/l, gases = 1 bar, water and solids = 1). The logarithmic term approaches zero. U^0 is the standard potential under standard conditions.

$\Delta U = 0$: corrosion comes to a standstill. Eq. (A.4.14) gives the associated concentration ratio of the dynamic Law of Mass Action $K = c_{\text{red}}/c_{\text{ox}} = \exp(U^0 nF/RT)$ in which the forward and backward reactions have equal rates. This is generally not achieved in practice..

$\Delta U < 0$: Does not occur. The chemical reaction would proceed in the opposite direction, and the anode and cathode would swap over. The deposition of metals on the cathode is known as electroplating..

It is expedient to calculate the value of ΔU indirectly with (Eq. A.4.9) by determining U_A and U_C. Eq.A.4.14 can be correspondingly applied to each half-potential. $U^0{}_A$ and $U^0{}_C$ are the standard potentials of metal dissolution or of the redox reaction. Tab. A.4.3 gives the standard potentials with respect to the standard hydrogen electrode for iron along with those of some other metals for comparison and for some frequently occurring redox potentials.

A.4.3 Chemical properties

Table A.4.3 Some standard potentials [volt] referenced to the standard hydrogen electrode

Metal electrode			Redox electrode		
Au	Au^{3+}	+ 1.42	$MnO_4^- + 8\,H^+$	$Mn^{2+} + 4\,H_2O$	+ 1.52
Ag	Ag^+	+ 0.80	Ce^{4+}	Ce^{3+}	+ 1.44
Cu	Cu^{2+}	+ 0.34	Cl_2	$2\,Cl^-$	+ 1.36
Pb	Pb^{2+}	- 0.13	$HCrO_4^- + 7\,H^+$	$Cr^{3+} + 4\,H_2O$	+ 1.36
Ni	Ni^{2+}	- 0.23	Fe^{3+}	Fe^{2+}	+ 0.77
Fe	Fe^{2+}	- 0.44	I_2	$2\,I^-$	+ 0.54
Cr	Cr^{2+}	- 0.56	$O_2 + 2\,H_2O$	$4\,OH^-$	+ 0.41
Zn	Zn^{2+}	- 0.76	Cu^{2+}	Cu^+	+ 0.17
Mg	Mg^{2+}	- 2.38	H_2	$2\,H^+$	+ 0.00
Na	Na^+	- 2.71	Cr^{3+}	Cr^{2+}	- 0.41

They provide an initial indication as to whether corrosion is possible. Whilst iron corrodes in aerated water under standard conditions because $\Delta U = 0.41 - (-0.44) = 0.85$ V, the more electrochemically noble gold is not attacked. On the other hand, iron is nobler than galvanised steel. It acts as the cathode and thus has cathodic protection, whereas the zinc anode can dissolve.

(b) Kinetics

If corrosion is possible from the thermodynamic perspective, the question arises as to how it proceeds with respect to time and location. This is the central issue of corrosion considerations for practical applications. An indicator of the chronological progression is the rate of removal in mol over time t in hours. This corresponds to the current $I = nF/t$ in amperes. A flow of current changes the potentials U_A and U_C of the half-reactions: they approach each other (polarisation) until $I_A = I_C$. At the corresponding resting potential U_R, the corrosion current $I_R = |I_A| = |I_C|$ (Figure A.4.19). An active external circuit can be used to increase the potential of the iron anode along with the corrosion current I until, at the passivation potential U_{Pa}, the current I decreases by several orders of magnitude. It only begins to increase again in the transpassive range above the breakdown potential U_B. This passivation of iron by the formation of a dense, firmly adhering and usually oxidic surface film plays an important role in corrosion protection. The formation and sustainment of this passivating layer (without the external circuit) requires that the redox potential U_{Red} of the cathodic half-reaction is $U_{\text{Act}} < U_{\text{Red}} < U_{\text{B}}$. Alloying the iron with e.g. chromium or adding inhibitors to the electrolyte influences this passivation process are used to reduce the corrosion rate. The corrosion sequence with respect to location depends on the position of the anodic and cathodic areas on a corroding surface. There are two limiting cases. In the first case, the anodes and cathodes continuously move about so that anodic dissolution of iron occurs uniformly over the surface. Such a homogeneous mixed electrode may be due to local polarisation. Areas being dissolved become more noble as the current flows causing the anode to jump to a nearby site. This uniform corrosion is evaluated using e.g. resistance diagrams. Naturally, it is undesirable and costly, but nevertheless still predictable. A loss in thickness of 1 mm/year with respect to the material density corresponds to a loss of e.g. $7.85 \cdot 10^3$ g of material per m² surface area or an hourly removal rate of $7.85/8.760 \approx 0.9\,\text{g/m}^2\text{h}$. From the technical perspective, a material is regarded as being resistant if its corrosion rate is less than $0.3\,\text{g/m}^2\text{h}$. Much more critical is the second case, which concerns localised corrosion of a heterogeneous mixed electrode with fixed anodic areas. A high local current density and a small amount of dissolved iron may cause a component to fail after a

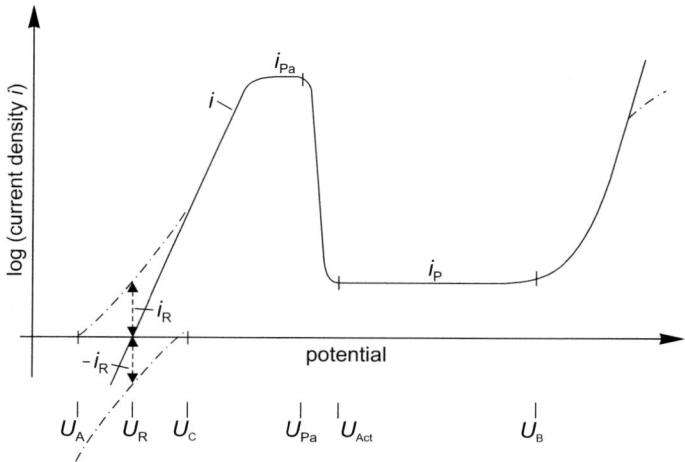

Fig. A.4.19 Current density/potential curve (iron in dilute sulphuric acid, schematic): **dash-dotted line**: polarisation of the anodic and cathodic potentials (U_{A}, U_{C}) towards the resting potential (U_{R}) at which the unmeasurable corrosion current (I_{R}) is flowing; **continuous line**: change in the potential due to an active external circuit and measurable current density i. Passive range between the passivation or activation and the breakdown potential (U_{Pa}, U_{Act}, U_{B}) with the passivating current density i_{Pa} and the passive current density i_{P} (for more information on the influence of chromium, see Figure B.6.1 a, p. 310)

surprisingly short time. Localised corrosion adversely affects the safety and service life of chemical equipment, machinery and civil engineering structures.

Deposition of corrosion products may produce loosely adhering surface films that inhibit the cathodic half-reaction. Faster electrolyte flow velocities clean the surface and bring in fresh medium, thus increasing the corrosion rate again. Particles entrained by the flow abrade the surface films and activate the surface (flow-assisted corrosion). Cast iron has a protective casting skin produced by the interaction of the melt with the sand mould. After initial corrosion of iron containing a coherent network of flake graphite, the surface is essentially covered with graphite and corrosion products that slow down further corrosion, particularly in static liquids. The carbide skeleton in white cast iron has a similar effect. The stable chemical bonds of graphite and carbide make them more resistant to wet corrosion compared to iron. In aqueous low-oxygen media, the potential difference of $\approx 1\text{V}$ between the more noble graphite and the matrix leads to selective corrosion of ferrite leaving behind the graphite skeleton filled with corrosion products (spongiosis).

(c) Localised corrosion

Localised corrosion of ferrous materials is caused by differences in the material or in the electrolyte and by damage of surface films. This damage may be chemical or mechanical.

If iron is in contact with a more noble metal, the resulting cell in the surrounding electrolyte causes anodic dissolution of iron, i.e. *galvanic corrosion*. Alloying differences in the microstructure may lead to selective corrosion of less noble areas. Thus, *intergranular corrosion* (IC) along grain boundaries is due to depletion of favourable elements in the grain boundary region or to the enrichment of grain boundaries with unfavourable elements or precipitates. Segregation-related alloying inhomogeneities lead to *segregation corrosion*.

There may be differences in the electrolyte composition e.g. in narrow crevices. Replenishment of the oxidising agent consumed by corrosion is insufficient to sustain a passivating layer in the tip of the crevice. Outward transport of the corrosion products is also impeded. Hydrolysis, e.g.

$$\text{Fe}^{++} + 2\text{ H}_2\text{O} = \text{Fe(OH)}_2 + 2\text{ H}^+ \tag{A.4.15}$$

increases the acidity within the crevice. Nickel and chromium in alloyed steels are precipitated as hydroxides whose solubility in an acidic medium is low so that the resulting depletion of OH^- in the crevice lowers the pH value below a value of 2 and *crevice corrosion* is intensified.

Chemical attack of a passivating layer by e.g. chloride ions, leads to local dissolution of iron and thus to *pitting corrosion*. The high anodic current density in the pit acts against a low cathodic current density in the comparatively large undamaged surface. Like crevice corrosion, acidification due to hydrolysis promotes the dissolution of iron.

A.4 Properties

Mechanical damage of a passivating layer by the movement of dislocations results in stress corrosion cracking or fatigue crack corrosion under critical system conditions. In anodic *stress corrosion cracking* (SCC), movement of dislocations due to unidirectional loading or residual stresses below the measurable yield strength can break through the passivating layer by forming a slip step. As a result, local dissolution of iron creates a notch that increases the tensile stress even further. The notch becomes a crack whose flanks are passivated but whose tip remains active due to hydrolysis and thus acidification. Examples of this are transgranular SCC of stainless austenitic steels in chloride-containing solutions and intergranular SCC of unalloyed steels in nitrate-containing media.

Corrosion fatigue (CF) occurs when the passivating layer is damaged by the emergence of slip bands as a result of cyclic loading. The formation mechanism of notches and cracks is similar to that of SCC; however, repassivation of active damage is hampered by cyclic stress. Unlike SCC, this does not require a specific corrosive medium. The endurance limit is lost and the fatigue strength increases with the frequency. This corrosion fatigue in the passive state differs from that in the active state.

In certain corrosion systems, a very low frequency or a low strain rate can prevent healing of the passivating layer so that *strain-induced cracking* occurs. This is a combination of classical SCC and corrosion fatigue. Another mechanism is *hydrogen embrittlement*. There are two types: 1. *Hydrogen embrittlement* caused by dissolved hydrogen atoms that lower the binding forces and thus the cleavage stress in the iron lattice. Recent investigations show that H enhances the mobility of dislocations. Their pile-up induces cracks at a lower stress: brittleness by microplasticity. 2. *Hydrogen-induced cracking* (HIC) caused by expelled hydrogen atoms recombining to molecules in defects, e.g. inclusions, thus producing a high internal pressure. If there is also tensile stress from an applied load or as residual stress, this is known as cathodic or *hydrogen-induced stress corrosion cracking* (HSCC). Such cracking starts after an incubation time during which the hydrogen diffuses into lattice defects and other defects. Subsequently, lattice expansion in front of the crack tip attracts hydrogen atoms so that the highest concentration of dissolved hydrogen coincides with the highest tensile stress. Propagation of the crack depends on subsequent diffusion of hydrogen to the crack tip. The hydrogen enters the material either during manufacturing (welding, pickling, electroplating) or is formed by corrosion directly at the crack tip, where it is adsorbed or dissolved and can thus promote crack propagation (Chapt. A.3, p. 58 and Chapt. B.2, p. 178).

A.4.3.2 High-temperature corrosion

In power engineering, petrochemistry, furnace construction, etc., the surface of the material is exposed to hot oxygen-containing gases such as air, steam, reaction mixtures, and flue gases. The oxygen partial pressure p_{O_2} in the gas that is able to oxidise iron according to:

$$2\ \text{FeO} = 2\ \text{Fe} + O_2 \qquad (A.4.16)$$

is given by the equilibrium constant K for a thermodynamic equilibrium ($\Delta G = 0$). The solid reaction partners - iron and scale - have an activity of one, thus $K = p_{O_2} = \exp\text{-}(\Delta G^0 / RT)$. The equilibrium pressure p_{O_2} of FeO increases from $\approx 10^{-25}$ at 600°C to 10^{-15} at 1000°C. Even minute amounts of oxygen will oxidise the iron.

(a) Kinetics

The key factor is thus the oxidation kinetics. In an ionic lattice, such as that of scale, ions diffuse rather than neutral atoms. The transport of positive charges in the form of Fe^{2+} cations is therefore associated with a flow of electrons or with the diffusion of O^{2-} anions in the opposite direction. This means that the composition of this semiconductor is no longer stoichiometric and there are vacancies in the sublattices of iron or oxygen. The resulting difference in charges is balanced by the change in valency of individual ions (e.g. $Fe^{2+} \rightarrow Fe^{3+}$). The most iron-rich oxide FeO (wustite) has a large metal deficit ($Fe_{0.85}O$ to $Fe_{0.95}O$). The iron ions have a high mobility owing to the large number of cationic vacancies. This p-type scale (positive charge carriers) grows quickly. Because this effect occurs above 570°C corrosive attack can be expected to accelerate above this temperature. The oxygen content decreases towards the outer surface from a thick layer of FeO to consecutive thin layers of Fe_3O_4 (magnetite) and Fe_2O_3 (haematite). The metal deficit is low in p-type Fe_3O_4. The diffusion of equal proportions of Fe^{2+} and Fe^{3+} ions via lattice vacancies is slower. Fe_2O_3 has an almost stoichiometric composition. This scale is considered to be n-type because the O^{2-} ions diffuse as negative charge carriers. However, migration of Fe^{3+} is also observed (Figure A.4.20). We have already pointed out in Chapt. A.3 that the increase in scale thickness s decreases over time or can be linear (Eq. A.3.4, p. 75). In the first case (q < 1), the diffusion of ions through the growing layer determines the rate of scale formation. In the second case (q → 1), cracks, pores and low adhesion allow the gas to penetrate directly to the surface of the material. These defects are produced by an increase in volume, temperature fluctuations and/or oxidation of the carbon in steel (CO/CO_2 pressure). The porosity increases with the amount of steam. The stresses within the scale layer govern adhesion and cracking. Tab. A.4.4 gives the stresses in the layer and the substrate that were measured at the same temperature. The compressive stresses prevailing in the layer generate

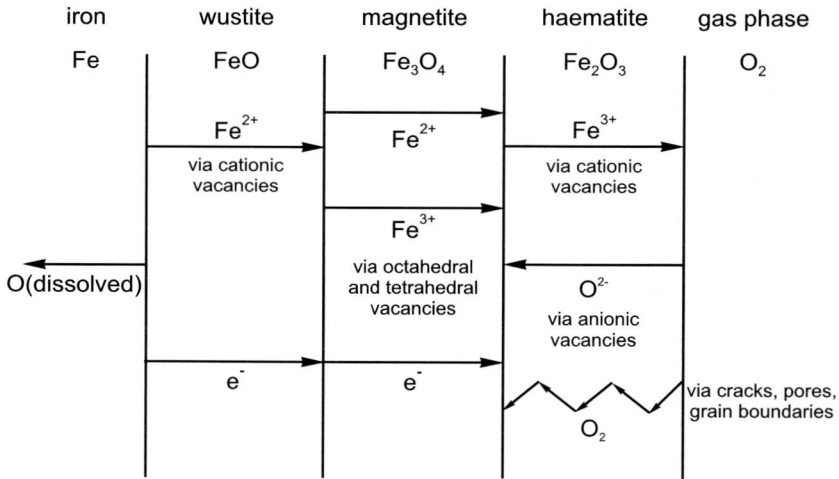

Fig. A.4.20 Structure of the layers and mass transport in iron scale (from A. Rahmel, schematic): Positive charge carriers (Fe cations) diffuse in p-type wustite and magnetite, whereas predominantly negative charge carriers (oxygen ions) diffuse in n-type haematite.

perpendicular tensile stresses in the transition region which support decohesion of the layer. If there is a temperature gradient, e. g. on rapid cooling, tensile stresses can build up in the layer. Cracks perpendicular and parallel to the surface not only increase the scale formation rate, they also contribute to loss of scale by flaking.

As Figure A.3.8 (p. 76) shows, iron is oxidised along with the carbon which is either dissolved in it or is present as carbide. The graphite in grey cast iron

Table A.4.4 Stresses in the oxide layer: Polished samples oxidised at 600°C in the air at atmospheric pressure for up to 100 h, measured by X-ray diffractometry and again after cooling to room temperature. It was not possible to measure the bottommost wustite layer; however, it should be under compressive stress (from S. Corkovic and A. R. Pyzalla)

Material	Temperature [°C]	Stresses in the layer [MPa]	
		Magnetite	Haematite
S355	600	100 → 200	-125
	20	135	-265
GJL 250	600	-120 / -400	-120 / -400
	20	-60	-420

→ increases or / fluctuates cyclically with the oxidation time

burns out on contact with oxygen. In continuous flake graphite, this results in deeply penetrating corrosion with liberation of a gaseous corrosion product. Cast iron with vermicular or spheroidal graphite should thus be given preference in high-temperature applications.

(b) Influence of alloying elements

Of the more noble metals, Ni suppresses formation of wustite in favour of the spinel $FeNi_2O_4$. This reduces the rate of scale formation. Less noble alloying elements, such as chromium, silicon and aluminium, form islands of spinel ($FeCr_2O_4$, $FeAl_2O_4$) or fayalite (Fe_2SiO_4) on the surface of the steel that reduce the rate of oxygen transport. These islands coalesce to an internal layer of scale as the degree of alloying increases. At high alloy contents, they form a layer of Cr_2O_3, Al_2O_3 or Fe_2SiO_4. These layers have a particularly small number of lattice defects and thus limit the diffusion of ions even further. Iron ions diffusing outwards form an external layer of scale that consists of iron oxides. Selective oxidation of the alloying elements can lower the concentration of these elements in the surface zone of the ferrous material. Damage or cracking of the protective inner layer of scale can then lead to local efflorescence of FeO.

Low alloy contents of elements with a high oxygen affinity, such as cerium or yttrium, segregate to the grain boundaries of the Cr_2O_3 or Al_2O_3 layer. This limits cation diffusion. The layer grows via diffusion of oxygen ions to the internal boundary surface which is significantly slower.

Oxygen can diffuse into the surface of the material where it oxidises elements such as Cr, Al, Si, and Y, whose oxygen affinity is greater than that of iron. This internal oxidation occurs preferentially on the grain boundaries. At very high temperatures, it proceeds with a straight front. Coalescence of internal and external oxides may improve the adhesion of the scale layer.

Not only oxygen, but also H_2O, CO / CO_2, SO_2, NH_3 and other gases as well as ash can attack the surface of the material. Various aspects of suitable alloy compositions are discussed in Chapt. B.6, p. 336

A.4.4 Special physical properties

Magnetism, thermal expansion and conductivity are important properties in a number of functional materials based on iron (Chapt. B.8, p. 369).

A.4.4.1 Magnetic properties

As a transition metal, every iron atom has a permanent magnetic moment. At high temperatures, these moments are randomly aligned (paramagnetism). Below the Curie temperature ($T_C = 769°C$ MO in Figure A.1.6, p. 10) small crystalline regions are formed that are delimited by Bloch walls (Weiss domains) in which the moments have a parallel alignment, i.e. they are polarised (ferromagnetism). However, owing to the random orientation of these

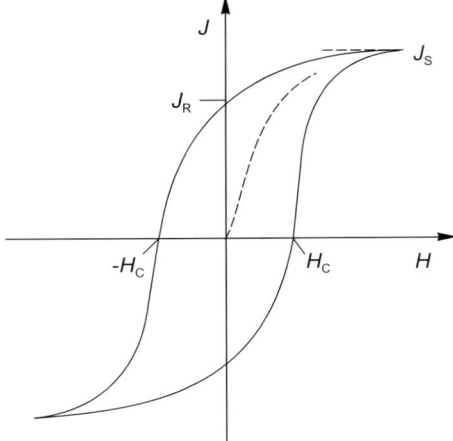

Fig. A.4.21 Magnetisation curve of ferromagnetic steels (schematic): As the field strength H increases, polarisation J follows the new dotted line until saturation at J_S. In the following remagnetisation cycles, J follows a hysteresis loop the area of which increases with increasing amount of dissipated energy. H_C and J_R are known as the coercitive field strength and remanence, respectively.

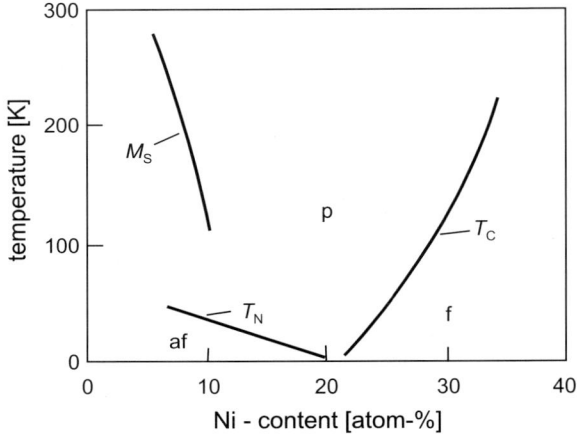

Fig. A.4.22 Magnetic transformation in Fe-Ni alloys containing 20 atom% Cr (from W. Bendick and W. Pepperhoff): M_s = martensite start temperature, T_N = Néel temperature, T_C = Curie temperature, p = paramagnetic, af = antiferromagnetic, f = ferromagnetic

A.4.4 Special physical properties

magnetic domains, there is no detectable external moment. Another possible orientation is an anti-parallel alignment of the moments in ferrous alloys below the Néel temperature T_N (antiferromagnetism). In this case, the moments compensate each other at the atomic level.

In contrast to the paramagnetic and antiferromagnetic states in which conventional external magnetic fields cannot produce significant polarisation, this can easily be achieved in the ferromagnetic state by migration of the Bloch walls and rotation of the polarisation direction of the domains. The relationship between field strength H and polarisation J is shown in Figure A.4.21. We differentiate between non-magnetisable (non-magnetic), magnetically soft and magnetically hard (permanently magnetic) ferrous alloys.

In *non-magnetisable* materials, the paramagnetism of austenite is stabilised down to room temperature and below by alloying with Mn, Mn + C or Cr + Ni. The application range of austenitic CrNi steels is governed by the decreasing M_s and (above 20 % Ni) the increasing Curie temperature as the nickel content increases (Figure A.4.22).

Soft magnetic materials respond to even a weak magnetic field with a high degree of polarisation. In this case, the individual domains rotate easily into alignment with the nearest bcc cube edge in the direction of the field. Positions not in this preferred orientation, such as the face or body diagonals, can only be aligned by inputting additional energy. This anistropic energy increases the amount of hysteresis. Hysteresis losses generate heat. Saturation polarisation is only reached at high field strengths. A high mobility of the Bloch walls lowers the anisotropic energy. This effect is increased by adding nickel and by lowering the content of precipitates containing carbon, nitrogen, oxygen and sulphur. Silicon also reduces hysteresis. It also lowers eddy current losses by increasing the electrical resistance. Crystalline anisotropy, in which the cube edges are preferentially oriented in the direction of the external field, is also advantageous. Once a magnetised state has been induced in a *hard magnetic* material, it should be maintained if possible. In addition to high saturation polarisation, there should be a large coercitive magnetic field strength. The hysteresis loop is then extremely widened. This requires obstruction of the movement of the Bloch walls, which can be achieved with precipitates such as carbides, that have the same thickness as the Bloch walls ($\approx 0.1\,\mu m$). An even more effective method is to avoid Bloch walls by embedding the individual domains in a non-magnetic matrix. Domains can be dispersed by segregation, precipitation or by powder metallurgy.

The graphite in grey cast iron is not involved in magnetisation. The iron carbides in white cast iron have a composition-dependent Curie temperature, e. g. 210°C for cementite, below which they are only weakly magnetisable. The above-described magnetic effects apply to the matrix of cast iron.

A.4 Properties

A.4.4.2 Thermal expansion

The thermal expansion of α-iron increases with temperature and is 12 μm/mK at ambient temperature. Within the stability field of γ-iron, the value remains more or less constant at 23 μm/mK. Thermal expansion at room temperature is 11, 10 and 9 μm/mK for spheroidal graphite cast iron, grey cast iron and white cast iron, respectively.

During cooling of paramagnetic iron-nickel austenite, ferromagnetic realignment just below the Curie temperature results in volume magnetostriction associated with a lattice expansion (Figure A.4.23). Over a certain temperature range, it compensates thermal contraction during cooling. This so-called Invar effect is most noticeable in iron with 36 %Ni (Invar alloy). It can be increased even further by adding cobalt (Superinvar). Because lattice expansion determines the elastic modulus, there is a temperature range with an almost constant Young's modulus (Elinvar). In spheroidal graphite cast iron containing 35 % Ni, thermal expansion at room temperature is about 5 μm/mK.

The desired thermal expansion coefficients can be obtained by varying the alloying content, e. g. as a sealing alloy compatible with glass. Joining of alloys with large and small coefficients of thermal expansion by roll bonding is used to make thermostatic bimetals (Chapt. B.8, p. 378).

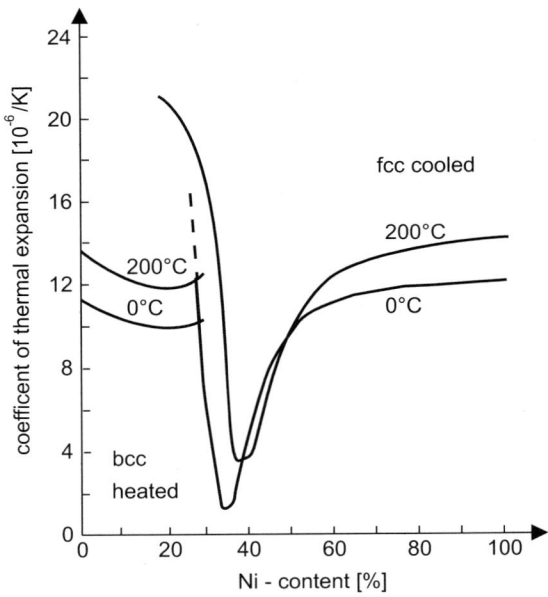

Fig. A.4.23 Thermal expansion of iron-nickel alloys (from P. Chevenard).

A.4.4.3 Conductivity

The freely mobile conduction electrons in metals are scattered by crystal defects. Steels with a good electrical conductivity σ are therefore be essentially free of alloying and tramp elements. Particularly carbon dissolved in interstices has an adverse effect on the conductivity and is thus precipitated as carbide. In contrast, work hardening is less important and is used e.g. in cold-drawn telephone wires. As the temperature increases, the electrons are decelerated by lattice vibrations, and the electrical resistance $\rho = 1/\sigma$ is raised. The resistance in the ferromagnetic state is lower than in the paramagnetic state. The thermal conductivity λ is proportional to σ. This is why e.g. a low-carbon steel is used for boiler tubes in steam generators. However, in spite of their lower thermal conductivity, more highly alloyed steels are often used in heat exchangers to ensure corrosion resistance.

A high electrical resistance paired with good scale resistance is expected in electrical resistance heating alloys. As discussed in Section A.4.3.2, chromium and aluminium lower high-temperature corrosion. At the same time, ρ may be increased by more than one order of magnitude. Addition of nickel produces austenitic alloys with a higher creep resistance.

The higher thermal conductivity of graphite means that λ is higher in grey cast iron compared to structural steel: by a factor of almost 1.2 in case of spheroidal graphite and by a factor of almost 3 in case of flake graphite. On the other hand, the electrical conductivity decreases from a factor of about 0.9 to a factor of 0.3. The Wiedemann-Franz Law which describes the proportionality of λ and σ and which is based on electron mobility, loses its validity, particularly in grey cast iron.

Machining of a free-cutting steel
(Steeltec AG, Emmenbrücke, Switzerland)

B
Ferrous materials and their applications

B.1 Materials for general applications

'Unalloyed' ferrous materials are often used for general engineering applications such as mechanical, automotive, plant and structural. Although they are called 'unalloyed', according to EN 10020, these materials do in fact contain certain amounts of manganese, silicon and other alloying elements. Carbon is not officially included with these elements; however, it determines the pearlite content in the microstructure and thus the mechanical properties of this group of materials, which can be divided into unalloyed structural steels and cast iron.

B.1.1 Unalloyed structural steels

Steels in this group contain either $< 0.9\,\%$ C and are identified according to their chemical composition (e.g. C45) or they are weldable grades containing $\leq 0.22\,\%$ C and are classified according to their yield point (e.g. S235). Their microstructure consists of a relatively soft ferrite and harder pearlite, whose proportion increases to $100\,\%$ for a eutectoid carbon content (Figure A.2.10, p. 30). As the example of a low-alloy steel given in Figure A.2.17 (p. 39) shows, an increase in the cooling rate increases the proportion of pearlite (from 60 to $70\,\%$ in the example) as well as the hardness (from 200 to 230 HV). This can be attributed not only to the higher proportion of harder pearlite grains, which now contain slightly less than the eutectoid carbon content, but also to the fact that they are more finely striped as a consequence of the faster cooling rate (Figure A.2.23, p. 46). Ideally, the pearlite grains in C10 should be dispersed in the ferrite similar to a dual-phase microstructure (Figure A.2.2 b, p. 22), C40 should have a duplex-type microstructure consisting of ferrite and pearlite, and C60 should have a network of ferrite around the pearlite grains. However, in reality, they contain rolling-induced banding (microstructural anisotropy, Figure A.2.6, p. 25) consisting of differing microsegregations of Mn and Si in which the Mn-enriched bands contain more C. The ferrite and pearlite bands lie side-by-side in the rolling direction (Figure B.1.1), and their delimitation is all the more pronounced as the degree of plastic deformation increases.

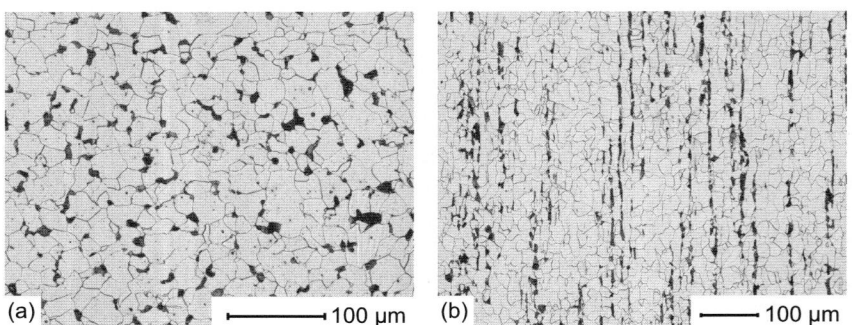

Fig. B.1.1 Microstructure of C15 steel consisting of ferrite (light) and pearlite (dark); ⌀30 mm: (a) transverse section, (b) longitudinal section

B.1.1.1 Properties

The service and manufacturing properties of unalloyed structural steels are closely linked to their carbon content. This low-cost element allows the properties to be tailored within a remarkable range.

(a) Mechanical properties

The proportion of pearlite increases with the carbon content and so does the tensile strength. In addition to this two-phase strengthening by the harder pearlite grains, there may also be some solid solution hardening by e. g. silicon and manganese and some grain refinement involved. The yield strength increases with the pearlite content to a lesser degree than the tensile strength because slip starts in the softer ferrite grains. Figure B.1.2 shows that alloying with approx. 1.5 % Mn leads to higher values for the weldable grades compared to the other steels. The post-rolling cooling rate decreases during air quenching as the thickness increases. This reduces the amount of pearlite (Figure A.2.17, p. 39), and the spacing between the carbide plates increases. Both lower the strength. At the same time, the ductility is reduced owing to the greater thickness of the carbide plates (Figure A.4.13, p. 96). The transition temperature T_T increases with the fraction of pearlite, the carbide plate thickness and the silicon content. This undesirable shift can be compensated by grain refinement (Chapt. A.4, p. 85 and p. 95) during a normalising heat treatment or normalising rolling.

(b) Wear resistance

Grooving abrasion as well as the tendency to adhesion decrease as the proportion of the harder pearlite grains increases. Thin cementite plates can also be grooved or broken by mineral grains that are softer. Pearlite with coarse lamellae is thus advantageous for this type of wear. In contrast, a high yield point is important for rolling wear in order to withstand rolling contact fatigue. This requirement is met by the fine lamellar microstructure of lower pearlite that does not contain any grains of ferrite. In applications with grooving abrasion by minerals, the surface can be wear-protected e. g. by hard-facing via flux-cored wire welding. Hard chromium and nickel coatings are used for machine components with narrow tolerances. This has the additional advantage of providing corrosion protection. Owing to the low load-bearing capacity of the substrate, high point loads must be avoided to prevent penetration of the thin hard coating.

(c) Corrosion resistance

Atmospheric corrosion in humid air produces a porous layer of rust consisting of Fe_3O_4 and $FeOOH$. Small additions of copper (as well as phosphorus, chromium and nickel) significantly increase the density and adhesion

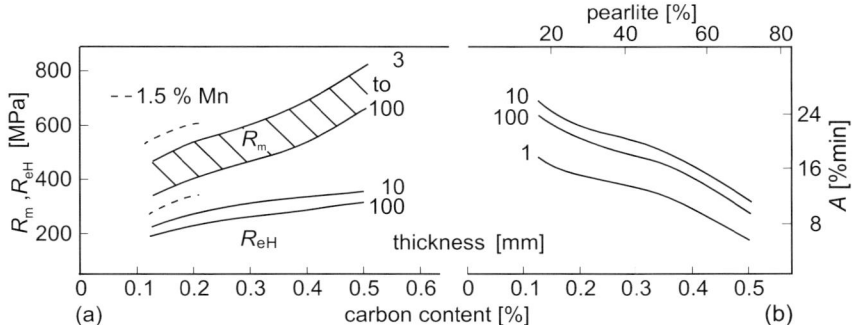

Fig. B.1.2 Mechanical properties of general structural steels according to EN 10025: (a) The proportion of harder pearlite grains increases with the carbon content. This is accompanied by an increase in the tensile strength R_m and, to a lesser extent, in the upper yield point R_{eH}. As the cross-section becomes thinner, the cooling rate from the normalising temperature increases. As a result, the proportion of pearlite increases slightly and also becomes more finely striped i. e. harder. In S235 to S355, added manganese causes solid-solution hardening, making the pearlite even more finely striped. (b) Starting from a thickness of 100 mm: as the cross-section becomes thinner, the elongation at fracture A initially increases on account of the thinner carbide lamellae, in spite of the increasing strength, and then decreases again with a further increase in strength.

of the layer and thus lower the corrosion rate (weather-resistant steels, e. g. S355J2WP as per EN 10025-5). Sulphur dioxide reacts with oxygen and water to sulphuric acid, which in turn reacts to give iron sulphate and is then regenerated by hydrolysis. Thus even the small amounts of SO_2 contained in industrial atmospheres can accelerate uniform corrosion or localised corrosion of structural steels. In drinking water pipes, a protective layer of lime scale forms if the water is neutral and sufficiently hard, i.e. contains Ca ions. Reinforcing steels exhibit passive behaviour in alkaline concrete (pH = 11 to 13). However, the steel begins to rust after CO_2 and SO_2 have consumed the depot of $Ca(OH)_2$ that formed as the cement was hardening. The associated increase in volume causes spalling of the outer layer of concrete and the whole process repeats itself.

In addition to painting and coating with polymers, the most widely used method of corrosion protection is coating with zinc. Zinc reacts with atmospheric humidity and carbon dioxide to produce a protective layer of basic zinc carbonate. Zinc also acts as a sacrificial anode, thus providing cathodic protection, because it is less noble than iron (Tab. A.4.3, p. 111) and dissolves first. This protection is able to bridge layer discontinuities of up to ≈ 1.5 mm, such as injuries, cut edges and laser welds.

Hot-dip coating with zinc at $\approx 450°C$ may alter the mechanical properties and induce distortion due to residual stresses, whereas electroplating at $< 80°C$ allows the uptake of hydrogen. The higher temperature produces an compound Fe/Zn zone between the steel and the layer of zinc. Enamelling at $\approx 800°C$ affords a glassy oxide-ceramic coating that is resistant to corrosion and wear. However, this involves heating the substrate into the austenite/ferrite range.

(d) Welding suitability

During fusion welding, the peak temperatures in the vicinity of the fusion line bring the heat-affected zone (HAZ) close to the melting temperature (Figure B.1.3 a). In spite of the high heating rate and short residence time, the TTA diagram indicates that grain growth increases with the peak temperature (Figure B.1.3 b). The high level of dissolution shifts the transformation lines of the TTT diagram to longer times during subsequent cooling (particularly for alloyed steels). Therefore, the microstructure in the HAZ is estimated on the basis of welding TTT diagrams (austenitising temperature $\geq 1300°C$). The cooling rate due to heat dissipation in steel (self-quenching) is particularly high during multi-pass welding of thick seams. This can produce some martensite and bainite and thus induce hardening compared to a normalised ferrite/pearlite microstructure (Figure B.1.3 c). Below M_f ($\approx 250°C$ for S355), the coarse needles of martensite produced from large grains of austenite are exposed to multiaxial shrinkage stresses on further cooling. This brings together all the prerequisites for brittle cleavage fracture, particularly if the

B.1.1 Unalloyed structural steels

seam is rigidly restrained and if there is an undercut. The tendency to such cold cracking can be estimated from empirical equations for the carbon equivalent value CEV. According to EN 10025,

$$CEV = C + \frac{Mn}{6} + \frac{Cr + Mo + V}{5} + \frac{Ni + Cu}{15} \quad (B.1.1)$$

and its permissible limit value increases with the yield point and the nominal thickness from 0.35 to 0.49.

Because a high carbon content is particularly crack-promoting, C is limited to $\leq 0.22\,\%$ in welding steels. Steels with higher carbon contents require special treatment if they are to be welded, e.g. preheating. This lowers the increase in hardness in the HAZ and reduces shrinkage stresses that also occur at higher temperatures, i.e. above T_T. Preheating also facilitates the effusion of embrittling hydrogen (Chapt. A.4, p. 114). The CEV of welding materials is adjusted to sufficiently low values by tailoring the composition of the weld filler. A low nitrogen content together with e.g. 0.02 % Ti can, under certain manufacturing conditions, lead to the precipitation of TiN particles of ≈ 20 nm that inhibit grain growth in the HAZ and thus allow a higher input of thermal energy. This means that higher welding speeds are possible.

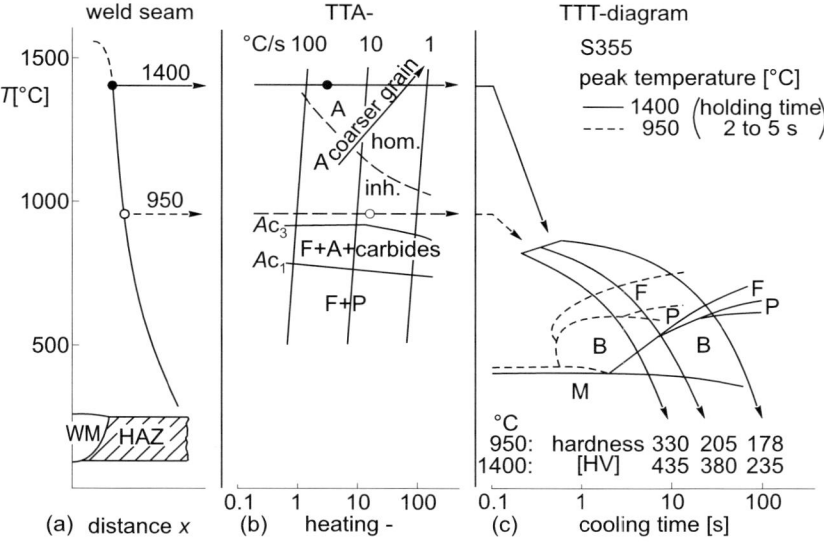

Fig. B.1.3 Hardening in the heat-affected zone (HAZ) during welding: (a) Temperature curve across the weld seam, WM = weld metal. (b) Position in the TTA diagram. (c) Transformation in the HAZ, solid line = close to the fusion line, dashed line = further away (TTT diagrams from J. Ruge, R.Müller and H.J. Peetz)

(e) Ageing

Dissolved carbon and nitrogen atoms collect in dislocation sinks. This segregation into defects (Figure A.2.7 c, p. 25) means that higher stresses are required in order to initiate the movement of dislocations. As a consequence, the upper yield point increases along with the transition temperature. This time-dependent change in the properties is known as ageing. At climatic temperatures, this diffusion requires months to years. Nitrogen is up to one order of magnitude faster and thus determines natural ageing. During ageing, $\alpha"$-nitride can precipitate from the Cottrell cloud of interstitial atoms around a dislocation (Tab. A.2.1, p. 52). Artificial ageing at elevated temperatures, e. g. 150 to 200°C, takes only minutes or hours. This induces precipitation of ε-carbide.

Here are two initial conditions for ageing: a homogenised distribution of interstitial atoms by solution-annealing of ferrite just below Ac_1 and quenching

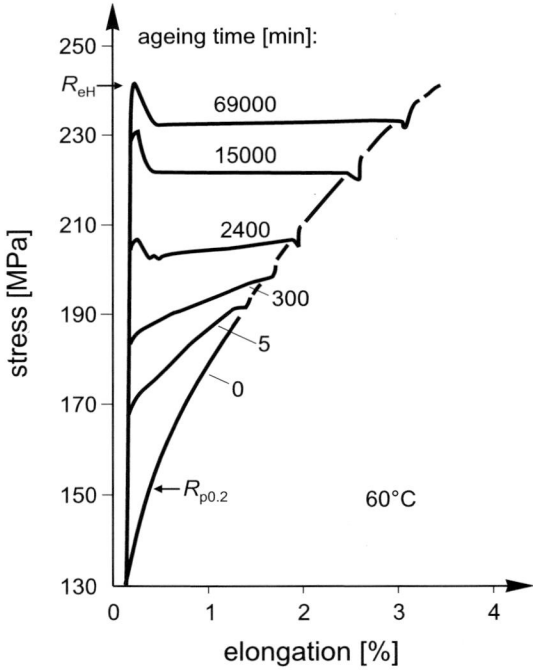

Fig. B.1.4 Ageing: Development of a discontinuous yield point in iron containing 0.01 % C after quenching from 720°C by ageing at 60°C (from W. Dahl and E. Lenz). $R_{p0.2}$ in the initial state increases over 48 days to R_{eH}. The initial curve is reached again after ≈ 3 % Lüders elongation.

(quench-ageing) or generation of new dislocations without C,N clouds by cold working (strain-ageing). In both cases, the materials do not have a discontinuous yield point; this develops on subsequent ageing (Figure B.1.4). The risk of brittle fracture during quench-ageing after welding can be eliminated by binding nitrogen with > 0.02 % Al to produce AlN. 'Bake-hardening' uses artificial strain-ageing by carbon to increase the strength of deep-drawn parts during paint baking.

(f) Anisotropy

Banding caused by elongation of microsegregations and non-metallic inclusions (Figure A.4.11, p. 93) has an adverse affect on the toughness transverse to the hot-working direction. In longitudinal specimens, these bands are parallel to each other (Figure B.1.1 b). Towards the end of the tensile test, short microcracks form in the harder pearlite bands; however, they can still be held in check by the softer ferrite. In transverse specimens cut through a series of bands, these microcracks in the pearlite bands can propagate (perpendicularly to the plane of the picture in Figure B.1.1 a) so that only a relatively small amount of applied strain is needed to fracture the specimen. There is a certain analogy to elongated inclusions (Figure A.4.6, p. 86). Manganese sulphides are very ductile at the hot-working temperature, e. g. they are elongated and widened during rolling of flat products (Figure A.4.11, p. 93). This may lower the notched impact bending energy in the through-thickness direction to only a tenth of that in the longitudinal direction, which thus increases the risk of lamellar tearing in butt-welded joints (Figure B.1.5) as well as cracking during bending of sheets if the fold line is parallel to the rolling direction or during heading of wire. Alloying with calcium, rare-earth metals, zirconium or titanium produces compact sulphides that are difficult to deform. In addition to a reduction in the sulphur content – which originates from the smelting coal – to a few thousandths of a percent, this influence on the sulphide shape contributes to decreasing the microstructural anisotropy. The shape of these sulphides can be determined according to Stahl-Eisen-Prüfblatt (Stahl-Eisen Testing Guideline) 1575. To avoid lamellar tearing, EN 10164 recommends a reduction in area in the through-thickness direction that can be agreed upon on ordering, e. g. Z35 for Z ≥ 35 %. In steel sheets, a preferred grain orientation of <111> rather than <100> in the plane of the sheet (Figure A.1.4, p. 8) has a favourable effect on the deep-drawing properties. The perpendicular anisotropy r is determined from the ratio of the width-to-thickness shape change, $r = \varphi_w/\varphi_s$ measured with flat tensile specimens. Averaging of results from specimens taken at differing angles with respect to the rolling direction gives

$$r_m = (r_{0°} + 2r_{45°} + r_{90°})/4 \qquad (B.1.2)$$

Precipitation of an AlN dispersion during recrystallising annealing after cold rolling can be used to obtain the favourable <111> texture with $r_m > 1$.

(g) Deep-drawability

Figure B.1.6 illustrates the deep-drawing process using the example of a simple round cup. The deep-drawability can be described by the limiting draw ratio $\beta_{max} = D/d$, which increases with the ductility of the sheet. Soft, low-carbon steels are thus often used. β_{max} increases linearly with the normal anisotropy r_m. A high strain-hardening exponent n counteracts a reduction in area of the sheet within the cup (Figure A.4.9 c, p. 90). Such is the case if the dislocations in the grain are highly mobile. This requires the lowest possible content of interstitial carbon and nitrogen atoms to lower their 'friction', with dislocations. The value of n increases as the friction stress R_{e0} (Figure A.4.12 c, p. 94) decreases. Utilisation of a coarse-grained structure to lower the yield point is out of the question owing to the possibility of producing a so-called 'orange peel' effect. Planar anisotropy Δr leads to undesirable earing.

$$\Delta r = (r_{0°}/2) + (r_{90°}/2) - r_{45°} = 2(r_m - r_{45°}) \tag{B.1.3}$$

If $\Delta r > 0$, ears are produced at the edge of the bowl at an angle of 0° and 90° to the rolling direction, or at an angle of 45° if $\Delta r < 0$. A value of $\Delta r = 0$ is preferable in the interests of an optimum exploitation of the material.

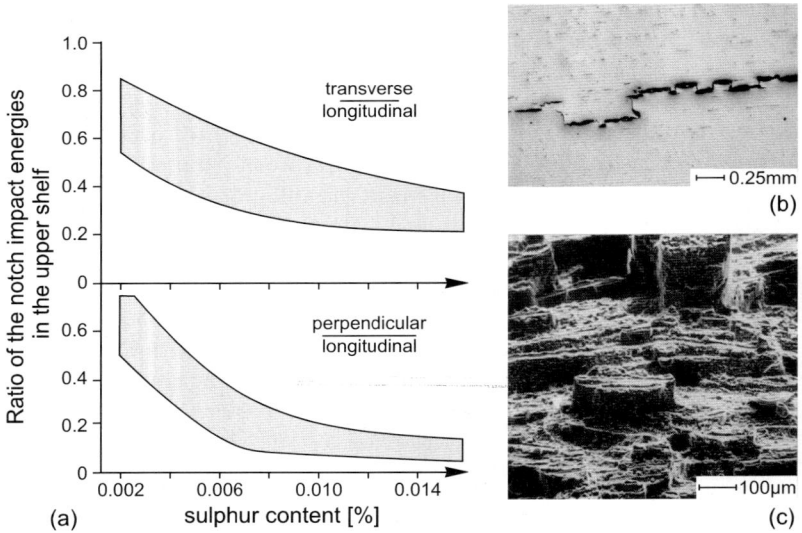

Fig. B.1.5 Influence of the sulphur content in flat steel: (a) Anisotropy of the toughness due to sulphides (from W. Haumann). (b) Lamellar tearing along the stretch-formed sulphides in a metallographic section. (c) Surface of a lamellar tear.

B.1.1 Unalloyed structural steels

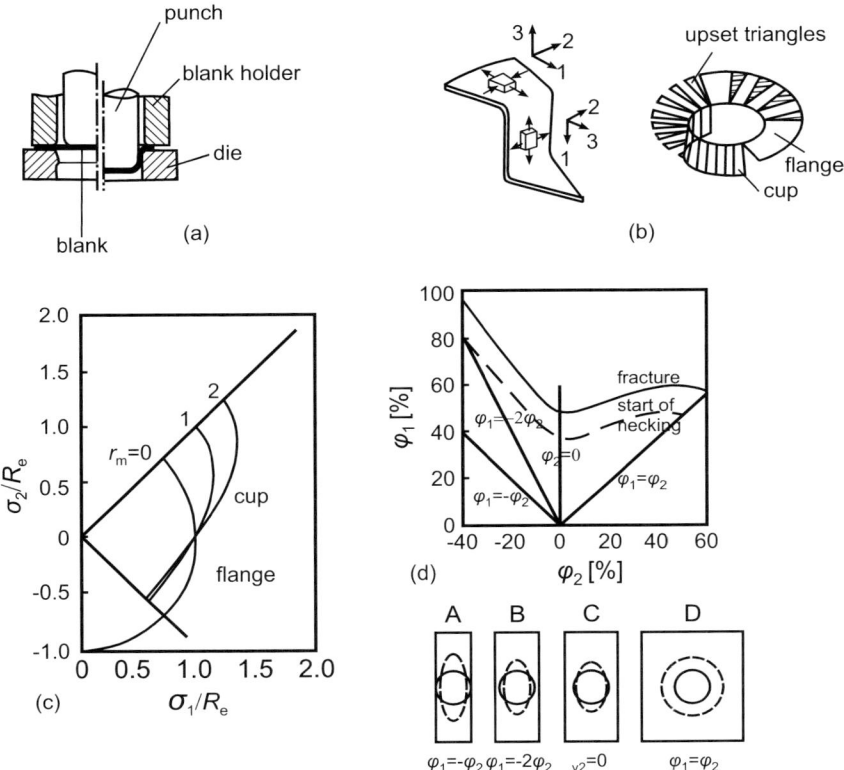

Fig. B.1.6 Deep drawing: (a) A circular blank with diameter D, clamped between the die and the blank holder, is cold-shaped to a cup by a punch with diameter d. (b) The plane stress condition ($\sigma_3 \approx 0$) induces shear stresses in the flange ($\sigma_1 \approx -\sigma_2$) and biaxial tensile stresses in the cup ($\sigma_2 \approx \nu\sigma_1$). The hatched triangles have to be upset and transformed into the length of the cup. (c) Start of yielding for plane stress is given by Figure A.4.1, p. 80. The elliptical yield locus curve (from R. Hill) is influenced by the normal anisotropy r_m, i.e. $R_e^2 = \sigma_{eq}^2 = \sigma_1^2 + \sigma_2^2 - \sigma_1 \cdot \sigma_2 \cdot 2r_m/(r_m+1)$. As r_m increases, the flange and the cup undergo textural softening and hardening, respectively. This latter counteracts an undesirable decrease in the sheet thickness caused by uniaxial stretch-forming in the cup. (d) The type of deformation can be demonstrated using a circular mark that changes shape during forming. We differentiate between A = forming under tension/compression, B = forming under tension, C uniaxial and D biaxial stretch-forming. C involves the lowest fracture strain and is the reason for tears in the base of the cup near the transition to the base.

(h) Machinability

Hard phases, such as oxides and carbides, wear down cutting tools; however, they crack and thus reduce the required cutting work as well as heating of the cutting tool. Wear can be reduced by lowering the carbon content and removal of oxides from the melt. At the same time, the addition of non-abrasive chip breakers reduces heating and produces a short-chipping grade (without continuous chips). The most widely used is manganese sulphide MnS. Even a few hundredths of a percent sulphur improve the machinability. For sulphur contents of up to 0.4 %, the manganese content must also be raised (Mn/S ≤ 4.5) to prevent red shortness during hot-working that is caused by low-melting iron sulphides. Selenium and tellurium can be used instead of manganese to reduce the anisotropy. However, its impact is usually less significant in e. g. steel bars for automatic lathes compared to sheets. Finely dispersed lead, which is insoluble in steel owing to its large atomic diameter (Tab. A.1.4, p. 15), has an effect similar to that of MnS. If Se, Te and Pb are used, their toxicity must be taken into consideration. Deoxidation of the melt with elements that have an affinity for oxygen may produce low-melting oxides belonging to the ternary CaO-Al_2O_3-SiO_2 system. Such oxides produce a glassy protective layer on titanium-containing hard-metal tools or hard coatings during machining at particular cutting speeds. Another development aims to transform cementite into graphite by annealing steels containing e. g. 1.5 % Si and 1.4 % Al at 680°C. Both alloying elements shift the constitution from a metastable to a stable system, and a fine dispersion of graphite precipitates improves the machinability with an acceptable loss of ductility (Figure A.2.25 d, p. 49).

B.1.1.2 Grades and applications

The huge range of applications for unalloyed steels will be discussed using various types of products. Particularly in volume production, manufacturing properties are at least as important as the service properties. The key development objectives are to reduce costs without losing quality or to increase the quality without increasing costs. Steel consumers are generally compelled to increase the quality and reduce the costs at the same time. Tab. B.1.1 summarises some examples.

(a) Flat products

Flat products in the form of plates or coiled strips are used in the hot-rolled (with and without scale) or cold-rolled (bright) state. EN 10079 classifies them according to their thickness (mm) as follows
< 0.5 very thin plate and strip (blackplate, e. g. for the packaging industry)
< 3.0 thin plate and strip (e. g. for the automotive industry)
$\geq$ 3.0 thick plate and strip (e. g. for structural engineering and shipbuilding)
EN 10029 covers thicknesses from 3 to 250 mm. Flat products thicker than

B.1.1 Unalloyed structural steels

Table B.1.1 Examples of unalloyed steels for general applications

Name	R_e [1] [MPa]	R_m [MPa]	A [%]	Content [%]	Application
TS230	≈ 230	≈ 325		≈ 0.06	blackplate for packaging EN 10202
TH620	≈ 620	≈ 625			S=batch-annealed, H=continuously annealed
DD14	> 170 [2]	< 380	> 31	< 0.08C	hot strip EN 10111
DC04LC	> 210	< 350	> 38	< 0.08C	cold strip EN 10139 skin-passed · for cold forming
DC06LC	> 180	< 350	> 38	< 0.02C < 0.3Ti	IF (interstitial-free) skin-passed
DX54D+Z	> 140	< 350	> 36		hot-dip zinc-coated strip EN 10142 for cold-working
DC04ED	< 210	< 350	> 38	< 0.004C	cold strip for direct-on enamelling EN 10209
S235	> 235	< 510	> 17	< 0.19C	hot-rolled products made
S275	> 275	< 580	> 14	< 0.21C	of unalloyed structural steels
S355	> 355	< 680	> 14	< 0.23C	EN 10025
S355WP	> 355	< 680	> 16	< 0.12C ≈ 0.1P	weather-resistant structural steel EN 10025-5
GS240	> 240	< 600	> 22	< 0.23C	cast steel EN 10293
B500	≈ 500	≈ 540	≈ 5	< 0.22C	concrete reinforcing steel EN 10080
Y1030H		≈ 1030		≈ 0.7C	prestressing rods EN 10138
R350HT	-	> 1175	> 9	≈ 0.76C	rails EN 13674-1, >350 HB
C22+N	> 240	> 430	> 24	≈ 0.22C	carbon steels
C45+N	> 340	> 620	> 14	≈ 0.45C	EN 10083-2
C60+N	> 380	> 710	> 10	≈ 0.60C	normalised
C2D1 to C92D		< 360 -		< 0.03C ≈ 0.93C	wire for cold-drawing regulated by EN 10016
11SMn37		> 370		< 0.14C ≈ 0.37S	free-cutting steel EN 10087 112-169 HB

[1] $>, <, \approx$: Lower limit, upper limit, reference value, for thin-gauge products

[2] Transverse specimens

25 mm are also referred to as very thick plate. They are used e. g. for thick-walled pressure vessels.

Very thin plate and strip (blackplate): EN 10205 regulates single cold-reduced strip of low-carbon steel for subsequent coating. It is subjected to continuous or batch annealing under an inert gas and then skin-passed with a $\leq 5\,\%$ reduction in thickness to obtain a continuous yield strength and to avoid flow lines caused by processing. The name of these grades is made up of the code letter T for blackplate and the hardness code, e.g. T61. Double-reduced blackplate undergoes a second reduction instead of a skin pass. It thus has a greater reduction in thickness and is strengthened to a 0.2 proof strength, which is given after the code letters DR (double reduced), e.g. DR620 with a hardness code of 76. EN 10202 regulates the use of this material as input stock to produce either tinplate by electrolytic coating with tin, including products with different thicknesses on the upper and lower sides (differential coating), or electrolytically chromium-coated steel (ECCS) by deposition of chromium with a hydrated top layer. The code letter T is followed by an S for batch annealing or an H for continuous annealing and then the 0.2 proof strength, which increases with the degree of cold reduction (example: TS230, TH620). These grades can be used to make packaging materials, e.g. food cans and biscuit tins.

Thin plate and strip: EN 10111 regulates hot-rolled strip of low-carbon steel with a thickness of 1.5 to 8 mm that is intended for further processing by cold-working. The code letters DD are followed by a code number (e.g. DD14). According to EN 10120, this can be used to produce e.g. cold-rolled thin sheets with a smooth surface and narrow tolerances. EN 10139 covers widths < 600 mm. This cold strip is designated DC, which is followed by code numbers (e.g. DC04). It is intended for cold-working. Therefore, this material is also subjected to inert gas annealing and skin-passing after cold rolling. Similar to blackplate, this skin-passing can also be used to increase the 0.2 proof strength (e.g. DC04 + C590). Fully killed hot- and cold-strip grades contain enough aluminium to bind most of the nitrogen and thus suppress strain-ageing (p. 130). The mechanical properties are guaranteed for up to 6 months, which means that the materials do not have to be processed further as quickly as semi-killed grades whose deadline may be as little as 1 month. The addition of < 0.3 % Ti or a corresponding amount of Nb binds C in addition to N. The ferrite thus contains almost no dissolved interstitial atoms. This not only completely excludes flow lines in the IF (interstitial-free) grade DC 06 (EN 10130), but also produces a lower yield strength ($\geq$ 120 MPa) and the highest strain-hardening exponent (n $\geq$ 0.22). The latter with its high uniform elongation is advantageous with respect to its stretch-forming properties. The first code letter D, which is used for both hot and cold strip, stands for ductility, i.e. its cold-workability. Applications of these products include deep-drawn lampshades, cold-profiled frames of automobile seats, embossed housing covers or stamped frames and covering panels.

Owing to its popularity in automotive manufacture, the proportion of coated cold strip has continuously increased. The aim is to provide corrosion protection by coating with metals (hot-dipped or electrolytic) and/or with polymers (liquid or powder coatings or foils) before the strip is processed by cold-working. Hot-dip coated, low-carbon strip steels (covered by EN 10142) are produced in a manner similar to that described for uncoated strip. The final treatment is dip-coating in a continuous process and an optional skin-pass. Their name thus also starts with D (e. g. DX54 D), followed by the composition of the coating (Z = zinc, ZF = zinc and iron, e. g. DX54 D + ZF). The proportion of iron increases if the material is heated to $\leq 550\,°C$ after treatment in the zinc bath at $\approx 450°C$ (galvannealed). The addition of only 0.2 % Al limits the formation of iron-zinc compounds by producing an extremely thin layer of Fe_2Al_5. For 5 % Al, the temperature of the zinc bath is reduced to $\approx 420°C$ and the resulting Zn-Al coating has a higher corrosion resistance in industrial atmospheres (Galfan). According to EN 10327, the designation for Zn-Al coatings is e. g. DX54 D + ZA. In addition to the thickness of the Zn or ZnAl coating (7 to 45 µm, equal or unequal on the upper and lower surfaces of the strip), the surface quality is also of importance, which increases from class A to class C. The formation of zinc spangles can be suppressed by reducing the amount of lead impurities. Hot-dip coatings of Zn - 55 % Al - 1.6 % Si (Galvalume) are more resistant to maritime climates. According to EN 10327, their designation is e. g. DX54 D + AZ. Aluminising with a near-eutectic Al - 10 % Si - alloy improves the service life of exhaust gas systems in vehicles (e. g. DX54 D + AS, as per EN 10327). For electrolytic zinc coatings, EN 10152 specifies a coating thickness of between 2.5 and 10 µm. Their designation is e. g. DC04 + ZE. Electrolytic coatings are applied at a lower temperature and no intermetallic phases between the steel substrate and the coating metal are formed, which improves the cold-processing properties. An electrolytic Pb-7 % Sn coating has been successful for fuel tanks.

If not only corrosion protection, but also a decorative appearance and colouring are important, thin sheets – even those with a metal coating – are given an organic coil coating. The general designation specified in EN 10169 is e. g. DC04 + OC (organic coating). A hot-dip galvanised strip with a polyester coating (SP) 25 µm-thick on both sides is designated e. g. DX 54 D + Z - SP 25/SP 25. The organic coating slows down corrosion of zinc, and the zinc prevents rusting underneath the coating. All in all, there are many types of flat coil-coated products available as wide strip, slit strip or plates. Applications include civil engineering, equipment, appliances, furniture and packaging. Another type of product is a sandwich construction consisting of two galvanised panels firmly bonded together by a thin layer of polymer to produce a sound-damping composite sheet that can be processed further like thin sheet. Such materials are used e. g. to encapsulate motors and compressors. Products with a thick polymer layer are used in lightweight engineering. In the automotive industry, the primer coat is often applied to a metal-coated strip in order to speed up the final painting step after processing.

If scratch and wear resistance are required in addition to corrosion protection, this can be achieved by enamelling at a temperature between 800 and 850°C to produce a ceramic coating on components made of cold-rolled strip, e. g. DC 04. According to EN 10209, the letters EK are to be used for materials with conventional two-layer enamelling consisting of a primer and a top coat, or ED for direct-on enamelling, e. g. DC 04 ED. The resistance of steel to the formation of fish scales is important here. These scales are caused by hydrogen, originating from moisture in the air or from the coating, that is reduced by Fe during enamelling to hydrogen atoms which then dissolve in the steel. The layer of enamel hinders their effusion during cooling. The H atoms from the resulting saturated solution recombine in the microvoids at the boundary interface. During service, the H_2 pressure slowly builds up and can lead to semicircular cracks in the coating known as fish scales. Cold-rolling produces tapered voids at either end of non-metallic inclusions in the direction of metal-working. A uniform distribution of these fine pores distributes the volume of H_2 more evenly and thus relieves the H_2 pressure in the boundary between the steel and the enamel. In addition to these low-carbon grades with a C content of a few thousandths of a percent, there are also IF grades with a C content of a few hundredths of a percent that also contain $< 0.3\,\%$ Ti (or a corresponding amount of Nb). This Ti binds the C and N as hard MX, some of which precipitates from the melt to form large particles $< 10\,\mu m$. Similarly large particles of hard oxide are produced if the melt is deoxidised with Ti. Some of these particles fracture during hot-working thus producing H_2 traps. In addition to household appliances with a coloured and decorative coating, enamels are also commonly used in industrial applications, such as tanks or heat exchangers, where they are in contact with erosive media. Examples include components of flue gas desulphurisation (FGD) plants that operate at temperatures of up to 260°C. Apart from the aforementioned thin sheet and strip used for cold-working and coating, there are also other unalloyed steels in this gauge range which are described below under thick plate.

Thick plate and strip: The unalloyed structural steels specified in EN 10025 are used in many types of products, including thick plate. Their steel designation starts with the letter S (structural steel) or E (engineering steel) followed by the minimum yield strength for a nominal thickness of $\leq 16\,mm$, which ranges from 235 to 360 MPa (e. g. S235, E360). The S grades guarantee a particular notch impact energy (J) at room temperature (R) or at 0°C (0) or at -20°C (2) (e. g. S 355 J2). Classes G1 to G4 are increasingly killed with Al, and the P and S contents decrease (e. g. S275 J2 G3). Further letters can be appended to indicate suitability for bending or roll-forming (C) or for normalising or normalising rolling (+N). The carbon equivalent (CEV) is designed to indicate weldability; however, there are limitations, such as the degree of kill i. e. the resistance to ageing. If the material is to be bent across the longitudinal direction, the amount of microstructural banding must be sufficiently low to avoid bending cracks. These steels are used e. g. in vehicle construction (lorries, trains), in structural engineering (high-rise buildings, bridges), in shipbuilding and in general mechanical and plant engineering. They are used not only as flat products, but are also bent into sections and welded into I-beams that are more lightweight than those made of roll-formed profiles. Similar steels are used in the construction of pressure vessels (e. g. P355 G H as per EN 10028-2), for which the notch impact energy in the transverse direction is guaranteed. The contents of P and S are thus limited to $\leq 0.025\,\%$ and $\leq 0.015\,\%$, respectively. Because the hoop stress of a pressure vessel is twice as high as the longitudinal stress, attention must to be paid to the direction of banding during roll bending. Lower requirements are specified in EN 10207 for steels used in simple pressure vessels.

Normalised (N) unalloyed steels with a C content of 0.22 to 0.60 % are covered by EN 10083-2 (e. g. C 45 E + N). E stands for a max. content of 0.035 %S. An R means 'resulphurised', with $0.02\,\% < S < 0.04\,\%$, to improve the machinability; however, this is usually specified for steel bars. The mechanical properties of these steels are determined by their pearlite content (Figure B.1.2). Steel C50 E containing 0.7 % Mn has a yield strength of 355 MPa, whereas this value is attained in S355 with only 0.2 % C and 1.6 % Mn. In contrast to the S grades, the C grades are not intended to be used for welding, but for screwed, bolted and riveted connections in structural and mechanical engineering, for flame-cut machine frames or for the construction of jigs and fixtures.

(b) Long products

Bars and sections: Hot-rolled steels are specified in EN 10058 to 10061 for flat, square, round and hexagonal cross-sections, e. g. made of unalloyed structural steels such as S355 (EN 10025) or C45 (EN 10083-2). They are used for a wide range of applications in mechanical engineering, particularly for machining. This is facilitated by using free-cutting steels as specified in EN 10087, e. g. 11SMn30 with 0.3 % S and 1.1 % Mn or 11SMnPb37 with 0.37 % S, 1.25 % Mn and 0.3 % Pb. Account must be taken of the welding suitability of ribbed concrete-reinforcing steel such as B500 A (regulated by EN 10080) for untensioned reinforcement. This is achieved by improving the yield strength by means of cold straining and artificial ageing. The ductility classes A, B and C in Parts 2 - 4 of this standard refer to the overall elongation at the maximum load. Welded mats and lattice girders are specified in Parts 5 and 6. The wire used for this is strengthened by cold drawing and subsequent cold-ribbing. It is also possible to partially quench the material in water from the rolling temperature so that only the surface zone has a martensitic microstructure, which is then tempered by the residual heat in the core of the wire. In contrast, prestressing steels are not intended for welding. Their fine pearlitic microstructure with $\approx 0.7\,\%$ C is subjected to cold straining and artificial ageing to obtain a nominal tensile strength of e. g. 1030 MPa (Y1030 H according to EN 10138-4). Rails have a similar microstructure with a hardness of e. g. ≥ 260 HB (R 260), which corresponds to a tensile strength of ≈ 900 MPa. The interlamellar spacing can be reduced even further by alloying with 1.2 % Mn and cooling of the rail head from the rolling temperature. These heat-treated rails attain values of e. g. 350 HB (R350 HT according to EN 13674-1), which corresponds to a tensile strength of ≈ 1200 MPa. Grain-boundary ferrite is undesirable because it forms a soft net (see Fig. A.2.2, p. 22)that is more strongly deformed by repeated rolling contact than the pearlite within, thus enhancing fatigue. This is also the reason why the decarburisation depth is limited because a loss of carbon would produce a network of ferrite. Pearlitic steels with coarse lamellae, such as 85Mn3, are suitable for components subject to wear. Steels specified by EN 10025 (e. g. S355) can be used for load-bearing elements, such as beams of I, T, U, L and Z shape, in structural steelwork. Alternatively, EN 10162 specifies profiles of cold-workable strip produced by roll-forming, e. g. a U section made of S235 or a Ω section made of galvanised DX52D+Z. I sections are also manufactured by welding together three strips for the upper flange, lower flange and web.

Some unalloyed steels, such as E335 and S355 (EN 10025) or C22 to C60 (EN 10083-2), are marketed as bright steel that has been peeled (+SH) or drawn (+C) with a smooth surface for components with narrow tolerances, such as shafts and locating pins, or for electroplating. Cold-drawing increases the yield strength, particularly for small diameters, e.g. $\varnothing\, 10$ mm to ≥ 520 MPa for S355. The surface quality and the resistance to ageing of the rolled steel are particularly important.

Wire: Rolled wire (D) that is intended for cold-processing (C) is made of carbon steel C4 D to C92 D (EN 10016-2). This includes shaping by cold-drawing and cold-extrusion to screws, nuts, bushings, etc., on automatic multistage metal-working machines. A soft initial condition is achieved by limiting the C content to 0.2 % and spheroidising the cementite (AC), e. g. C15C + AC (EN 10263-2). The S content is kept at ≤ 0.025 % to limit the number of elongated sulphides and thus prevent fine microcracking of the surface during cold-heading. Other surface defects, such as laps and seams, must also be avoided. A sharply defined core segregation can also lead to tearing. Link chains are made of round wire according to EN 10025 (e. g. S355) by cold-bending followed by resistance pressure welding of the individual links. Wire with a fine pearlitic microstructure and ≈ 0.7 % can be used to prestress concrete. This is produced according to EN 10138-2 by cold-drawing (C), which increases the nominal tensile strength as the diameter decreases to values exceeding 1800 MPa (Y1850 C). Such wires, smooth or profiled, can also be used to make stranded wire for prestressing purposes (EN 10138-3). A similar fine pearlitic microstructure is suitable for wire ropes and is achieved by patenting (Chapt. A.3, p. 63). According to EN 10264-2, cold-drawing can increase the tensile strength up to 2160 MPa. However, this decreases the attainable number of bends or twists to fracture. Wires coated with Zn or ZnAl are also made into wire ropes. The same applies to spring wire, which is patented and drawn and then cold-coiled to spiral springs. EN 10270-1 regulates diameters down to 0.05 mm that have a tensile strength of up to 3500 MPa. A steel cord with a diameter of e. g. 0.2 mm has a tensile strength of about 3000 MPa. A good cleanness with only very small oxide inclusions is a prerequisite for a high fatigue strength and thus also e. g. for a long service life and good safety of tyres. A brass or bronze coating improves the adhesion between the

B.1.1 Unalloyed structural steels 141

reinforcement and the rubber in a tyre, conveyor belt or driving belt. These and other metallic coatings on steel wires are regulated by EN 10244. Organic coatings, such as PVC and PE, are regulated by EN 10245.

Pipes and tubes: This group of products is classified according to the method of manufacture (welded/seamless, hot-finished/cold-drawn), the shape (round, square and other profiles), the application (pressure-retaining for pipelines, non-pressurised for mechanical engineering and structural steelwork) and the coating (internal/external). The C content is limited to $\leq 0.22\,\%$ in tubes with a longitudinal or spiral weld seam or to $0.6\,\%$ in seamless tubes. Seamless (EN 10216-1) and welded (EN 10217-1) steel tubes for pressurised pipelines have yield strengths that increase up to grade P265 TR 2, where P indicates the pressure and TR 2 the Al content and notch impact energy. Line pipes (L) for water pipelines (e.g. made of L355 according to EN 10224) or gas pipelines (e.g. made of 360 GA according to EN 10208-1) are usually welded, but seamless varieties are also available. If a protective layer of lime scale cannot form in contaminated water, the pipes can be lined with cement mortar (EN 10298). Pipelines buried in the ground or passing through water require corrosion protection of their external surface by coating with zinc (EN 10240), polymers (EN 10288 to 10290) or tar/bitumen (EN 10300). Tubes are used in structural steelwork as linear members, supports and frameworks. EN 10210-1 regulates the production of seamless or welded hot-finished hollow sections (H) with a circular, square or rectangular cross-section made of e.g. S355 J2 H. Cold-shaped, welded hollow sections are regulated by EN 10219-1. Mechanical and vehicle engineering applications use welded tubes (EN 10296-1) to reduce weight (e.g. hollow shafts made of E355) or seamless tubes (EN 10297-1) for surface hardening (e.g. runners made of C60 E). Sizing or cold-drawing is used to limit the surface roughness and tolerances and to increase the yield strength. These precision steel pipes (EN 10305; e.g. E335) are also available as tubular steel sections with a square or rectangular cross-section and are widely used in lightweight construction of vehicles and mobile equipment.

(c) Other products

In addition to the aforementioned flat and long products, EN 10079 also regulates other groups, namely free-form and drop-forged parts as well as castings and powder metallurgy products.

Forgings: Parts produced by open-die forging include large disks, rings, bushings, shafts and rolls for mechanical engineering applications. The Stahl-Eisen Materials Guideline SEW 550 specifies steels C22 to C60 for a diameter of up to $d=1\,m$ and alloyed steels for up to $d=2\,m$. For products deviating from a cylindrical shape, EN 10222-1 defines a ruling section with comparable cooling conditions during heat treatment. The starting point is a reference cross-section of a plate with a ruling thickness t_R, a width of $>2t_R$ and a length of $> 4\,t_R$. For other cross-sectional shapes, an equivalent thickness t_{eq} is specified, e. g. for a round bar $t_{eq} = d \approx 1.5 t_R$ or for a square bar $t_{eq} \approx 1.2\,t_R$. The mechanical properties specified in this standard depend on t_R. Their values generally decrease in thicker cross-sections owing to the slower cooling rates. Taking C45 N as an example, for an increase in the diameter from ⌀ 250 to ⌀ 1000 mm, the minimum values of the yield strength drop from 325 to 295 MPa and the elongation at fracture in the longitudinal direction decreases from 18 to 15 %. Thick-walled steel forgings for pressure vessels are manufactured according to EN 10222-2. They can be longer than 5 m. In the example steel P245 GH, P stands for pressure and H for heat resistance up to 400°C. For $t_R = 150$ mm, the yield strength drops from e. g. ≥ 220 MPa (20°C) to 120 MPa (400°C).

Unalloyed drop-forged parts can be made of C22 to C60 according to EN 10083-2 or from billets, such as S355 or E360, according to EN 10025. Examples include fasteners used in construction, pedal cranks for bicycles or the connecting rod in a lawn mower.

Steel castings: For general applications, EN 10293 regulates weldable steels with a minimum yield strength of 200 to 300 MPa. The code letters S or E for structural steelwork or mechanical engineering is preceded by a G for a casting (e. g. GS200 or GE300). In contrast to rolling steel, an elevated Mn content is included in the name (17Mn5). Similar to rolling steel, steel castings for pressure vessels are regulated by EN 10213, e.g. as GP280 GH, with a guaranteed yield point at elevated temperature, which is designated as G280 in ISO 4991. Examples of large steel castings include frames for presses and rolling stands, mounts for large stamping tools and pressure-retaining machine housings for large compressors. Smaller series parts include fixtures and fasteners. A high polarisation or induction is important in housing components and magnet poles for DC motors (Figure A.4.21, p. 118). Low-carbon grades are suitable for such applications.

Sintered steel: Steel powders can be obtained in high outputs by atomisation of molten steel, reduction of powdered iron ore or vapour deposition (carbonyl iron). The powder can be shaped into green bodies by cold-pressing or by mixing with a polymer and then injection-moulding. Cost-effective protective gas sintering produces a porosity (in volume%) that can be adjusted within a wide range by optimising the compaction and sintering conditions and which can be lowered by sinter-forging. DIN 30910 regulates the following material classes and applications:

Sint-AF	$> 27\,\%$	filters
Sint-A	$25 \pm 2.5\,\%$	plain bearings, self-lubricating
Sint-B	$20 \pm 2.5\,\%$	plain bearings and shaped parts with anti-friction characteristics

Sint-C	15±2.5 %	plain bearings and shaped parts
Sint-D	10±2.5 %	shaped parts
Sint-E	6±1.5 %	shaped parts
Sint-F	< 4.5 %	shaped parts, sinter-forged

The first appended number characterises the chemical composition, the second is used for differentiation (e. g. Sint-D35). Classes AF to C have an open porosity. In plain bearings, the pores are filled with lubricants that exude when heated during service thus lubricating the bearing surface. Filters are manufactured from powdered stainless steel or bronze. Soft and easily compacted powders with < 0.33 % C are suitable for moulding. Alloying with 1 to 5 %Cu powder decreases sintering shrinkage. The copper completely dissolves in the iron at the usual sintering temperatures of 1100 to 1300°C. Nickel powder only partially dissolves on account of its low coefficient of diffusion and thus increases shrinkage. Copper and nickel improve the mechanical properties. Application fields include mass-produced parts with a complex shape, such as shock absorber pistons of Sint-C10, synchroniser rings of Sint-F31, sprockets of Sint-C21, internally and externally toothed pump wheels, toothed belt pulleys as well as parts for hydraulic machinery, domestic equipment and locks.

(d) Variety of grades

The large number of different grades and designations is rather confusing at first; however, they are necessary because the service properties for the intended application and the manufacturing properties during steel processing are both of crucial importance. This is also the reason for the extensive set of standards. For example, EN 10025 specifies not only the mechanical service properties but also the engineering properties that are important for processing and the surface finish. The former are covered by the basic designation e. g. S355, with additional codes for size- and temperature-dependent strength and toughness properties. The latter properties are usually more difficult to define, which is reflected by the use of terms indicating suitability for welding, cold-heading, bending or enamelling. Some of these specifications are not relevant, e.g. for equipment machined in the works, and this is expressed by E360. In contrast, enhanced testing or a greater hydrogen resistance can be stipulated for pressure-vessel steel P355. For line pipes, L355 must permit e. g. dilation of the sockets. Steel 355 is thus available in a wide range of varieties that are related to its processing as sheet, wire, pipes etc. Owing to this end-product dependence, the respective extent of testing and the associated costs, it is not useful to combine all the different attributes into a single 355 grade. Companies processing steel are not usually affected by the overall complexity of the steel standards because screw manufacturers are specialised in working with wire, packaging manufacturers with tinplate and pipeworks with hot strip.

B.1.2 Cast iron

B.1.2.1 Composition of grey cast iron

In many applications, a near-net shape is particularly important so that grey iron castings are used. The aim is to obtain the desired properties by cooling from the casting temperature without a heat treatment and thus save costs. Nevertheless, in some circumstances, thermal treatments such as soft annealing or normalising (Chapt. A3) are used to produce particular manufacturing or service properties. These are determined not only by the microstructure of the steel-like metal matrix but also by the solidification microstructure (shape, quantity and distribution of graphite precipitates as well as the number and size of the metal cells). A key parameter that describes the microstructure of cast iron is the degree of saturation S_c (s. Eq. A.2.2, p. 34). It indicates how far away an alloy concentration is from the eutectic composition (4.3 %C in the Fe-C system). This composition is shifted to a lower C content by elements such as silicon, phosphorus and sulphur that increase the activity of carbon (Chapt. A.2, p. 31). A degree of saturation of $S_c = 1.05$ thus represents a slightly hypereutectic composition, whereas a value of $S_c = 0.7$ indicates a significantly hypoeutectic composition. It is therefore a criterion that can be used to estimate both the technical casting properties as well as those relating to application engineering.

Unalloyed grey cast iron contains 2.5 to 4 %C as well as the tramp elements silicon (0.5 - 3.5 %Si), manganese (0.3 - 1.5 %Mn), phosphorus (<0.2 %P) and sulphur (<0.2 %S). The most important alloying elements are nickel and copper (along with Si, P and S to promote grey solidification) as well as the carbide-stabilising elements manganese, chromium, molybdenum, niobium, vanadium and titanium.

These elements not only influence the solidification microstructure, but also the microstructure of the metal matrix. For example, austenite-stabilising copper (0.4 - 1.8 %Cu) has a strong pearlitising effect, particularly because it hinders the deposition of carbon on graphite that is necessary for ferritisation. Nickel (0.5 - 3 %Ni) expands the γ field and promotes the growth of pearlite by suppressing ferritisation. Nickel also increases the solubility of Cu in ferrite and thus prevents the precipitation of embrittling copper-rich phases. Alloying with less than 0.3 %Mo stabilises the ferrite, but contents between 0.3 and 1 % delay the growth of pearlite. Chromium is a significantly more effective carbide stabiliser than molybdenum, and the addition of > 0.3 %Cr already results in the formation of carbides. Although these carbides increase the hardness and the wear resistance, they also lower the tensile strength and toughness. Tin is a strong pearlite promoter and even very low concentrations of 0.1 % produce a predominantly pearlitic metal matrix in walls up to 50 mm thick.

B.1.2 Cast iron

A particular group of elements affects graphite inclusions, even at very low concentrations. In contrast to magnesium and cerium, which are added with the intention of precipitating graphite spherulites, the addition of other elements such as titanium, antimony, lead, bismuth, selenium and tellurium at levels of $< 0.01\,\%$ results in degeneration of the spheroidal graphite, which is only desirable in cast iron with vermicular graphite. The shape of the graphite precipitates is used in European standards as the basis to classify the different types of grey cast iron:

- cast iron with flake graphite, GJL (EN 1561)
- cast iron with spheroidal graphite (ductile cast iron), GJS (EN 1563)
- cast iron with vermicular graphite, GJV (VDG Specification W50)
- blackheart malleable cast iron, GJMB (EN 1562)
- whiteheart malleable cast iron, GJMW (EN 1562)

Tab. B.1.2 summarises the most important properties of the various groups of materials for selected alloys. A more detailed commentary on the material designations used in the European standards is given in Appendix C1.

Table B.1.2 Properties of grey cast iron at room temperature: Selected grey cast irons for general applications

Name	Tensile strength [MPa]	0.2 % Proof strength [MPa]	Elongation at fracture [%]	Impact energy[7] [J]	Notched impact energy[8] [J]	Young's modulus [GPa]	Density [g/cm³]	Thermal conductivity [W/(m·K)]	Graphite shape	Matrix
GJL-150[1]	150 - 250	98 - 165[2]	0.3 - 0.8	4	—	78 - 103	7.1	53[6]	flake	F / P
GJL-250[1]	250 - 350	165 - 228[2]	0.3 - 0.8	5	—	103 - 118	7.2	49[6]	flake	P / F
GJL-350[1]	350 - 450	228 - 285[2]	0.3 - 0.8		—	123 - 143	7.3	46[6]	flake	P
GJS-400	370 - 400	250	15 - 30	98 - 196	10 - 19	160 - 180	6.9	38.5	spheroidal	F
GJS-600	550 - 600	380	3 - 8	39 - 78	3.5 - 10	170 - 180	7.0	32.9	spheroidal	P / F
GJS-800	800	500	2 - 4	9 - 29		170 - 180	7.1	32	spheroidal	P
GJV-300[3]	300 - 375	220 - 295	1.5	15 - 35	3 - 6	140 - 160	7.0	44	vermicular	F
GJV-400[3]	400 - 475	300 - 375	1.0	6 - 10	≤ 3	150 - 170	7.0	41	vermicular	P / F
GJV-500[3]	500 - 575	380 - 455	0.5		—	160 - 180	7.1	38.5	vermicular	P
GJMB-350	350	200	10	90 - 130	≥ 14	170	7.2 - 7.5	45 - 63	temper carbon	F
GJMB-600	600	390	3	25	3 - 5	175 - 195	7.2 - 7.5	45 - 63	temper carbon	P
GJMB-800	800	600	1	—	30 - 70	—	7.2 - 7.5	45 - 63	temper carbon	M
GJMW-360	280 - 370	170 - 200	7 - 16	130 - 180	14	175 - 195	7.8	42 - 63	no graphite	F
GJMW-550	490 - 570	310 - 350	3 - 5	30 - 80	6	175 - 195	7.3 - 7.8[5]	42 - 63	temper carbon[4]	F / P

[1]: separately cast specimens with ⌀ 30 mm
[2]: 0.1 % proof strength
[3]: VDG Specification W 50, not European standard
[4]: in the core, no graphite in the surface zone
[5]: depending on the degree of decarburisation
[6]: at 100°C
[7,8]: unnotched, notched, DVM specimen

F = ferrite
P = pearlite
M = martensite

B.1.2.2 Cast iron with flake graphite

Flake (or lamellar) graphite is the natural precipitate shape in hypoeutectic Fe-C-Si alloys (Figure A.2.12 a). As discussed in Chapter A.4, p. 79, the graphite flakes act as internal notches and thus have an unfavourable effect on the tensile strength and elongation at fracture. Thus cast iron with flake graphite (GJL) has a strength range of between 100 MPa (GJL-100) and 350 MPa (GJL-350), whereby the elongation at fracture drops from 0.8 to 0.3 % as the strength increases. High strengths require a low degree of saturation (Figure B.1.7) and a predominantly pearlitic metal matrix, which can be achieved by carefully selecting the appropriate combination of the above-mentioned elements. GJL exhibits a pronounced dependence of the mechanical properties on the wall thickness. The cooling rate in thin walls is high so that dendrites and graphite precipitates remain small and the matrix is pearlitic (Figure A.2.18). On slow cooling (e. g. large wall thickness), the carbon can diffuse to the graphite flakes leaving a ferritic matrix that is strengthened by solid-solution hardening with Si and Mn. The European standard takes account of the effects of wall thickness and cooling conditions by specifying the properties for various cross-sections of specimens that have been cast separately, cast on or taken from a casting. Inoculation of the melt with e. g. FeSi produces finer metal dendrites and graphite precipitates, which increases the tensile strength and lowers the hardness. The microstructural refinement

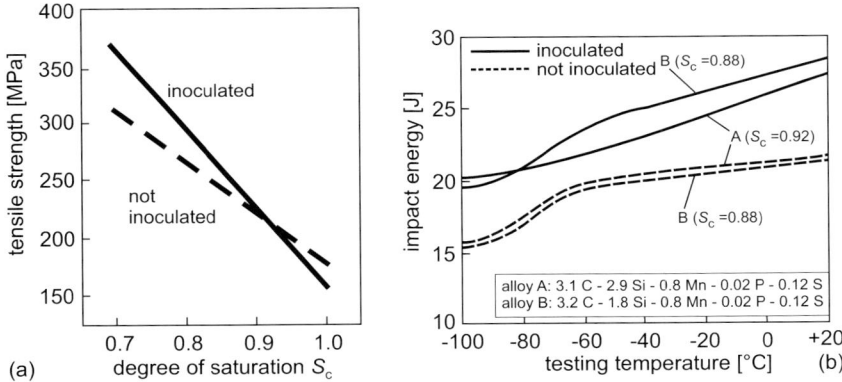

Fig. B.1.7 Inoculation of cast iron: The addition of 0.2 - 0.3% FeSi or CaSi (0.5 - 3 mm mesh size) just before casting increases the number of crystallisation nuclei in the melt. They produce a larger number of small eutectic metal cells that promote the formation of flake graphite, thus improving the mechanical properties and also reducing the likelihood of chilling. (a) The relative gain in tensile strength increases as the degree of saturation S_c decreases, i. e. as the solidification interval increases. (b) The development of a fine microstructure has a favourable effect on the impact energy (from R. Deike et al.)

resulting from the inoculation also increases the impact toughness, which is determined using unnotched samples of GJL (Figure B.1.7 b). A network of ternary phosphide eutectic (steadite) forms if the phosphorus content is > 0.5 %. Cementite precipitates on the grain boundaries and steadite have a strong embrittling effect. Manganese sulphides do not significantly lower the toughness because they are much smaller than the graphite flakes.

Unlike steel, the Brinell hardness of cast iron with flake graphite cannot be related to the tensile strength because the hardness and strength are frequently inversely affected by the microstructure. However, the European standard includes casting alloys classified - with a view to their construction properties - according to their tensile strength (e. g. GJL-150) and - with respect to their machinability - according to their hardness (e. g. GJL-HB175).

The microstructure of grey cast iron has a number of advantages and disadvantages with respect to applications at elevated temperatures. The strength and hardness drop steeply above 400°C because the pearlite decomposes to ferrite and small graphite particles. Above 600°C, they redissolve and the carbon is precipitated on the graphite flakes (Chapt. A.2, p. 51). The castings expand owing to the higher specific volume of the graphite (factor of 3 with respect to the iron phases), and the length may increase by up to 0.5 % for a fully pearlitic initial microstructure. Whereas silicon accelerates this process, the carbide-stabilising elements manganese and chromium slow it down. Because even a small amount of added molybdenum noticeably increases the creep resistance of the ferrite due to solid solution hardening, cast irons containing molybdenum and chromium exhibit a greater creep strength at 500°C compared to unalloyed variants.

Below 500°C, the growth of scale on unalloyed and low-alloy GJL is essentially determined by the oxidation of iron and silicon. Small amounts of chromium ($< 1.2 \%$) improve the oxidation behaviour by reducing the growth rate of the oxide film. Above 500°C, uniform surface oxidation is accompanied by an additional internal oxidation along the graphite flakes that increases with the proportion of graphite and the flake size.

Cast iron with flake graphite is often used because of its high thermal conductivity. In ferritic grades, this increases with the degree of saturation to values of up to $\lambda = 52$ W/mK and is attributed to the high thermal conductivity of graphite (80 - 85 W/mK along the longitudinal flake axis), which effectively dissipates the heat via the three-dimensional network of flakes in GJL. The value of λ decreases with increasing pearlite fraction in the metal matrix because the thermal conductivity of Fe_3C is only 7 W/mK. The high thermal conductivity is also responsible for a good thermal shock resistance. Thermal stresses (simplified to $\sigma_{th} = E \cdot \alpha \cdot \Delta T$) remain relatively low because the Young's modulus (E) and the thermal coefficient of expansion (α) are lowered by graphite and the high thermal conductivity keeps ΔT small.

Cast iron with flake graphite exhibits a remarkable wear behaviour, particularly under sliding wear conditions in a lubricated metal-metal contact. The graphite flakes provide an internal reservoir of lubricant that guarantees certain emergency service characteristics if external lubrication is insufficient. A homogeneous distribution of fine graphite flakes (length 100 - 250 µm, thickness 2 - 3 µm) is particularly favourable because the 'lubrication paths' are short. Even if the subsurface graphite has been consumed, the resulting voids form pockets that can be filled with an external lubricant. The microstructure of the matrix also plays an important role. Whereas the soft ferritic matrix offers little resistance to abrasion and wear, the wear rate decreases with increasing pearlite fraction and decreasing distance between pearlite lamellae.

Phosphorus has a similar effect; above 0.5 % it precipitates as a network of eutectic steadite that is significantly harder than pearlite (Figure B.1.8). The high thermal conductivity of GJL also lowers wear because it is responsible for rapid dissipation of the frictional heat and thus counteracts the temperature-related decrease in yield strength.

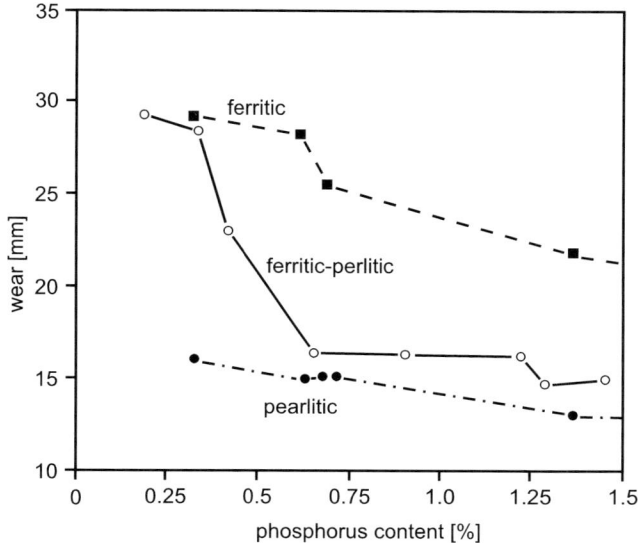

Fig. B.1.8 Sliding wear of GJL (from E. Piwowarsky): Influence of the pearlite and phosphorus contents on the loss of material due to sliding wear.

B.1.2.3 Cast iron with spheroidal graphite

Cast iron with spheroidal graphite (Figure A.2.12 c, p. 32.) was produced for the first time in 1948 by two groups – P. Gagnebin et. al. in the USA and H. Morrogh and W.J. Williams in England – on the basis of a patent held by C. Adey. This was achieved by treating an iron-carbon-silicon melt with magnesium and cerium. In the meantime, the good price/performance ratio and a controllable casting technology has led to a very high growth rate for the use of cast iron with spheroidal graphite (also known as ductile cast iron, nodular cast iron and spheroidal cast iron). The current global production capacity is about 20 million tons per year. European standard EN 1563 defines this group of materials as GJS (Appendix C.1, p. 383).

As discussed in Chapt. A.4, p. 99, the internal notch effect of graphite nodules is significantly lower than that of flakes, so that the mechanical properties of cast iron with spheroidal graphite are superior to those containing flake graphite. The spherical shape is obtained by treating the melt with magnesium, calcium or cerium. However, the melt must be previously desulphurised to avoid growth of embrittling sulphides. Because magnesium evaporates easily on account of its high vapour pressure, it is more effective if added to the melt in the form of a Mg prealloy (NiMg or FeSiMg). Subsequent inoculation with 0.4 to 0.7 % FeSi and other elements with an oxygen affinity (Al, Zr) increases the number of spherulites, produces an ideal graphite shape and counteracts chilling in small wall thicknesses. The effect of magnesium and inoculation gradually weakens so that large graphite nodules are often found in thicker walls.

Apart from limitations in the phosphorus (max. 0.08 %) and sulphur contents (max. 0.02 %), the basic composition of ductile cast iron (GJS) corresponds to alloy concentrations in GJL with Si contents of 1.7 - 2.8 %.

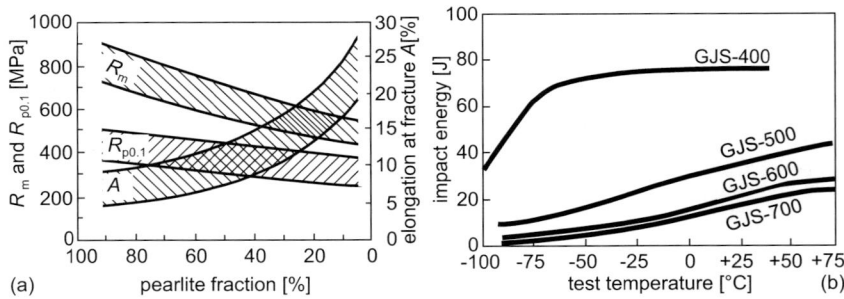

Fig. B.1.9 Mechanical properties of GJS: (a) Tensile strength, 0.1 % proof strength and elongation at fracture of unalloyed GJS as a function of the pearlite fraction in the matrix (from K. Peukert). (b) Impact energy as a function of the test temperature for selected GJS alloys (from W. Siefer)

A saturation level close to unity is desirable to obtain a melt with good flow characteristics. The properties of GJS are essentially determined by the metal matrix. Its ferrite/pearlite ratio can be altered with specific alloying elements and by adjusting the cooling rate. Slow cooling increases the ferrite fraction, which is frequently found as a shell around the graphite spherulites. This type of structure is known as a bull's eye.

EN 1563 divides GJS into strength classes GJS-350 to GJS-900. Even though the dependency on the wall thickness is lower than in GJL, there are specifications for different wall thicknesses and cooling conditions. Tab. B.1.2 lists the most important properties of a few selected GJS alloys. This shows that GJS has a higher strength than GJL. The ferritic grade GJS-400 has a good strength and an elongation at fracture of 18 %. Its low hardness confers good machinability. Grades with a higher strength contain increasingly more pearlite, which lowers the elongation at fracture and the toughness, although they are still higher than those of GJL (Figure B.1.9). The increase in the proportion of pearlite from GJS-400 to GJS-700 lowers the impact toughness in the upper shelf region and shifts the transition temperature from -80°C to room temperature. The lower notch effect of the graphite nodules results in an acceptable notched impact toughness, even for unalloyed GJS. The standard recommends two ferritic grades with a guaranteed notched impact toughness (e.g. of 15 - 17 J at room temperature for GJS-350-22U-RT) whose phosphorus and silicon contents must be kept below 0.1 % and 2.6 %, respectively. The influence of silicon and phosphorus on the notched impact toughness is shown in Figure B.1.10.

The elements copper, nickel and molybdenum increase the pearlite fraction and thus strengthen alloyed GJS. This is accompanied by a decrease in the notched impact toughness and an increase in the transition temperature.

The tensile strength and proof strength of ferritic grades (e.g. GJS-400) increase with decreasing temperature, and the yield ratio $R_{p0.2} / R_m$ increases from 0.75 at room temperature to ≈ 1 in the cryogenic range. The increasing

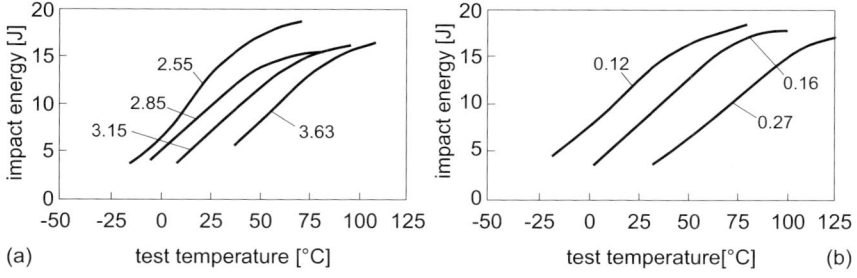

Fig. B.1.10 Notch impact energy of GJS (from the Metals Handbook): Influence of (a) mass% Si and (b) mass% P on the notched impact strength (ISO-V specimens) of GJS. Silicon and phosphorus increase the transition temperature

pearlite fraction lowers the linear increase in the tensile strength with a drop of the temperature and becomes degressive for grades with a pearlite content > 50 volume-% (e. g. GJS-600) because the high-strength grades become progressively sensitive to internal notches.

Unlike GJL, pearlite in GJS is destabilised at lower temperatures. It already exhibits a large decrease in the strength (tensile strength, yield strength and hardness) at temperatures below 300°C. The associated expansion is significantly smaller than in GJL. The creep resistance can be increased by alloying with 6 %Si, 3 %Ni and 1 %Mo. The oxidation resistance is greater than that of GJL because there is less internal oxidation of nodular graphite. Ferritic GJS with > 4 %Si is resistant to scaling until it transforms into austenite and is thus discussed in greater detail in Chapt. B.6, p. 341.

Owing to the higher Young's modulus and the lower thermal conductivity compared to GJL, rapid temperature changes with high cooling rates lead to higher thermal stresses and thus more quickly to thermally induced cracks. If the cooling rate is lowered, GJS can withstand – as a consequence of its greater strength and ductility – many more thermocycles than GJL until it fails.

B.1.2.4 Cast iron with vermicular graphite

Cast iron with vermicular graphite (GJV) was developed in Europe around 1960 based on the fact that a low concentration of certain elements, namely titanium, antimony, lead, bismuth, selenium and tellurium, are able to modify the shape of graphite nodules. It is manufactured by adding magnesium or a MgTiCeCa prealloy or a mixed metal compound containing cerium. The latter requires desulphurisation of the melt, which is not necessary for treatment with magnesium as long as a narrow Mg/S ratio is maintained. All three methods produce a cast iron with compacted graphite precipitates (also known as compacted graphite cast iron) whose shape lies between a nodule and a flake (Figure A.2.12 b.). Unlike the three-dimensional spatial coherence of flake graphite in GJL, GJV contains graphite in the form of sparsely branched, compact graphite lamellae with rounded ends along with a small amount of graphite nodules. At saturation levels just below 1, the properties are determined by the ferrite/pearlite ratio, similar to GJL and GJS, and by the proportion of vermicular graphite (Figure B.1.11).

Overall, the properties of cast iron with vermicular graphite represent a compromise between those of GJL and GJS (Tab. B.1.2). Tab. B.1.3 compares the properties of grey cast iron and summarises the advantages and disadvantages with respect to GJL and GJS.

The creep resistance and frequently the thermal conductivity are of interest for applications at elevated temperatures. The tensile strength of GJV with 90 % vermicular graphite lies half way between that of GJL and GJS; however, its thermal conductivity is closer to GJL than to GJS owing to the degree of spatial coherence (Figure B.1.12).

In oxygen-containing media, oxidative attack may occur above 550°C along the interface between graphite and the metal matrix; however, it cannot penetrate as far into the interior as in GJL. A GJV with 4 % Si and 0.5 % Mo has proven suitable for applications at elevated temperatures.

Owing to its good ductility and endurance limit, cast iron with vermicular graphite has about the same resistance to thermal shock cracks as cast iron with spheroidal graphite. As a consequence of the greater thermal conductivity

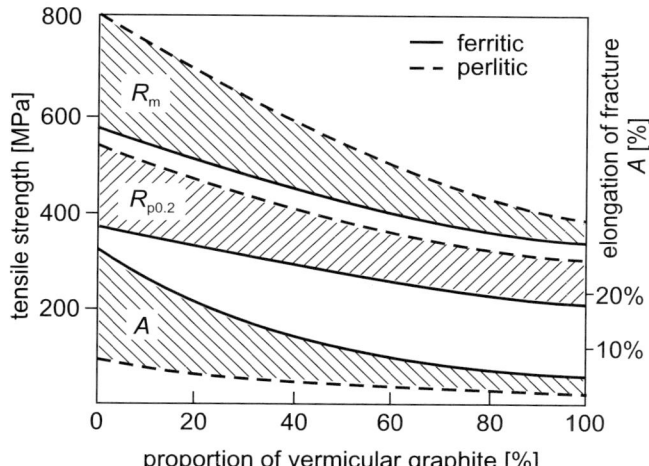

Fig. B.1.11 Influence of the graphite shape on the mechanical properties (from K. Röhrig): Starting from 100 % graphite nodules, the tensile strength, yield strength and elongation at fracture decrease as the proportion of vermicular graphite increases. In the pearlitic state, the reduction in the properties is noticeable up to 50 % vermicular graphite, but is only slight above this value.

Table B.1.3 Properties of GJV compared to those of GJL and GJS

GJV versus GJL	GJV versus GJS
- higher strength	- improved thermal shock resistance due
- higher ductility and toughness	to higher thermal conductivity and
- lower dependence of the wall thickness on the properties	lower coefficients of expansion
	- lower Young's modulus
- higher oxidation resistance	- lower distortion tendency
- lower tendency to expand at elevated temperatures	- better damping capacity
	- better technical casting processability

B.1 Materials for general applications

and lower Young's modulus of GJV, the thermal stresses are lower thus also reducing the risk of thermal shock cracks and distortion (Tab. B.1.3).

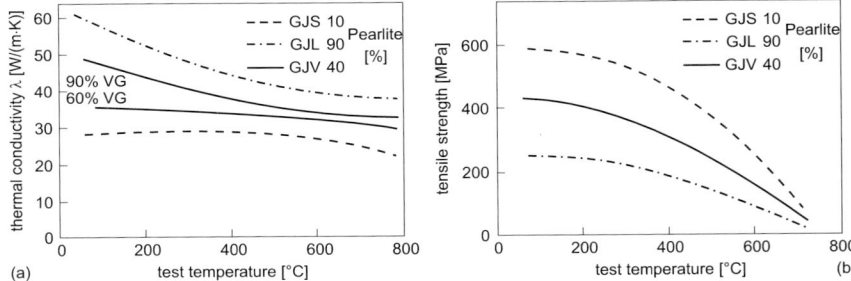

Fig. B.1.12 Thermal conductivity and tensile strength of grey cast iron (from W. Renfang): Although it has a significantly higher tensile strength, the thermal conductivity of GJV-300 with 90 % vermicular graphite (VG) is only slightly less than that of GJL-200.

B.1.2.5 Malleable cast iron

Malleable cast irons are materials that have originally solidified from a hypoeutectic Fe-C-Si melt as white cast iron (Figure A.2.13) that is then subjected to temper annealing (Chapt. A.2, p. 34). This annealing treatment brings the material back towards the equilibrium state and the cementite decomposes. If annealing is carried out in an inert atmosphere, the cementite transforms into graphite aggregates and blackheart malleable cast iron is produced. However, if the white cast iron is annealed in an oxygen-containing atmosphere, the carbon is oxidised and forced out of the surface zone (decarburisation) to produce whiteheart malleable cast iron. The principle of decarburisation was already known in the 18th Century from the transformation of pig iron in open-hearth processes. In the middle of the 19th Century, industrial development focused on whiteheart malleable cast iron in Germany and on blackheart malleable cast iron in America. Both groups of materials are now regulated by EN 1562, which designates blackheart malleable cast iron as GJMB (B: black) and whiteheart malleable cast iron as GJMW (W: white; see Appendix C.1, p. 390).

(a) blackheart malleable cast iron

The basis for blackheart malleable cast iron are Fe-C-Si melts containing 2.4 to 3 %C whose silicon content lies between 1.2 and 1.5 %, with respect to the white solidification of the cast iron. The annealing process to produce blackheart malleable cast iron is carried out in two stages in an inert atmosphere.

In the first annealing stage (900 - 970°C), the Fe_3C carbide decomposes. A fine initial microstructure, a silicon content at the upper alloying limit and addition of 0.002 to 0.003 % B accelerate this carbide decomposition. The liberated carbon is precipitated on the γ-Fe/Fe_3C interface as temper carbon clusters. The shape of these clusters varies from aggregates of flakes to compact clusters and spheroidals with corresponding changes in the properties (Figure B.1.13 a.).

The microstructure of the metal matrix depends on the temperature profile in the second annealing stage and on the subsequent cooling rate. Rapid cooling to 740 - 760°C followed by slow cooling at a rate of 3 - 10°C/h transforms the austenite into a completely ferritic matrix accompanied by precipitation of more carbon onto the existing graphite particles. The properties of the ferritic blackheart malleable cast iron are determined by the graphite fraction (6 - 8 volume%) and its distribution as well as by solid-solution hardening of the matrix by Mn and Si. The ferritic grade GJMB-350 has a tensile strength of 350 MPa and an elongation at fracture of 10 %. It is thus a ductile cast iron. Slow cooling from the first annealing stage to approx. 870°C reduces the C content in the austenite (to approx. 0.75 %C), which transforms into finely striped pearlite on cooling in moving air. This produces a tensile strength of 600 MPa for the grade GJMB-600. A martensitic or martensitic/bainitic metal matrix is obtained if the material is quenched in oil from 870°C. The subsequent tempering treatment allows the microstructure of the resulting blackheart malleable cast iron to be tailored to give a maximum strength of 800 MPa (Tab. B.1.3) and the necessary hardness for wear applications.

The impact toughness of unnotched and notched specimens lies just below that of GJS due to the jagged shape of the temper carbon clusters, although the dependencies on the Si and P contents are similar. As in GJS, there is a noticeable decrease in the tensile strength at temperatures above 350°C because the pearlite or martensite decomposes with precipitation of ternary

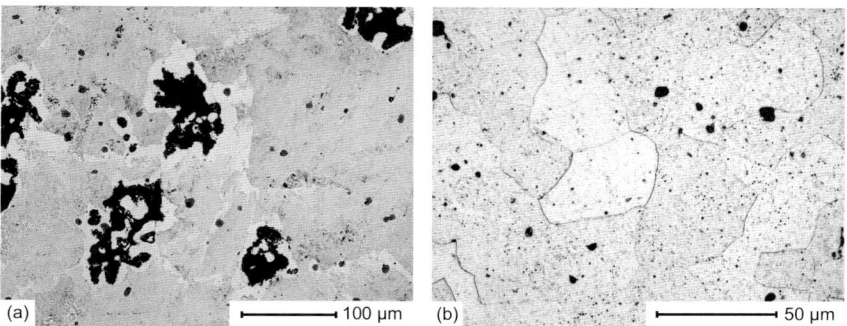

Fig. B.1.13 Microstructure of malleable cast iron: (a) blackheart malleable cast iron (GJMB), (b) whiteheart malleable cast iron (GJMW)

graphite. The damping capacity of blackheart malleable cast iron is approximately twice that of GJS.

(b) whiteheart malleable cast iron

Whiteheart malleable cast iron contains between 3.1 and 3.4 %C and is produced by annealing white cast iron for several hours between 1000 at 1050°C in a weakly oxidising atmosphere. The cementite in the surface that is accessible to the oxygen thus decomposes according to the following equation:

$$Fe_3C + O_2 \rightarrow 3Fe + CO_2 \quad (B.1.4)$$

This surface zone is thus decarburised whereas the Fe_3C in the remaining cross-section is transformed to temper carbon. In order to equalise the concentration gradient between the surface zone and the core, carbon diffuses towards the surface thus allowing further decarburisation. Because a high silicon content promotes graphitisation and slows down the diffusion of carbon in austenite, only 0.4 to 0.8 %Si is added so that whiteheart malleable cast iron can be produced within acceptable annealing times. If the annealing time is increased, the surface zone becomes increasingly decarburised until ultimately, at annealing times of several days, castings with small wall thicknesses (<15 mm) exhibit a completely ferritic microstructure (Figure B.1.13 b.). For larger wall thicknesses, the core consists of temper carbon in a predominantly pearlitic matrix. Therefore, the properties of such components are extremely dependent on the wall thickness. After holding at the elevated temperature, accelerated cooling in air with subsequent soft annealing at 700°C has proven successful for large wall thicknesses. This spheroidises the cementite in the pearlitic core regions and thus significantly improves the ductility.

Whiteheart malleable cast iron covers a tensile strength range of between 280 MPa (GJMW-360-12, completely ferritic) and 570 MPa (GJMW-550-4, with a pearlitic core); the elongation at fracture lies between 12 and 4 % (Tab. B.1.2). Whiteheart malleable cast iron with a purely ferritic matrix and without graphite precipitates has an impact toughness of between 150 and 200 J, which is comparable to that of unalloyed cast steel.

B.1.2.6 Processing and applications of cast iron

(a) Technical casting properties

Its proximity to the eutectic means that grey cast iron has a small solidification interval and a low melting or solidification temperature. This is the reason why cast iron components can be produced much more cheaply than steel castings, which frequently require a complicated feeder system with a large amount of unused melt in the sprues. The good mould filling capacity of cast iron melts with a saturation level close to $S_c = 1$ allows complex filigree shapes and thin walls. The addition of up to about 1 %P improves the flowability even further by forming a low-melting ternary Fe-C-P eutectic, and is thus sometimes used for grey cast iron with flake graphite.

The apparent detour via initial white solidification of white cast iron followed by lengthy temper annealing does not represent a cost disadvantage with respect to cast steel because small thin-walled parts of whiteheart malleable cast iron, in particular, could only be made as a low-carbon steel casting using a complicated feeder system.

The volume reduction during solidification is small owing to the small solidification interval and is decreased even further by the large specific volume of the precipitating graphite. If the amount of graphite is large, contraction during solidification of the metallic phases can even be completely compensated. Shrinkage in the solid state is less than that of cast steel because the influence of the graphite inclusions lowers the coefficient of thermal expansion so that there is less overall distortion.

(b) Machining

In spite of the near-net shape of castings, the functional surfaces and mounting dimensions have to be adjusted by machining. Grey cast iron is very suitable for machining because the graphite precipitates produce chips that break into small pieces and also act as a lubricant on the working surfaces of the cutting tools. As a consequence of the lower three-dimensional spatial coherence of the graphite precipitates, short chipping decreases in the order GJL→GJV→GJS. Cast iron with fine metal dendrites and graphite precipitates, as generated by inoculation, can be machined more easily than coarse-grained microstructures. Similar to steels and steel castings, machining becomes more difficult as the strength of the metal matrix increases. Because the tensile strength is dependent on the type of graphite inclusion, the hardness is a more suitable parameter to evaluate the machinability of cast iron. Like steels, the hardness, which increases with the pearlite fraction, has an adverse effect on the wear of the cutting tool. Cast iron with spheroidal graphite can be machined more easily than grey cast iron with flake graphite of the same hardness and steel of the same strength. Cast iron alloyed with chromium contains carbides that become more difficult to machine as their size increases. Similar effects are exhibited by titanium carbonitrides that can form in cast iron with vermicular graphite if the melt was treated with MgTi-CeCa prealloys.

The machinability of whiteheart malleable cast iron is excellent because the ferritic matrix is soft and the inclusions of manganese sulphide and temper carbon produce short chips that are responsible for the low tool wear and low cutting forces. This results in an approximately 25 % improvement in the machinability compared to free-cutting steels.

A comparison of the cutting force F_c in a lathe trial with various grey cast irons and steels showed that this force increases with the strength. The forces measured for grey cast iron were always lower than those for a comparable steel. The surface quality increases with the strength; however, it is slightly lower than that of steel or cast steel because the graphite precipitates in the grey cast iron detach during plastic deformation leaving behind open cavities.

Recent studies have reported that ageing has a positive effect on the machinability of cast iron. In spite of the various factors that mask ageing phenomena, such as dependency on the wall thickness, degree of saturation, solidification rate, chemical analysis etc., a statistically substantiated influence of ageing on the machinability was observed for clutch and brake discs made of GJL. A significant improvement in the machinability was found after natural ageing for up to 1000 hours (Figure B.1.14 a). This improvement is associated with an increase in the tensile strength that is dependent on the nitrogen content (Figure B.1.14 b), whereby the increase is proportional to the content of free nitrogen that is not bound to titanium and which accumulates at dislocation sinks. Neutron diffraction and calorimetric studies of samples aged for 30 days at room temperature detected the presence of α''-nitride precipitates with a size of 2-4 nm (p. 130). They increase the strength by up to 13 %. This effect can also be observed in the ferritic shell around the graphite nodules in GJS by means of microhardness

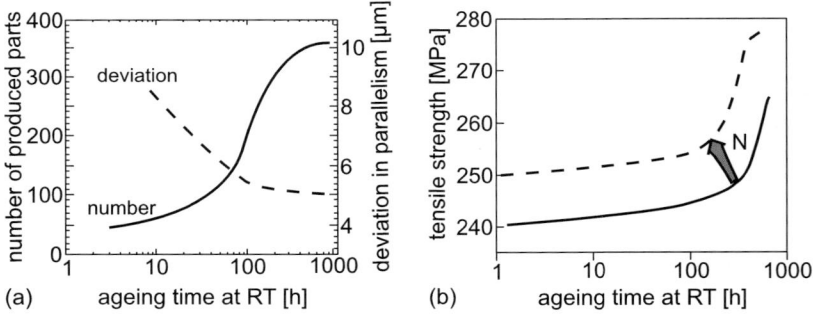

Fig. B.1.14 Ageing phenomena in cast iron (from L. Richards): (a) The number of brake disks made of GJL processed with one tool increases significantly if the parts are previously aged at room temperature. The deviation from parallelism also improves. (b) Tensile strength of GJL versus the ageing time at RT. Nitrogen shifts the curve to higher values.

measurements (190 to 260 HV in 30 days). The embrittlement associated with the increasing strength is assumed to be the reason for the shorter chips that lower the wear of cutting tools.

(c) Applications

The range of applications for cast iron is very diversified and includes the fields of structural, domestic, general mechanical and automotive engineering.

The construction industry uses predominantly GJL and GJS. Whereas the mechanical properties of GJL are sufficient for road manhole covers, gullies and tree grates, GJS is used for water and gas pipes with diameters of up to 1000 mm, which have to withstand pressure surges and earth movements. In contrast to GJL pipes that used to be cast in sand so that their casting skin thus contained a corrosion-inhibiting silicate layer, modern pipes are cast in moulds and have to be protected against corrosion by a cement lining and a coating of paint on the outside.

In recent years, the amount of cast iron products used in mechanical engineering has been continuously increasing. This is chiefly due to advancements in production techniques in the foundries. Modern casting facilities can now produce complex castings whose properties can be precisely tailored with a high degree of reproducibility. Arguments for the use of cast iron include its near-net shape, good castability, low manufacturing costs as well as its special service properties, such as high thermal conductivity, thermal shock resistance, its extraordinary damping capacity and its low relative density that is particularly important in lightweight engineering of vehicles (Table B.1.2). The diverse microstructural features of the aforementioned grades offer some unique combinations of properties that cannot be achieved with other metallic materials.

A large field of application for grey cast iron with flake graphite is housings for gears, fittings, pumps and compressors as well as their impellers, which are manufactured with piece weights of up to several tons. This material is used because complex three-dimensional structures are possible and it has a good damping capacity. Furthermore, the damping capacity of GJL is exploited in a number of applications, including machine beds or frames for cutting tools and for metal-working machines with piece weights of up to 200 tons. If the strength of GJL is insufficient, GJS is used instead, although it has a lower damping capacity. In modern wind power facilities with a capacity of up to 5 MW, ferritic GJS-400-18 is used for the rotor hubs, which have piece weights of up to 50 t.

Its low alloying and production costs as well as its low density makes cast iron an interesting material for automotive series parts. They now make up 40 % of the total amount of cast iron produced. Engine blocks are traditionally made of GJL. However, the continuous demand for weight savings has led to the use of aluminium alloys, even though they are twice as expensive. They are capturing the European market because of the possibility of weight savings of up to 40 %. Nevertheless, in more highly loaded diesel engines, cast iron still holds its ground on account of its higher strength, higher damping capacity and higher stiffness. Here, GJL is in competition with GJV. Developments of GJL are aimed at increasing the strength by alloying with Cu, Ni, Cr, Ti, Mo and V, as well as reduction of P and S with simultaneous limitations in the scattering of strength values by means of precise quality control measures. Casting developments are focussing on making the ends of the graphite flakes less sharp-edged so that these GJL grades achieve tensile strengths of about 300 MPa, thus approaching those of vermicular grades.

The high requirements demanded of a tribosystem consisting of the cylinder wall/piston/piston ring with respect to a low oil consumption as well as low wear and friction loss can be met by grey cast iron in combination with a special surface finish. The aim is to create isolated pockets that assist hydrodynamic lubrication. The lubricant is pressed into these pockets instead of into the honed cross-hatching in conventional systems. If a low amount of Ti ($\approx 0.04\,\%$) is added to the GJL, nitrides and carbides form with a size of approx. 5 µm that lower the coefficient of friction of the material pair along with the tendency to adhere. During machining, they are removed together with the surrounding matrix down to the next graphite flake. This produces large craters of $\approx 40\,\mu m$ that are uniformly distributed over the surface as isolated lubricating pockets and which thus represent a 'micro pressure-chamber system'. A similar effect can be achieved by microstructuring with a laser. The cylinder walls of GJV in large automobile diesel engines are honed with a pulsed UV laser ($\lambda = 300\,nm$). As the surface is remelted to a depth of $\approx 2\,\mu m$, surface tension causes the melt to expose the graphite precipitates, which leave behind cavities that function as lubrication pockets later on. Furthermore, a metal vapour plasma forms during melting that allows approx. 18 % N to penetrate into the molten zone so that a nanocrystalline ceramic layer forms when the laser is switched off. Finally, the surface conditions itself during service: the high combustion temperature produces a 200 nm-thick layer of iron nitrides and carbides that protect against wear.

GJV-300-4 is used for the engine block in large engines for cars and commercial vehicles owing to its greater strength, ductility and resistance to expansion. Its poor thermal conductivity is compensated by special cooling techniques. Cylinder heads of modern four-valve engines are exposed to higher mechanical vibrational stresses and cyclic thermal loads than the crankcase. Fully pearlitic GJV-450 alloyed with 0.8 % Cu and 0.06 % Sn has been proposed for this. When exposed to cycling temperatures of 50 - 420°C and completely restrained thermal expansion, it withstood twice as many cycles to failure as the predominantly ferritic GJV-350. A low-alloy grade, such as GJL-250, was exceeded by a factor of 5.

Ductile cast iron (GJS) is not an alternative for such applications on account of its lower damping capacity, high tendency to distortion and poor thermal conductivity. However, it is used in automotive components that are subjected to predominantly mechanical loads. These include housings for disc brakes, steering gear, rear axles (particularly engine brackets), steering knuckles and brake pads for commercial vehicles. A further development of GJS400-15 is used for transverse links and swivel bearings. Its elevated Si content and alloying with boron increases the yield strength to 320 MPa and it thus has a 25 % higher fatigue strength. GJS is a key material for crankshafts and camshafts, which are made of GJS-600 not only because of its low notch sensitivity under cyclic loading conditions. The running surfaces are additionally surface-hardened (Chapt. B.3, p. 222). The first hollow camshafts are now being used as castings in diesel engines in some premium class automobiles. These are produced by chill-casting the cams against the mould so that the surface layer undergoes white solidification, which means that the casting does not require any further heat treatment. This natural hardness of the surface layer of 50 to 55 HRC has proven favourable because this hardness is maintained even at the highest oil temperatures.

Grey cast iron (GJL-250 to GJS-600) is used as near-net-shaped continuously cast semi-finished products for the cost-effective manufacture of control blocks for hydraulic systems. Materials produced by continuous casting have a higher solidification rate compared to sand-casting and thus have a more finely grained and denser microstructure that provides the high pressure tightness required for this application. Furthermore, the continuous casting diameter of up to ≈ 450 mm means that these semi-finished products can be used to manufacture toothed wheels and ring gears as well as glass-moulding tools.

The only cast iron suitable for structural welding is whiteheart malleable cast iron. GJMW-380-12, which has been specially optimised for welding, has a carbon-depleted surface layer ($< 0.3\,\%$C) that allows castings with complex geometries to be welded to shaped steel components without preheating. Examples of welded cast iron/steel assemblies include semitrailing arms on rear axles of high-performance automobiles as well as stabilisers and spring seats for lorries. In these cases, GJMW-380-12 is welded to a weldable structural steel. Furthermore, GJMW-400-5 is used for small parts such as fittings, wing nuts, pipe unions and hooks as well as connecting elements for scaffolding and formwork. They are often galvanised to protect them against corrosion.

Blackheart malleable cast iron is also more suitable for small parts with piece weights of up to about 2 kg. For example, heavy-duty expansion anchors as well as door and furniture keys are made of GJMB-350-10. Grades with a higher strength, e.g. GJMB-650-2, are used for hand tools such as pliers, adjustable wrenches and screw clamps. The working surfaces can be surface-hardened if necessary. These few selected examples demonstrate that cast irons are an effective and highly variable group of materials that can be precisely tailored to their ultimate application by exploiting their wide range of production and finishing methods. In analogy to the tailored blanks, tailored strips and tailored tubes described in Chapter B.2 (p. 180), they certainly earn the term 'tailored castings'.

Drop-forged part of quenched and tempered steel
(SCHMOLZ + BICKENBACH Distributions GmbH, Düsseldorf, Germany)

B.2 High-strength materials

Alloying and heat treatment can be used individually or in combination to increase the strength of ferrous materials. Steels can also be subjected to controlled hot-working and cooling, which is known as thermomechanical rolling. The purpose of this is to achieve a low final hot-working temperature, which is out of the question for die forging because it would overload the tool. However, the materials are often quenched or precipitation-hardened from the forging temperature, thus saving the subsequent austenitising step. In cast irons, the main focus is on transformation in the bainite range.

B.2.1 Weldable rolled steels

B.2.1.1 Fine-grain steels

The most cost-effective method of increasing the strength using carbon (Figure B.1.2, p. 127) usually ends at S355 or at 0.22 %C because of restrictions with respect to the weldability and the transition temperature. Normalising or thermomechanical working can be used to harden by grain refinement, which not only increases the yield point, but also decreases the transition temperature T_T (Chap. A.4, p. 85 and p. 95). Lower carbon contents decrease the CEV during welding (Eq. B.1.1, p. 129) to such an extent that small amounts of vanadium, niobium and titanium (between 0.03 and 0.3 % in total) are tolerated. During hot rolling and further cooling, this microalloying produces a fine dispersion of carbide/nitride precipitates of type MX (see Tab. A.2.1 p.52). In the austenite field, these precipitates inhibit grain growth and recrystallisation. This increases the likelihood of fine-grained, non-recrystallised austenite undergoing the γ/α transformation. The resulting ferrite grain diameter is < 10 µm, so that T_T sinks. Within the transformation front, but also during subsequent slow cooling, MX particles precipitate again and now even more finely and thus precipitation-harden the ferrite. T_T thus increases slightly again. These grades are known as low-pearlite fine-grained microalloyed structural steels or just as HSLA (high-strength low-alloy) . Rapid cooling into the bainite temperature range directly after rolling produces fine-grain bainitic steels.

Figure B.2.1 illustrates various methods of producing hot strip. Whereas conventional rolling requires a subsequent normalising step, grain refinement is achieved during normalising rolling (N) by lowering the final-rolling temperature. The austenite recrystallises before it is able to transform. No recrystallisation of austenite occurs during thermomechanical rolling (M). To precipitate small MX particles, most of the microalloying elements have to be dissolved beforehand at the hot-working temperature. The temperatures required for this increase in the order V, Nb, Ti, and increase even further with the nitrogen content. Complete dissolution of MX would induce rapid grain growth in the slab and thus slightly coarsen the final grain size in the

B.2 High-strength materials

Table B.2.1 Comparison of the yield point and susceptibility to cold-cracking: For the same CEV, the yield point of the thermomechanically rolled steel (M) is double that of the normalised steel (N). For the same yield point, the CEV of the normalised steel increases on welding into the range where the material is susceptible to cracking (from L. Meyer).

Chemical composition [%]	$CEV = 0.3$		$R_e = 400\,\text{MPa}$	
	N	M	N	M
C	0.16	0.06	0.18	0.08
Si	0.30	0.30	0.30	0.30
Mn	0.85	1.40	1.40	1.20
Al	0.03	0.03	0.03	0.03
N	-	0.04	0.03	0.03
V	-	0.08	-	-
	Yield point in MPa		CEV equivalent	
	240	480	0.44	0.28

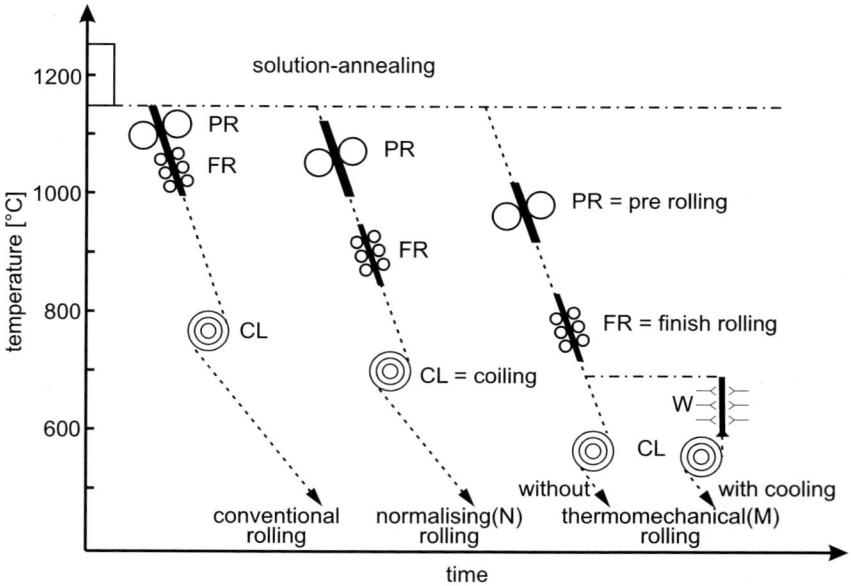

Fig. B.2.1 Manufacture of hot-rolled strip: The diagram shows the approximate final-rolling and coiling temperatures, which decrease from conventional rolling through N-type to M-type manufacture. Water cooling is used to suppress the growth of pearlite in favour of bainite (dashed line = air-cooling; W = water cooling).

coiled strip. Titanium reacts with oxygen, nitrogen, carbon and sulphur, in this order if they are simultaneously present. This example demonstrates how difficult it can be to accurately control the amount of precipitatable titanium.

Poorly soluble compounds are particularly effective austenite grain refiners after they reprecipitate (Figure B.2.2). Small amounts of Nb and Ti are thus more effective than V. Because this is dependent on the atomic concentration, the heavier niobium (Tab. A.1.4, p. 15) is at a disadvantage, but it is nevertheless effective. The binding of nitrogen to titanium or zirconium keeps boron in solution so that it can exert its hardening effect in rapidly cooled steels (p. 192). As the hot-working temperature decreases, the load applied to the rolling mills increases, but then so does the strength of the strip. Metal-working facilitates nucleation and thus accelerates the transformation. At final-rolling temperatures close to Ar_1 ($\approx 700°C$), hot-working also affects the transformed ferrite/carbide microstructure. If it no longer recrystallises, the elongated grain shape can lead to separations e.g. in notched impact bending specimens. To achieve the same strength at higher final-rolling temperatures ($\approx 800°C$), a larger amount of microalloying elements is necessary or accelerated cooling rates from the rolling to the coiling temperature. In the first case air-cooling at a rate of e.g. 0.5°C/s leads to a ferrite/pearlite transformation. In the second case, spray-cooling at a rate of e.g. 15°C/s shifts the transformation to the bainite range. In both cases, the coiling temperature lies between 600 and 550°C. Subsequent slow cooling in the coil allows the steel (Figure B.2.2 b) to harden by MX precipitates (Figure B.2.3). The

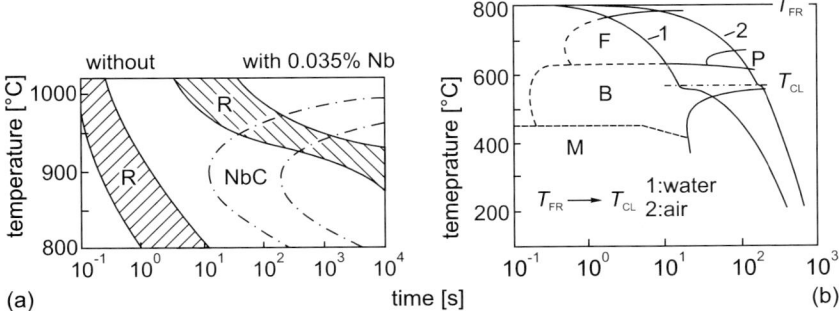

Fig. B.2.2 Metallurgical processes during manufacture of hot strip: (a) Influence of $\approx 0.035\%$ Nb in steels containing $\approx 0.1\%$ C and $\approx 1.3\%$ Mn. According to M.G. Akben et al., recrystallisation (R) during hot working is retarded by the precipitation of NbC. (b) After hot-working of the microalloyed steel, air-cooling from the finish-rolling temperature (T_{FR}) to the coiling temperature (T_{CL}) produces a fine-grained ferrite-pearlite microstructure. Water-cooling suppresses the growth of pearlite in favour of bainite. Precipitation hardening can take place in the more slowly cooling coil at T_{CL}.

bainitic microstructure is usually tempered, the ferritic/pearlitic structure is not.

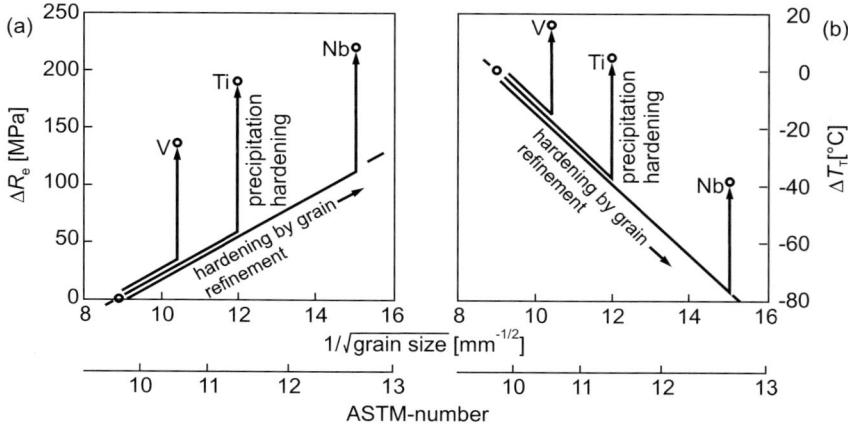

Fig. B.2.3 Strengthening by microalloying: (a) Proportions of hardening by grain refinement and by precipitation-hardening versus the increase in the yield point ΔR_e and (b) versus the change in the transition temperature ΔT_T (from C. Straßburger).

B.2.1.2 Multi-phase steels

Whereas the hardness of microalloyed steels depends on lattice defects due to solid solutions, fine grains and fine precipitates, multi-phase steels can also undergo coarse two-phase strengthening, i.e. from a mixture of two or more different types of grains, such as ferrite, bainite, martensite or retained austenite. This is similar to the strengthening effect associated with an increasing proportion of pearlite grains (Figure B.1.2). However, pearlite contains almost 0.8 %C in the form of brittle carbide plates. The higher strength is thus associated with a loss of ductility and weldability. High-strength multi-phase steels are frequently used as sheet materials in car body manufacture, where they have to withstand deep drawing, stretch-forming, hydroforming and welding. This is why fine-grain mixtures are produced from the aforementioned tougher phases. Microstructural development can be controlled by means of two supplements to the rolling procedures shown in Figure B.2.1. The first is applied at the end of the rolling process, the second during cooling.

(a) Intercritically rolled steels

If the temperature after pre-rolling and finish rolling lies below Ac_3, brief holding in the two-phase ferrite/austenite field may lead to segregation of the alloying elements. Owing to its high diffusion velocity, carbon is rapidly enriched in the austenite according to the Lever Rule (Figure B.2.4 and Figure A.1.8, p. 16). However, the substituted elements are also mobile, e. g. manganese diffuses into the austenite, and Si into the ferrite. This intercritical rolling or holding produces up to 90 % ferrite, and the amount of C remaining in the austenite increases with the proportion of ferrite. The thus enriched austenite transforms during subsequent cooling. For a ferrite/austenite ratio of e. g. 80/20, water cooling produces a dual-phase microstructure (DP) of ferrite and martensite (Figure A.2.2 b, p. 22). If this ratio is e. g. 55/45 after intercritical rolling, then cooling and isothermal holding between 500 and 350°C leads to a bainite transformation in which the carbon can be enriched in the untransformed austenite to such an extent that retained austenite (RA) remains after cooling to room temperature. This is promoted by silicon, which suppresses carbide precipitation in the bainite and forces carbon into the austenite. Because small amounts of austenite are not sufficiently enriched, they transform to martensite on cooling from the isothermal holding temperature. The microstructure contains ferrite from the intercritical transformation and ferrite from bainite transformation along with martensite

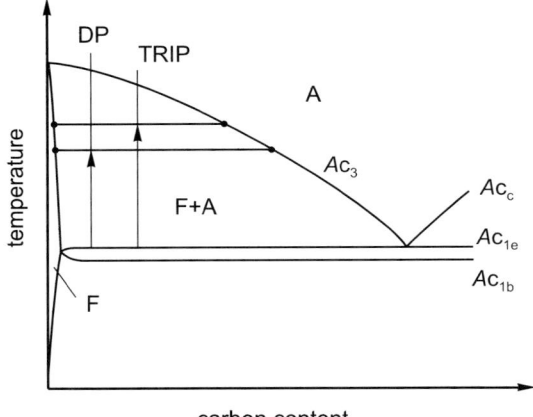

Fig. B.2.4 Intercritical microstructural tailoring (schematic): Holding in the two-phase field leads to near-equilibrium proportions of ferrite F and austenite A according to the Lever Rule. Their amounts can be varied by changing the alloy content and the temperature. In a dual-phase steel with ≈ 0.1 % C, the major phase is ferrite, whereas only half is present as ferrite in a TRIP steel with ≈ 0.2 C. Austenite becomes enriched with carbon because it is poorly soluble in ferrite.

from subsequent cooling and the aforementioned RA, which bestows the designation 'TRIP' to these multi-phase steels. Transformation-induced plasticity (TRIP, Chap. A.2, p. 41) arises from the partial transformation of RA during cold deformation to produce martensite with its greater hardness and specific volume at locations where transformation is starting. This increases the resistance to deformation at these points. As a result, the surrounding regions of the microstructure undergo slip and strengthening by RA transformation, and so on. This TRIP effect thus starts with a low yield point and causes uniform deformation and strengthening. This is very desirable during sheet processing. In addition to this beneficial TRIP effect during manufacturing, it is also advantageous in the event of a crash or during general service of a component. The latter also applies to the TRIP effect in austempered ductile iron (ADI, Chap. A.3, p. 71 and p. 208).

In addition to DP and TRIP steels with $>50\%$ ferrite, there are also PM (partially martensitic) steels containing $>50\%$ martensite. Common to all variants is the fact that the microstructural components have to be as small and as homogeneously distributed as possible to obtain a high product of the strength multiplied by the ductility. This is achieved by grain refinement and, to some extent, by microalloying with niobium, as described in Section B.2.1.1. Another method is to exploit carbon partition between ferrite and austenite at two different temperatures: (a) at a high intercritical temperature in the two-phase field between Ac_3 and Ac_1, (b) at a low temperature in the region of the bainite nose. Partition of type (a) is near-equilibrium, type (b) affects supercooled austenite. Type (a) is used in DP and PM steels, whereas (a) and (b) are used in TRIP steels.

The implementation of a processing sequence consisting of rolling, holding and cooling steps into continuous production of hot strip makes extreme demands on the process engineering. Further factors have to be considered: firstly, the surface quality – because a multi-phase microstructure should not be subsequently cold-rolled – and secondly, the type of intended coating, e.g. a hot-dip zinc-coating subjects the steel to a heat treatment. In addition to hot-strip production, which combines rolling and heat treatment, there is also cold-strip production in which rolling and heat treatment are carried out separately: the hot strip is pickled, cold-rolled, and then heated to the intercritical temperature under a protective gas, which induces grain refinement and allows to adjust the ferrite and austenite phases. This is followed by rapid cooling to below the M_s temperature (DP and PM steels) or an isothermal transformation in the bainite range (TRIP steels). The different production methods for hot and cold strip are outlined in Figure B.2.5. The schematic TTT diagrams of hot strip are based on 100% austenite. As the ferrite forms, the M_s temperature drops due to enrichment of the remaining austenite with carbon. In cold strip, the transformation refers only to the austenite fraction after intercritical annealing. Its M_s temperature also decreases due to partitioning of carbon in the bainite temperature range of TRIP steels.

Dual-phase steels: The microstructure of dual-phase steels consists of ferrite containing a dispersion of 10 to 25 % martensite with a grain size of between 1 and 4 µm. The volume increase associated with the austenite/martensite transformation leads to local deformation of the surrounding ferrite so that the yield point is continuous rather than discontinuous. An optimum combination of tensile strength and elongation at fracture is found close to 15 % martensite. At this content, Figure B.2.4 indicates a considerable enrichment of C in the initial austenitic phase, which, however, should not lead to the more brittle plate martensite with a greater hardness (Chap. A.2, p. 42). This can be countered by keeping C < 0.1 % and raising the intercritical temperature. Alloying with Mn ≤ 1.5 % and Si ≤ 0.5 % shifts the transformation noses for pearlite and bainite (Figure B.2.5 a, c) to times that are long enough

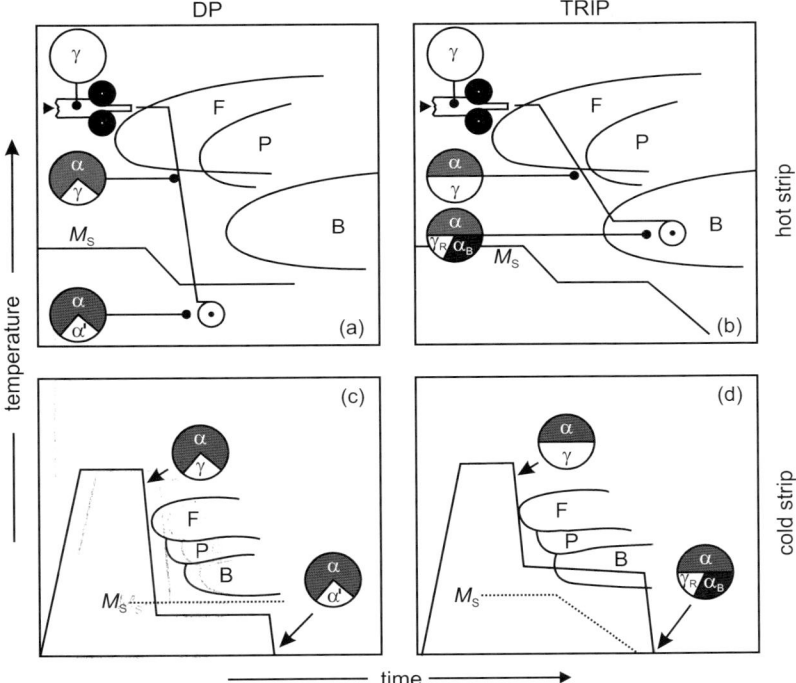

Fig. B.2.5 Manufacture of multi-phase steels (schematic, from W. Bleck): (a) and (b) manufacture of hot strip by rolling and microstructural tailoring from the rolling temperature; (c) and (d) manufacture of cold strip by cold rolling of descaled hot strip and microstructural tailoring by cooling from the intercritical annealing temperature; (a) and (c) dual-phase steel; (b) and (d) TRIP steel. The approximate proportions of the phases are represented by pie charts: γ, γ_R =austenite, retained austenite, α, α_B, α' = ferrite, bainitic ferrite, martensite.

to allow martensite to form. Owing to the greater thicknesses and more challenging temperature control associated with hot strip, its alloying content is generally slightly higher than that of cold strip and may also include small amounts of Cr and Mo. Because Si diffuses to the surface and its oxide has an adverse effect on hot-dip zinc-coating, its content is reduced if this post-treatment is to be carried out and may be partially replaced by $P < 0.1\%$. Additions of $Cr + Mo < 1\%$ partly reverse the loss of hardness of martensite in the zinc bath. Adding $\approx 0.05\%$ Al improves the resistance to ageing, whereas $\geq 0.03\%$ Nb reduces the grain size. If the level of alloying is insufficient and the bainite nose is entered, a little bainite and even retained austenite may form; however, it is less stable than that intentionally produced in TRIP steels. Paint baking exploits the effect of bake-hardening (BH, Chap. B.1 p. 131).

TRIP steels: These steels contain 50 to 60 % ferrite and 25 to 35 % bainitic ferrite as well as 5 to 15 % retained austenite and $< 5\%$ martensite. In a steel with 0.2 %C, the C content of austenite increases during intercritical holding to e. g. 0.5 % and to $\geq 1.5\%$ during bainitic transformation with an alloying content of 1.5 % Si. Addition of 1.5 % Mn retards the growth of pearlite and increases the stability of the retained austenite, which is also dependent on its grain size and distribution. Isothermal holding in the bainite range can be combined with hot-dip coating in a Zn or Zn-Al bath between 460 and 420°C. In this case, alloying with Al instead of Si helps to avoid disruptive oxides in the surface of the steel. Steels containing $\approx 0.2\%$ Si and 1.8 % Al may also contain 0.1 % P because, like Si, it contributes to delaying carbide precipitation in bainite. A few tenths of a percent of Cr and Mo allow a slower cooling rate and facilitate process control. Compared to dual-phase steels, TRIP steels have a higher product of the tensile strength multiplied by the elongation at fracture, but their 0.2 proof strength is slightly higher.

(b) Complex-phase steels

As for microalloyed fine-grain structural steels, the austenite is rolled to the Ac_3 temperature or slightly below it and then cooled with the intention of producing more than one type of grain. The first cooling step is rapid water-cooling of the hot strip as it exits the rolling mill, the second is coiling at the selected temperature with slow cooling of the coil. Figure B.2.6 illustrates the influence of the coiling temperature on the microstructural composition and on the mechanical properties of a weakly alloyed steel for cooling rates of 30°C/s between the final-rolling and coiling temperatures and 50°C/h for cooling in the coil. If soluble boron is to be used to delay transformation, this requires binding of N by Ti to prevent precipitation of BN (p. 192). Any excess Ti remains in solution at the start of rolling and, together with a little added Nb, it precipitates as MX thus contributing to grain refinement and hardening of the ferrite at a higher coiling temperature. As the coiling temperature decreases, martensite growth promotes strengthening of the structure and the development of a continuous yield point. The retained austenite is less stable than in TRIP steels. To save on alloying content, the cooling rate to the coiling temperature can be increased by integrating a high-speed cooling line with rates of e.g. 300°C/s (for 4 mm-thick strip). This two-stage cooling strategy provides a further opportunity to influence the microstructural composition by selecting the intermediate temperature for the start of the fast stage. Production trials with an unalloyed steel gave the range of property combinations shown in Figure B.2.7. Complex-phase steels (CP) are essentially composed of different types of bainite and martensite with small proportions of ferrite and

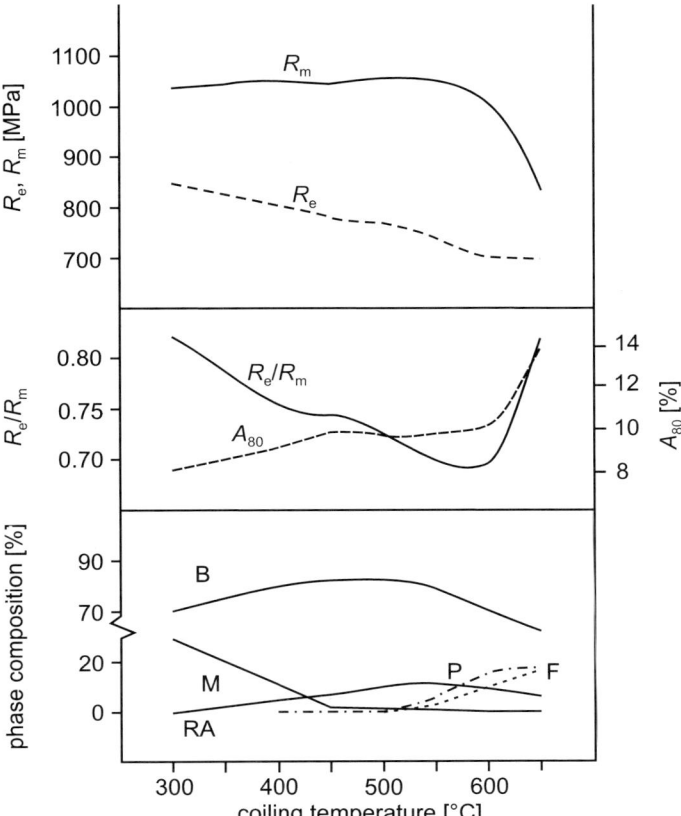

Fig. B.2.6 Influence of the coiling temperature: on the microstructure and mechanical properties of a steel containing (%) 0.16 C, 1.5 Mn, 0.2 Mo, 0.43 Cr, 0.002 B, 0.06 Nb, 0.04 Ti. F, P, B, M, RA = ferrite, pearlite, bainite, martensite, metastable retained austenite (from Y. Pyshmintsev et al.).

unstable retained austenite. The martensite fraction increases with decreasing coiling temperature, and thus leads into the group of martensite-phase steels (MS).

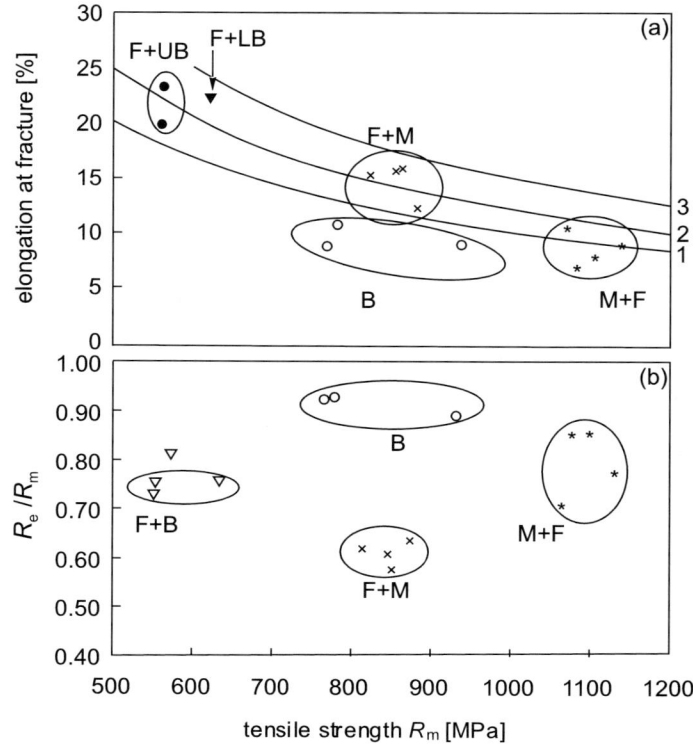

Fig. B.2.7 High-speed cooling and the coiling temperature: exert a considerable influence on the microstructure and mechanical properties of an unalloyed steel containing 0.14 % C und 1.5 % Mn. The curves in (a) represent the product of the yield point [MPa] multiplied by the elongation at fracture [%]; 1 = 10000, 2 = 12500, 3 = 15000; where F, UB, LB, B, M = ferrite, upper bainite, lower bainite, almost entirely bainitic, martensite (from A. Lucas et.al.).

B.2.1.3 Applications of weldable steels

It is not easy to keep track of all the high-strength, weldable rolling steels because their properties are determined by rolling and cooling rather than by their chemical composition, indeed, this applies all the more, the thinner the nominal cross-section. Historically, they were developed from microalloyed low-pearlite fine-grained structural steels (HSLA), whose ferrite was strengthened by precipitation hardening. One prerequisite was the construction of rolling stands that could exert the higher rolling forces required for thermomechanical working of austenite. Because this production process achieves good properties even with mild cooling, it is still being used for e.g. thicker nominal cross-sections. Installation of a water-cooling system then allowed transformation to bainite or, for a corresponding alloy content, to martensite. Continuation of this development can be seen in the newer complex-phase (CP) steels: thermomechanical working of austenite with very fast cooling of the strip and slower cooling in the coil produces a pearlite-free phase mixture that is particularly suitable for high-strength grades. Between the low-pearlite and the CP steels are the DP and TRIP steels in which the final rolling temperature is reduced into the intercritical two-phase field consisting of ferrite and austenite. Ferrite forms first and the remainder transforms so that it is pearlite-free. This also applies to PM (partially martensitic) steels, whereas MS (martensite-phase) steels remain ferrite-free by quenching of the initial austenitic state. The boundaries between the individual groups of steels cannot be sharply defined, indeed they overlap somewhat and we must wait and see how they will be regulated by the set of standards.

The effect of subsequent heating during hot-dip zinc-coating or welding increases as the strength increases. This can be countered as follows: the low heat input during laser welding keeps the heat-affected zone narrow, the flow constraint increases and a local drop in hardness has a lower impact under loading conditions. A narrow weld experiences better cathodic protection by the intact zinc coating of its surroundings. Uptake of hydrogen can be avoided during welding. If electrodeposition is used instead of hot-dip zinc-coating, the processing temperature can be reduced from $\approx 450°C$ to $< 100°C$ so that even a strength of 1500 MPa is maintained. However, the uptake of hydrogen is a drawback because it cannot outgas through the electroplated layer during baking. This problem can be solved by electrodeposition of a $\approx 10\,\mu m$-thick AlMg layer from a mixture of water- and halogen-free organometallic compounds dissolved in aromatic hydrocarbons such as toluene or xylene. This process does not include any heavy-metal ions, avoids hydrogen uptake and thus embrittlement of high-strength steels, and their strength is unaffected by the process temperature of 95°C. However, at present, it is only used to protect rack goods (sheet panels). If hydrogen has been excluded during manufacturing, the formation and uptake of hydrogen produced by corrosion must be strictly excluded if the material is to be used in high-strength applications. A zinc coating helps to protect against atmospheric corrosion. In a more aggressive environment, caution should be exercised with respect to high-strength grades (see Figure B.2.10, discussed later on). In high-strength steels used for automotive manufacture, corrosion protection against industrial atmospheres, deicing salt and encrustation does not only prevent rust, but also hydrogen embrittlement. In view of this, improvement of the protecting layers is especially important.

A new strip-coating system consisting of a 3.5 µm-thick electrolytic zinc coating and vapour deposition of an equally thick layer of magnesium is used to avoid excessive heating, improve the corrosion resistance and increase cathodic protection of bare cut edges. Ablation craters are avoided during laser welding. These are caused by vapourisation of a thick Zn layer. An intact coating and a retained strength as well as production aspects and costs are the reasons behind further developments in joining by riveting techniques, chiefly self-piercing riveting and impulse riveting. This is associated with a decline in the popularity of spot welding. Compared to spot joins, adhesive bonding increases the stiffness and is successful with thinner materials. New adhesives extend the range of applications and, in addition to sealed joints, they also save weight and costs.

As already discussed in previous sections, the objective of lowering the fuel consumption of vehicles by means of lightweight constructions using high-strength steels requires consideration of not only the mechanical and chemical properties but also their relationship with processing techniques such as coating, cold-working and joining. The possibilities must explored as to which level of strength in combination with which production strategy at which costs a component can be successfully manufactured.

Fine-grain structural steels: Normalised or normalising rolled (N) fine-grain structural steels produced according to EN 10113-2 range from S275N to S460N (Figure B.2.8). They are available with the same range of yield strength as pressure vessel steels (EN 10028-3) but with a defined hot strength or low-temperature toughness (Figure B.2.9). The product thickness is up to 250 mm. For thinner flat products for cold-working, EN 10149-2 specifies thermomechanically (M) rolled steels S315MC to S700MC, and EN 10219-1 covers tubes of S275M to S460M. As shown by the example in Tab. B.2.1, changing from N to M doubles the yield point for the same weldability/CEV or lowers the CEV for the same yield point. High-strength tubes are quenched and tempered up to P690Q (EN 10216-3), and steels up to S960Q (EN 10137-2) are available as sheet and wide flats (Tab. B.2.2).

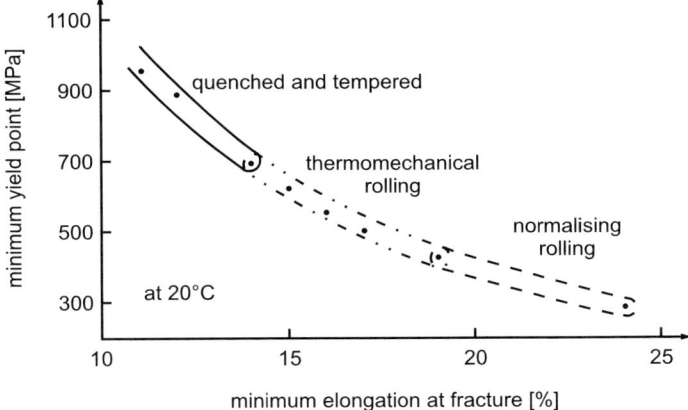

Fig. B.2.8 High-strength fine-grained steels: Minimum values of the yield point and elongation at fracture of steels S275 to S960 according to EN 10113-2, 10149-2 and 10137-2.

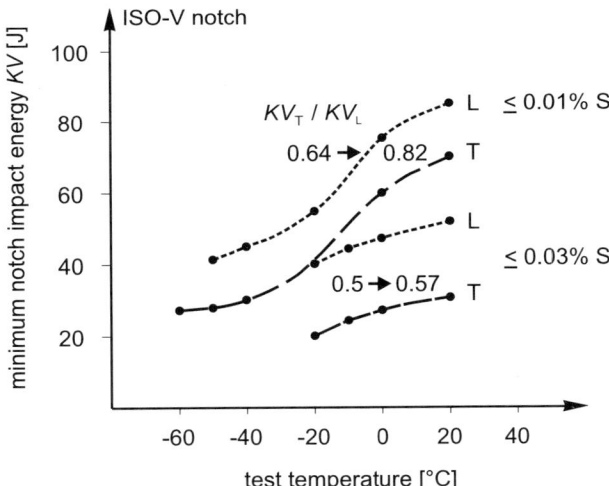

Fig. B.2.9 Toughness of fine-grained steels: Dependence of the test temperature and sulphur content exemplified by steel S355N (EN 10113-2) and the less sulphurised steel P355NL2 (EN 10028-3) L, T = longitudinal, transverse specimens

B.2 High-strength materials

Table B.2.2 Examples of higher-strength, weldable steels: Pressure vessel steels (P) contain less P and S, particularly class L2 (low). The minimum yield strength of 355 - 960 MPa is achieved by thermomechanical rolling (M), normalising rolling or normalising (N) or QT treatment (Q). Microalloying (m_1) is used for grain refinement and precipitation-hardening. The suppression of pearlite in favour of higher-strength bainite/martensite is controlled by tailoring the alloy content (m_2) to the product thickness. An Al content of $> 0.02\,\%$ together with Ti contribute to killing and resistance to ageing (see Figure B.2.8).

Name	Chemical composition (upper limit in %)							
	C	Si	Mn	P	S	N	m_1	m_2
S355M	0.14	0.5	1.6	0.035	0.030	0.015	0.20	0.90
P355NL2	0.18	0.5	1.7	0.020	0.010	0.012	0.12	0.95
S460 M	0.16	0.6	1.7	0.035	0.030	0.025	0.22	1.05
P460NL2	0.20	0.6	1.7	0.020	0.010	0.025	0.22	1.90
P690Q	0.20	0.8	1.7	0.025	0.015	0.015	0.23	5.0
P960Q	0.20	0.8	1.7	0.025	0.015	0.015	0.23	4.7*

$m_1 = $ Nb+Ti+V, $m_2 = $ Cr+Cu+Mo+Ni, * <0.005B

The tensile strength of boron-alloyed QT sheets used in automotive manufacturing is already as high as 1500 MPa. In addition to their use as thin sheets, one of the key applications of fine-grain structural steels is as thick and very heavy plates in structural, mechanical and plant engineering. Examples include buildings, bridges, lock gates, water tanks, storage containers, and mobile cranes.

Corrosion resistance: One of the major problems associated with the use of high-strength steels is hydrogen embrittlement (Chap. A.3 p. 58 and Chap. A.4, p. 114). Detrimental quantities of atomic hydrogen can only be expected to be adsorbed and absorbed by the steel in the presence of promotors such as H_2S, acidic aqueous solutions (pH $<$ 3) or clean active surfaces that facilitate hydrogen uptake. Active surfaces are created by plastic deformations due to cyclic loading or slow strain rates e. g. at notches.

The absorption of hydrogen produced by corrosion can induce cracking in the absence of externally applied tensile stresses by recombining to H_2 at e. g. band-shaped sulphides, thus causing a local build-up of pressure. These cracks can unite under an external load to produce lamellar tearing (Figure B.1.5 c, p. 132). Hydrogen-induced cracking (HIC) can be successfully combatted by lowering the sulphur content and influencing the sulphide shape. The resistance to HIC is evaluated according to EN 10229. Hydrogen generated during welding causes delayed cracking in the coarse-grained and hardened HAZ, and residual welding stresses also contribute here. Suitable precautions must be taken during welding to limit the uptake of hydrogen and to allow its effusion e.g. by preheating.

External tensile stresses in the presence of a H_2S promotor leads to hydrogen-induced stress corrosion cracking (HSCC). The ratio between the critical tensile stress that initiates HSCC and the yield point (Figure B.2.10) shows that the high yield point of high-strength fine-grained steels, in particular, cannot be exploited under these conditions. In high-strength steels with e.g. $R_{p0.2} > 1200$ MPa, HSCC also occurs in the absence of promotors. Susceptibility to strain-induced cracking can be decreased by design measures that reduce vibrational and notch stresses and by avoiding corrosion notches.

Natural gas and oil pipelines are preferentially made of microalloyed low-pearlite steels of type X70 ($\approx$ S460M). Sour natural gas containing H_2S may cause HIC. This can be alleviated by reducing the amount of sulphides and spheroidising the remainder as well as by reducing the amount of oxides. Rapidly cooled bainitic steels of type X80 ($\approx$ S550M) are designed for pipelines with 100 bar internal pressure. Carbon contents of < 0.03 % and sulphur contents of about 0.001 % are desirable here. However, if the applied stress increases, even steels with a high cleanness may exhibit a special type of HIC known as SOHIC (stress-oriented hydrogen-induced cracking). Boron and other elements are used to increase the hardenability, depending on the plate thickness. The suppression of harder pearlite bands lowers the materials's susceptibility to HIC. The bainitic microstructure remains without a discontinuous yield point, which means that during cold bending to a tube, no Lüders

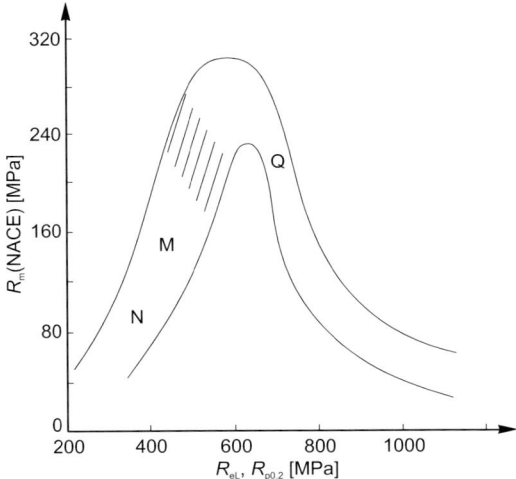

Fig. B.2.10 Hydrogen-induced stress corrosion cracking (HSCC): Weldable structural steels; N, M, Q = normalised, thermomechanically rolled, quenched and tempered. For specimens exposed to an aqueous NACE test solution (2.5 g H_2S/l, pH = 3) for 1000 h, a plot of the yield strength (determined in air) versus the elongation at fracture R_m exhibits a maximum. (from W. Haumann et al.).

deformation occurs and the Bauschinger effect decreases. This effect means that the bending-induced residual tangential compressive stresses in the inside surface layer of the sheet lower the yield point in this area when strained at the service pressure. Cold expansion can increase the load-bearing capacity and improve the roundness of longitudinally welded large-diameter tubes.

Multi-phase steels: Multi-phase steels are chiefly used as thin sheets for cold working in automotive manufacture. A low yield point coupled with a high elongation and strengthening is advantageous not only during processing, but also in the event of a crash because this requires the absorption of a large amount of energy in the crumple zone (crashworthiness). However, such a zone must be supported by a stiff, high-strength substructure. Steels with a strength of up to 1500 MPa are used for this, although their cold-workability is, of course, limited. It is thus frequently necessary to carry out hot-shaping with hardening from the hot-working temperature, followed by pickling, zinc-coating and straightening (e.g. side-impact beams). Cold-shaping of zinc-coated strip is cheaper, and is also already possible with some complex-phase steels (CP) and martensite-phase steels (MS), which contain a mixture of extremely fine-grained phases. They undergo significant bake-hardening during paint baking. The high deformation velocity of a crash does not usually cause embrittlement.

In addition to microalloyed fine-grain steels (HSLA) and fine-grain steels quenched and tempered from the hot-rolling temperature, these multi-phase steels represent a new group of materials for automotive manufacture, some of which are still being tested. This demonstrates that sophisticated rolling and cooling techniques can produce a combination of all strengthening mechanims and still achieve sufficient ductility for cold-forming and crashes. These high-strength materials are used to lower the weight of vehicles and thus reduce fuel consumption. Their elongation at fracture and tensile strength are compared to other types of steel in Figure B.2.11.

A further possibility of saving weight involves selecting the sheet thickness and strength according to the local loads applied to the component. For example, a number of different pieces of sheet are joined by high-speed laser welding to a blank that is then processed by deep drawing, bending, trimming, etc. An example of this is a car door with reinforcements around the lock and hinges. Such tailored blanks are also convenient for cold-shaping, e.g. a piece of highly deformable IF steel is inserted at a point where a high local draw depth is required (patchwork blanks). Tailored tubes, e.g. with a conical shape, are subjected to internal high-pressure hydroforming to produce lightweight components, e.g. automobile side-members. This component is also produced by another method starting from wide zinc-coated strip that is cold-rolled to different thicknesses at periodic intervals and then cut to size (tailor-rolled blanks). Each of these periodic lengths is then bent to a tunnel profile. As a further alternative, two or three steel strips of different thickness and/or strength are placed next to one another and then laser-welded to a wider strip (tailored strips), which is then subjected to roll-profiling or die-bending to produce the component. Roll-forming is used for small radii and sheets with a high strength, e. g. for the torsion profile of a twist-beam rear axle made of DP or CP steel with a tensile strength of 600 or 800 MPa, respectively. The high strength and the design of the roll-formed cross-section provides considerable weight savings. This is particularly important for car wheels because it relieves the recoil on the suspension. By use of a roll-profiled wheel rim, a weight-reduced wheel disc and a DP steel with a tensile strength of 600 MPa, the weight can be brought close to that of a cast aluminium wheel, and indeed at lower costs.

Facilities to transport and store mineral materials are made of weldable sheet. Because the depth of grooving by hard particles decreases with the sheet hardness, fine-grain high-strength QT steels are more resistant (Chap. A.4, p. 106). However, dual-phase steels are also suitable, particularly if the carbon content is increased to 0.2 % because the hard martensitic islands within the ferritic matrix resist grooving.

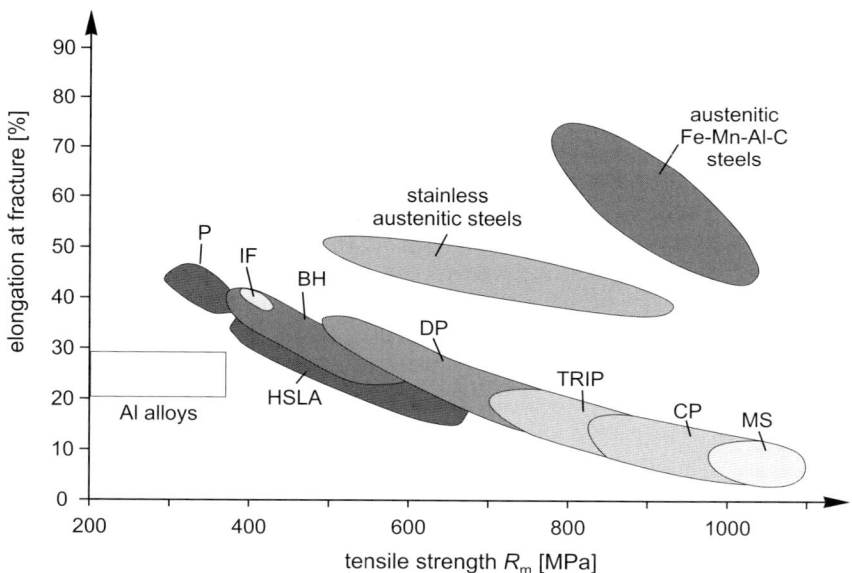

Fig. B.2.11 **Properties of high-strength weldable steels compared to austenitic steels and aluminium alloys:** P = phosphorus-alloyed, IF = interstitial-free, HSLA = high-strength, low-alloy, BH = bake hardening, DP = dual phase, TRIP = transition-induced plasticity, CP = complex phases, MS = martensite phases (from U. Brüx and G. Frommeyer).

B.2.1.4 Lightweight steels

In addition to the use of high-strength steels as well as tailored blanks and tubes, another way of reducing the weight of vehicles is to use lightweight steels, which are produced by alloying with lighter elements. Al is used for ferritic steels, whereas Al and Si are used for austenitic grades.

Iron can hold up to 6.5 % Al in solution at room temperature, which makes it about 9 % lighter. A carbon content of about 0.01 % and binding of the remainder with 0.05 % each of Ti and Nb keeps the grain boundaries free of precipitates. An additional 1.5 % Mn and optionally 0.002 % B leads to grain refinement and improvement of the rolled texture. The resulting tensile properties are significantly better than those of a pure Fe-Al alloy. Above 8 % Al, the iron is completely passivated due to the formation of an Al_2O_3 top layer. However, even 6 % Al increases the corrosion resistance.

If the manganese content is increased from 15 to 20 % in an iron containing 3 % Al and 3 % Si, the initial duplex microstructure consisting of approx. half and half ferrite and austenite changes to a microstructure with 95 % of austenite, which has a tendency to form twins, and ε-martensite during cold-working on account of its low stacking-fault energy. α-Martensite grows at the intersections of the differently oriented ε bands or directly from the austenite. Deformation twins predominate above 25 % Mn, and at 30 % Mn there is no longer any forming-induced transformation to ε- or α-martensite, i.e. no TRIP effect. It changes to a TWIP effect (twinning-induced plasticity). Tensile strengths of 600 to 1100 MPa and an elongation at fracture of 95 to 50 % have been measured, depending on the deformation mechanism, for a specific weight of $7.2\,\text{g/cm}^3$ (-8.5 % less than iron). This high elongation is associated with the fact that at the start of necking the local increase in deformation induces increased twinning and thus strengthening, which hinders further necking, and the uniform elongation A_u increases. The product of the tensile strength (MPa) multiplied by the elongation at fracture (%) reaches values of up to 50000, which are twice as high as those given for the TRIP steels discussed on p. 172. Figure B.2.12 shows that the ductility of a TWIP steel is retained, even at the highest deformation velocity, which is important with respect to the crumple zone in the event of crash

Carbon increases the solubility of Al. Solution-annealing and quenching by artificial ageing causes spinodal decomposition into C-depleted and C-enriched zones, from which 30 nm κ-carbides precipitate with the composition $(FeMn)_3AlC$ which considerably increase the proof strength (Figure B.2.13). This triplex steel (austenitic matrix with carbide and δ-ferrite) saves almost 15 % weight due not only to the lower atomic mass of Al and Mn, but also to their larger atomic diameters (Tab. A.1.4, p. 15). With its $R_\text{m} \cdot A \approx 60000$ and a specific fracture work in the tensile test of $450\,\text{J/cm}^3$, it is far superior to low-alloy high-strength steels. However, contact with oxidising media must be avoided during melting and casting of Al-rich steels to prevent burn-off and oxide inclusions. These high-alloy steels lead on to stainless austenitic steels, which are discussed in Chap. B.6, p. 322. Lightweight steels are still in the development phase.

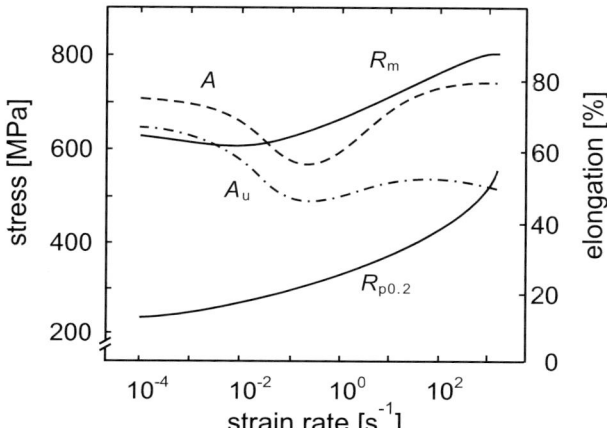

Fig. B.2.12 Influence of the strain rate on the mechanical properties of the lightweight austenitic TWIP structural steel X3MnSiAl25-3-3 in a tensile test (from O. Grässel et al.)

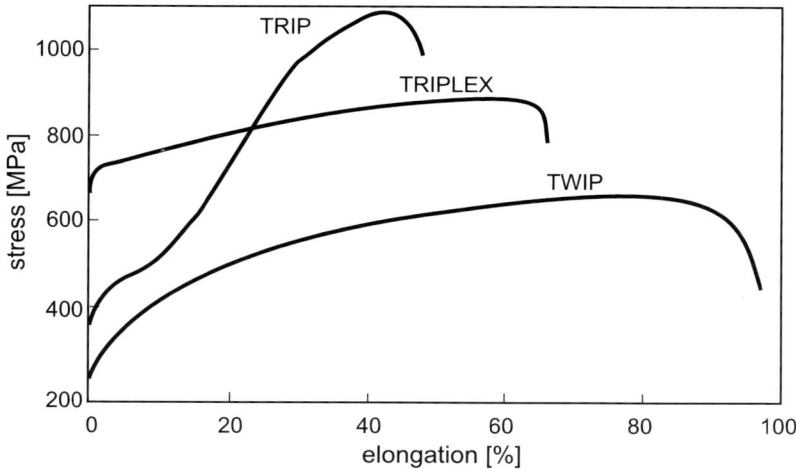

Fig. B.2.13 Typical curves for the tensile test of high-Mn TRIP, TRIPLEX and TWIP steels (from U. Brüx and G. Frommeyer)

B.2.1.5 Pearlitic rolled steels

For weldable steels, the effect of microalloying with V, Nb, and Ti is particularly effective because the low C content ($\leq 0.22\,\%$) facilitates dissolution of MC carbides at the hot-working temperature. This dissolution is a precondition for their finely dispersed reprecipitation in austenite (grain refinement) and ferrite (precipitation-hardening). The development of precipitation-hardened ferritic/pearlitic steels (FP-PH, p. 188) demonstrated that steels containing up to $0.5\,\%$ C can also be strengthened by microalloying. This effect was eventually extended to include fully pearlitic steels with $\approx 0.7\,\%$ C. Rails, prestressing rods and high-strength steels rely on a pearlitic lamellar microstructure (Chap. B.1, p. 139).

Strengthening by microalloying is also possible in these application examples (Figure A.4.13, p. 96). Niobium, in addition to vanadium, is used for grain refinement of rail steels. Both contribute to precipitation hardening of the ferrite lamellae by forming extremely small MC particles. Additional alloying with $\approx 1\,\%$ chromium and aluminium is used for heavy-duty grades having an extremely fine lamellar microstructure with strengths of up to $> 1400\,\mathrm{MPa}$. Microalloyed steels for prestressing concrete attain similarly high values (Y1230), or even higher (Y1860) after cold-drawing and tempering. Strengths of approx. $4000\,\mathrm{MPa}$ are reached in microalloyed cold-drawn steel cord.

B.2.2 Steels treated from the forging temperature

The successful combination of hot-working and heat treatment used for rolled steel is also exploited with forging steel. The objective is to tailor suitable properties in the forgings with the lowest possible costs. In contrast to the emphasis placed on the weldability of rolled flat products, the key factors for near-net-shaped drop-forged parts tend to be distortion, machinability or their suitability for surface-hardening. Compared to general hardening, hardening from the forging temperature is associated not only with an increase in grain growth as a consequence of the higher temperature, but also with a greater hardenability on account of the higher degree of dissolution: residual alloyed carbides and AlN dissolve and segregation-related inhomogenities are reduced. In contrast to rolled flat or round steel, there are differing degrees of deformation within a forging. For this reason and because of tool loading, the final forging temperature cannot usually be lowered as uniformly and as far. Grain refinement by recrystallisation is thus less significant compared to thermomechanical rolling. Microalloying with Nb/Ti leads to the formation of MC carbides that slow down austenite grain growth during heating to the forging temperature. At the same time, some of these carbides dissolve and subsequently hinder grain growth by reprecipitating during forging. However, this effect decreases as the C content increases because either undesirable coarse MC precipitates from the melt or because too little dissolves.

Two final microstructures are desirable in forgings obtained by cooling from the forging temperature: martensite and ferrite/pearlite. A homogeneous multi-phase or complex-phase microstructure is very difficult to achieve owing to differences in the cross-section, and the associated advantages could only be exploited if the grain size is extremely small. The type of cooling has to be adjusted to the capabilities of the forging facility with its frequent change of procedure. A water bath is easier to integrate than an oil bath plus cooler; however, it is more liable to cause distortion. Both have shorter cooling and overall times compared to air cooling.

In addition to tailoring particular service properties, cooling from the forging temperature can also be used to obtain certain manufacturing properties. In low-carburised steels, the machinability can be improved by a microstructure consisting of ferrite and lamellar pearlite, which acts as a chip breaker thus avoiding continuous chips. In hypereutectoid steels, Ac_c increases on alloying with Cr, Mo, and V. If the final forging temperature lies below Ac_c, spheroidal carbides precipitate within the grains and continue to grow as the temperature falls. This facilitates soft annealing to spheroidal cementite (AC) and also improves the machinability.

B.2.2.1 Martensitic steels

Bending usually requires less pressing force than deforming the bulk. Therefore, the temperature for hot bending of the individual leaves of laminated leaf springs for commercial vehicles can be reduced to the hardening temperature of spring steels. Bending is carried out in a lamellar press-quenching die that is kept closed during hardening in oil; however, it allows oil to penetrate between the lamellae to reach the spring leaf, which is tempered to the spring hardness after cooling. Whereas thin coil springs are cold-coiled from patented and drawn wire or QT wire (EN 10270), thicker wires or rods are coiled around a mandrel at the hardening temperature, cooled in oil and tempered. If the rods have already been ground, the amount of undesirable decarburisation is limited and thus the fatigue strength remains high. In addition, the dimensional accuracy is improved by quenching in the bending or coiling tool.

Going from multi-leaf trapezoidal springs to two or three-leaf parabolic springs, the individual leaves are rolled out at both ends and thus tapered. The final hot-working temperature is kept close to the hardening temperature to reduce decarburisation and also to obtain a finely grained microstructure. This is followed by oil-hardening and tempering. Such thermomechanical working appears to decrease temper embrittlement in high-strength steel 50CrV4.

Whereas bending and oil-hardening of leaf springs can be integrated into a continuous line with a hardening and tempering furnace, water-hardening from the forging temperature is more convenient in a drop-forging facility. Surface-hardenable QT steels, such as 41Cr4, may crack if they are cooled too fast in water. As this problem decreases with the C content, steels such as 10MnB6 have been developed that form high-strength, ductile lath martensite in water and which no longer require tempering (Figure B.2.14 b). This represents a further cost advantage in addition to the savings on a separate hardening treatment. However, the yield point increases by approx. 30 MPa with each hundredth of a percent C, so that the strength parameters are noticeably more melt-dependent compared to highly tempered QT steels. For reasons of weight reduction and also tool loading, drop-forged parts for automobiles are often sufficiently thin-walled to guarantee sufficient hardening by 1.5 % Mn and 0.002 to 0.005 % B. The effect of dissolved B on the suppression of ferrite can develop fully at a low C content; however, it presupposes microalloying with Ti to bind N and thus prevent precipitation of BN (p. 192). A secondary effect is a decrease in grain growth at the forging temperature. Owing to the high M_s-temperature, auto-tempering becomes increasingly important as the cooling rate decreases (Figure A.2.22, p. 45) resulting in a lower hardness in thicker regions of the cross-section. The point at which bainite starts to contribute is difficult to judge and requires a TEM study.

Partial replacement of Mn by Cr reduces segregation and improves the cold shearability. Steel 5CrB4 with 0.9 % Mn is also suitable for cold working. Steel 7CrMoBS5 with 0.25 % Mo and 0.09 % S is used for thicker cross-sections and improves the machinability. Low-C lath martensites do not contain suitable chip breakers and may thus produce tough continuous chips that hinder some processing steps. Water cooling from the final forging temperature at a rate of < 250°C/s is reported to produce a pure bainitic microstructure in 8MnCrMoB5-4 with 0.1 % Mo; however, there is no TEM evidence of this.

Semi-hot forging exploits the lower tendency to scale growth and decarburisation at lower temperatures (usually close to Ac_1) in order to produce near-net-shaped parts with narrower tolerances and a better surface quality compared to conventional drop-forging. One field of development is directed at raising the semi-hot temperature to just above Ac_3 to combine thermomechanical working with subsequent water hardening whilst maintaining the surface quality as far as possible (Figure B.2.14 c). The low hot strength of a low-C and low-alloy austenite (e. g. 5CrB4) and the lower distortion on quenching should be advantageous compared to conventional forging.

Depending on the component's shape, martensitic forging steels can be problematic with respect to distortion. This can be countered by increasing the substitutional alloying content so that a good combination of strength/toughness of lath martensite can be achieved by less harsh cooling (example: X3CrMnNiMo2-2-1-1). The low C content avoids embrittling grain boundary carbides. Thermomechanical working at 750°C was found to reduce the lath length from 10 to 0.2 µm, which not only considerably increased the strength but also significantly lowered the transition temperature. The highly dislocated microstructure with its thermal stability is suitable for precipitation hardening by the G phase (Tab. A.2.1, p. 52). This is achieved by alloying with Si and Ti, and the resulting ferrite stabilisation is compensated by adding

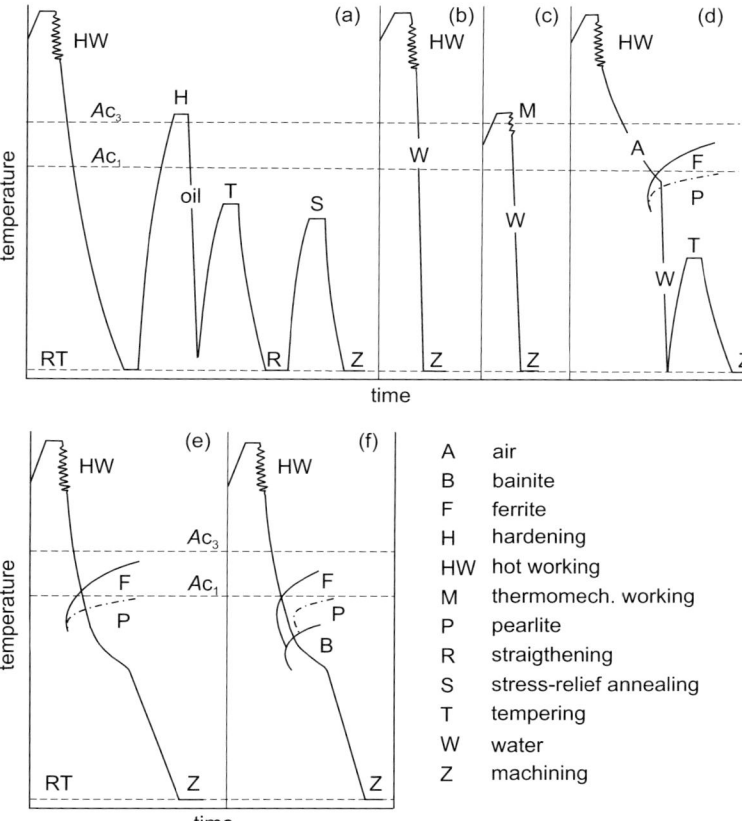

Fig. B.2.14 Relationship between forging and heat treatment. Concepts as a schematic representation: (a) Forging and subsequent QT treatment of slender components, e.g. of 42CrMo4, with straightening and stress-relief annealing. (b) Hardening in water of a forged martensitic steel, e.g. 10MnB6, from the hot-working temperature. This steel can be machined without tempering if the shape and machining allowances are appropriate. (c) As for (b), however, with thermomechanical working after the forging temperature has been lowered into the semi-hot range - provided that the deformation force and tool service life permit this (from V. Ollilainen, E. Hocksell). (d) Interrupted hardening of e.g. 30MnVS6 with air cooling until ferrite starts to grow, with subsequent hardening in water and tempering (from I. Gonzalez-Baquet et al.). This leads up to (e) precipitation-hardened ferritic-pearlitic (FP-PH) steels, e.g. 30MnVS6, produced with controlled air cooling, which is delayed after the ferrite/pearlite transformation to allow precipitation-hardening of the ferrite (from S. Engineer). (f) Similar to (e), however, with added Mo to suppress the growth of pearlite (38MnSiMoVS5 from K.W.Wegner).

more Mn and Ni (X2MnNiSiTi4-3-2-1). Slight overageing can reduce the cuttability of these coherent precipitates and thus lower brittleness due to the coarse distribution of slip bands.

A simultaneous increase in hardening capacity and hardness penetration is achieved in martensitic forging steel 21CrMnCu8-6 with 0.5 % Cu: microalloying with Ti/Nb counteracts grain coarsening and also makes it suitable as a case hardening steel. Tempering is no longer necessary after air-hardening from the forging temperature. Its mechanical properties are superior to those of precipitation-hardenable ferritic-pearlitic steels (FP-PH, see the next section); however, they remain below those obtained after conventional QT treatment at 950°C/water + 190°C/air during which martensite forms with 5 % retained austenite as thin films. Finally, there are also air-hardenable steels, such as 40SiMnCrMo7-6-6-4, that can be highly tempered. The higher C content allows surface-hardening without a water spray. Their Cr and Mo content generally increases the hardness after gas nitriding to > 700 HV without affecting the core strength. These steels can also be subjected to subsequent quenching and tempering as an alternative to treatment from the forging temperature.

B.2.2.2 Ferritic-pearlitic steels

Figure B.1.2, p. 127 shows that faster air-cooling of a thin cross-section improves the tensile strength and elongation of C45, although it still has a low yield stress/tensile strength ratio. This is due to the softer ferrite in the ferrite/pearlite mixture. Microalloying makes this grade precipitation-hardenable, thus raising the yield strength of the ferrite. The solubility of their carbides at the forging temperature decreases in the order V, Nb, Ti, so that V is chosen for 0.45 % C. The formation of ferrite with finely striped pearlite is complete after accelerated air-cooling from the hot-working temperature to about 650°C, and VC is then precipitated by delayed air-cooling between 600 and 500°C. Finely striped pearlite can also be produced in thicker cross-sections by cost-effective alloying with Mn (and Si) (Figure B.2.14 e). Sulphur is added to the melt to improve the machinability. This leads to steels 46MnVS3 and 46MnVS6 (EN 10267) that can be surface-hardened to about 55 HRC. Lowering the C or pearlite content increases the ductility in three stages up to 19MnVS6, but with a trade off in strength (Tab. B.2.3). The lower the C content is, the more effective is microalloying with Nb/Ti to limit grain growth at the forging temperature. The influence of e.g. 0.02 % N is also more noticeable, which is regarded as having a strengthening effect due to precipitation of MX. V-alloyed steels are included in the group of precipitation-hardenable ferritic-pearlitic (FP-PH) steels, and their condition after controlled cooling from the hot-working temperature is designated as BY. V-free forged pearlitic steels also belong to this group; this includes ferritic-pearlitic steel 38MnS6, which has a lower yield point than 38MnVS6 but costs less. The fully pearlitic steel C70S6 does not contain the softer ferrite grains; however, the yield point increases at the expense of the ductility. In the BY condition, it is fracture-splittable. As a consequence, the big end of a connecting rod can be cost-effectively divided at notches produced during forging by splitting at room temperature. Air-cooled pearlitic forged steels are superior to water-cooled martensitic forged steels on account of their lower distortion; however, their yield point is lower. A method of compensating this has been proposed: addition of Mo to 38MnVS6 and rapid air-cooling into the bainite range to produce a low-distortion hardened microstructure without soft ferrite grains and which also has a greater creep resistance (Figure B.2.14 f). Another proposal is to suppress pearlite and thus increase the ductility. To this end, 30MnVNbTi6 was air-cooled from the forging temperature into the ferrite nose. On reaching 660°C, approx. 10 % ferrite had formed, and subsequent quenching in water produced martensite that hardened during tempering at 420°C, to result in a high product of $Z \approx 50\%$ multiplied by $R_{\mathrm{p0.2}} \approx 1000\,\mathrm{MPa}$ (Figure B.2.14 d). Because harsh cooling conditions were only used in the lower half of the temperature range, i. e. at a greater thermal stability, a lower distortion can be expected compared to direct water-hardening from the forging temperature.

FP-PH steels are widely used in automotive manufacture. The cost reduction is most obvious for crankshafts, which were originally cooled to room temperature after forging and then subjected to hardening, tempering, straightening and stress-relief annealing (Figure B.2.14 a). These four energy- and time-consuming production steps are no longer necessary if FP-PH steels are used instead of QT steels; however, the subsequent surface-hardening step requires a higher carbon content. Grades with lower carbon contents, which are thus more ductile, are used for wheel suspension systems.

Table B.2.3 Mechanical properties of precipitation-hardened ferritic-pearlitic steels (FP-PH): The approximate values apply to forgings after precipitation-hardening. Steels regulated by EN 10267 contain C and Mn as well as [%] 0.15 - 0.80 Si, $\leq$ 0.025 P, 0.02 - 0.06 S, 0.01 - 0.02 N, $\leq$ 0.3 Cr, $\leq$ 0.08 Mo, 0.08 - 0.20 V, the latter can be partially replaced by Nb.

Name	R_e [1] [MPa]	R_m [MPa]	A [1] [%]	Z [1] [%]
19MnVS6	420	650-850	16	32
30MnVS6	470	750-950	14	30
38MnVS6	520	800-1000	12	25
46MnVS6	570	900-1100	8	20
46MnVS3	470	750-950	10	20

[1] minimum values

B.2.3 Structural steels for full heat treatment

The combination of hot-working and heat treatment in Sections B.2.1 and B.2.2 does not apply to cast irons and cast steels. A steel with a soft-annealed microstructure is often required for cutting and non-cutting shaping operations so that the actual service properties are achieved by heat treatment of near-net-shaped components. Therefore treatments are discussed below that lead to the most uniform strength increase in the entire component made of steel or cast iron (next section). In contrast, treatments of the surface layer are aimed at producing differences between the surface layer and the core. This is discussed in Chapter B.3.

B.2.3.1 QT steels

(a) Hardenability and properties

The goal is to produce tough components that do not readily fracture when subjected to severe dynamic or cyclic loads. Resistance to failure means that, on the one hand, high mechanical stresses are absorbed elastically by the component and, on the other, that an overload causes only permanent deformation without a fracture: when driving over a pothole, the steering knuckle may bend at most, but not break. Tab. B.2.4 summarises the mechanical properties of selected steels.

The aim is to produce a large product of the macroscopic yield point multiplied by the ductility ($R_{p0.2} \cdot Z$) or, for notched components, of the yield point multiplied by the toughness ($R_{p0.2} \cdot KV$). For components subjected to cyclic loads, the stress amplitude σ_a must remain below the microscopic yield point. Such loads are then permanently elastically tolerable: the component is fatigue-resistant. If some of the stress amplitudes do exceed this value, the material should be able to withstand small plastic deformations for as long as possible without cracking: ($R_{p0.1} \cdot Z$)↑. Once a fatigue crack has developed, the time until the critical crack length is reached increases with the fracture toughness. A large product of the microscopic yield point multiplied by the fracture toughness ($R_{p0.01} \cdot K_{Ic}$) thus also has a favourable effect on the fatigue strength. Based on the discussions in Chapter A.4.1, these combinations of properties are achieved by avoiding coarse hard phases, by means of a good cleanness and refinement of the carbides. This slows down the formation of crack nuclei both under unidirectional and cyclic loading conditions. The transition temperature decreases. A good cleanness becomes increasingly important as the hardness increases (Figure A.4.15, p. 97). Non-metallic inclusions are removed during post-treatment of steel prior to solidification. Coarser carbides of pearlite or upper bainite are avoided by less segregation and sufficient hardenability. The required alloying levels also depend on the costs. The ratio between the price of the alloying elements and their efficiency in the Jominy test can be used as a rough guide (Figure B.2.15).

B.2.3 Structural steels for full heat treatment

Table B.2.4 Mechanical properties of some QT steels: The relationship between alloy content and ruling section of the product is demonstrated by these examples

Name	Standard	Thickness [mm]	$R_{p0.2}$[1] [MPa]	R_m [MPa]	A [1] [%]	Application
G17CrMo5-5	EN 10293	100	315	490-690	20	welded components
20MnMoNi5-5	SEW 640	400	390	560-680	18	
26NiCrMoV14-5	SEW 555	1000	850	950-1100	15	heavy forgings
C35E	∧	8	430	630-780	17	unalloyed
30MnB5	EN	20	650	800-950	13	for cold heading
34CrNiMo6	10083	100	700	900-1100	12	high hardness penetration
42CrMo4	∨	60	650	900-1100	12	moderate hardness penetration
51CrV4	EN 10089	15	1200	1350-1650	6	spring steel

[1] minimum values

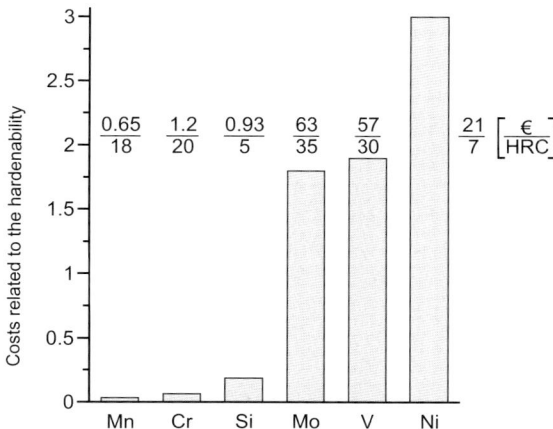

Fig. B.2.15 Costs related to the hardenability: Costs in Euro per kg alloying element (average price) or per percent alloy content in 100 kg steel with respect to the increase in hardness for J20 in HRC per percent alloy content (see E. Just).

Boron has a special status because even 0.0008 to 0.005 % contributes to increasing the hardenability at very little cost. According to Tab. A.1.4 (p. 15), a maximum of 0.005 % B dissolves in austenite, namely, as a borderline case between interstition and substitution. The available space on the austenite grain boundaries attracts the boron atoms. This grain boundary segregation lowers the interfacial energy and thus delays nucleation of ferrite on cooling: The ferrite/pearlite transformation is shifted to longer times in the TTT diagram, and the hardenability increases. In order for boron to be effective, a sufficient amount must dissolve during austenitisation, and this is hindered by its bonding to carbon – or even more strongly – to nitrogen. Boron-alloyed steels thus usually contain <0.35 % C. The increase in hardenability due to boron is particularly effective in weldable steels. Owing to the lower solubility of boron nitride, nitrogen has to be bound by pretreating the melt with Al and Ti before boron is added.

The paramount importance of the degree of hardening (Eq. A.3.2, p. 67) can be seen in Figure B.2.16. Grain refinement during quenching and tempering increases the yield point and lowers T_T. Higher nickel contents improve the low-temperature toughness even further. The strength is increased by increasing the carbon content to ≈ 0.6 % and decreasing the tempering temperature. At the same time, the region of temper embrittlement should be avoided as far as possible (Figure A.2.25 a, p. 49). Molybdenum and traces of dissolved boron decrease the sensitivity to temper embrittlement, whereas manganese and silicon increase it.

The manufacturing properties may impose conflicting requirements on the steel cleanness. For example, the sulphur content is placed in the upper half of the permissible range in order to improve the machinability (e. g. 41CrS4 with 0.02 to 0.04 % S as opposed to 41Cr4 with < 0.035 % S). Calcium silicate inclusions form a protective deposit during turning (Chap. B.1, p. 134). Preference should be given to low-Ni and -Si steels for soft annealing with respect to cold-working (Chap. A.3, p. 59). Hardening in water increases not only the hardness penetration, but also the distortion and, as the carbon content increases, the cracking susceptibility (Chap. A.3, p. 74), and is suitable for compact components with ample machining allowances. Slender, asymmetrical components should be alloyed so that a milder cooling rate is sufficient (Figure A.3.4 p. 66). The removal of mass e. g. by pre-drilling of hollow parts, improves hardness penetration.

(b) Applications

Steels with <0.25 % C: A low carbon content favours welding suitability. High-strength conveyor chains for mining equipment are made of round bars of 23MnNiMoCr6-4. The chain is made up of successively cold-bent links that are then flash butt-welded and deburred. This is followed by water-quenching to $R_m > 1100$ MPa and tempering. Off-shore engineering uses QT castings (e. g. nodes) of G13MnNi6-4 that are welded with load-bearing elements of

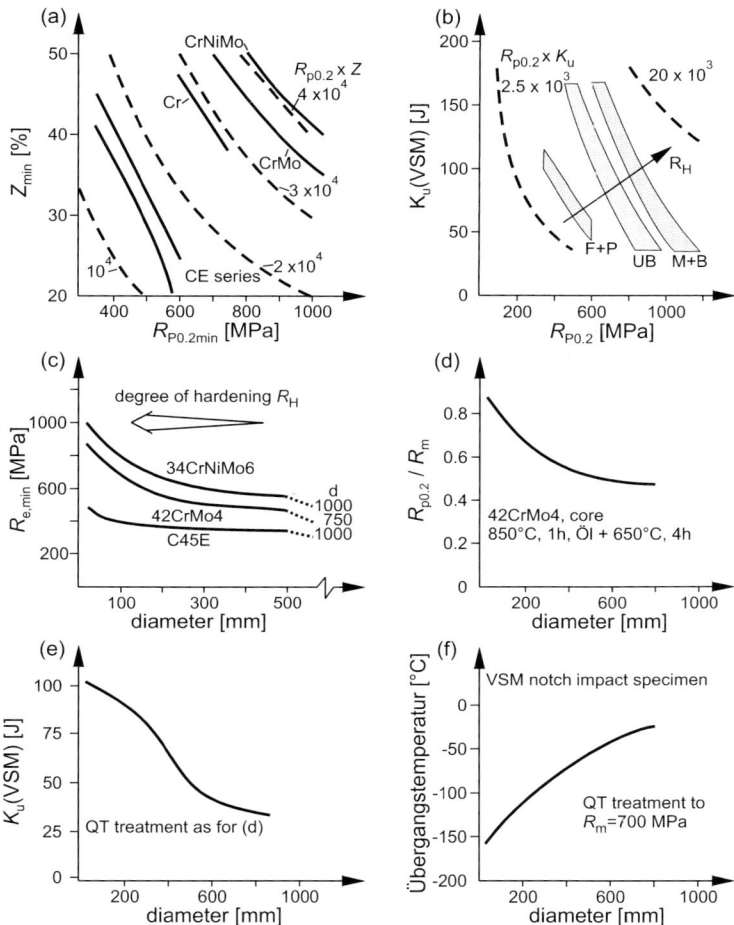

Fig. B.2.16 Mechanical properties of QT steels: (a) Influence of alloying according to EN 10083 up to 16 mm diameter (CrNiMo steels up to 40 mm diameter) with respect to the minimum values of the proof strength $R_{p0.2}$ and reduction of area Z. Dashed curves of the same product of $R_{p0.2} \cdot Z$ are given for comparison. (b) Influence of the microstructure and degree of hardness R_H on the notch impact energy K_u for a given proof strength $R_{p0.2}$ of CrMo and CrNiMo steels with 0.25 to 0.35 % C, where F + P = ferrite + pearlite, UB = upper bainite, M + B = martensite + bainite. (c) Influence of the diameter d on the minimum yield strength R_e according to EN 10083 and SEW 550. (d-f) Properties in the core of steel 42CrMo4, which are not affected by segregation and hot-working because all specimens were taken from rods with a 30 mm diameter that were cooled to correspond to oil-hardening of other diameters. The ratio between the product of the proof strength $R_{p0.2}$ multiplied by the notch impact energy K_u for diameters of 30 mm versus 800 mm is 5:1 (b, d, e and f were evaluated according to F. Hengerer et al.).

S355 for oil drilling platforms. According to the Stahl-Eisen Material Guideline 520, G12MnMo7-4 is suitable for joining to S500 owing to a certain degree of precipitation-hardening by molybdenum carbides during tempering so that its yield point corresponds to that of the rolled steel, even in thicker cross-sections. Thick-walled pressure vessels for light-water nuclear reactors must have a high initial toughness because they are embrittled by neutron radiation during service. Grades such as 20MnMoNi5-5 are used in these applications. They are protected against corrosion by a two-layer lining produced by submerged-arc strip cladding with a stainless austenitic CrNi steel as the weld filler.

As the carbon content decreases, the amount and size of the temper carbides as well as the transition temperature decrease (Figure B.2.17 a). They decrease even further due to grain refinement during quenching and tempering (Figure B.2.17 b). Grades alloyed with nickel are used in cryogenic engineering (Figure B.2.17 c). Further ways of improving the low-temperature toughness of steels include a good cleanness and thermomechanical rolling - also in combination with microalloying - to refine the austenite grain size. The service temperatures of QT steels containing 3.5 to 9 % Ni (EN 10222-3) lie between those of the low-temperature fine-grained low-pearlite steels according to EN 10025-4 (e. g. S 355 ML) and steels with stable austenite (Figure B.2.17 d). One key application is for equipment used to produce and transport liquified

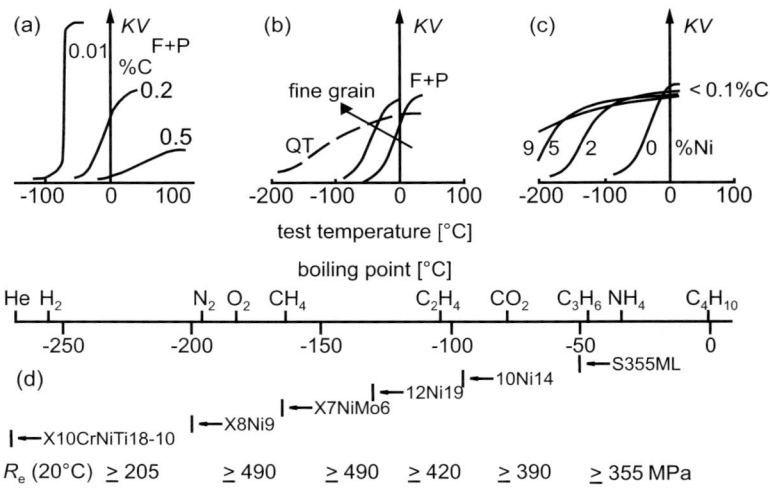

Fig. B.2.17 Low-temperature steels: (a-c) Influence of carbide content, grain refinement, QT treatment and alloying with nickel on the notch impact energy KV and the transition temperature. (d) Boiling points of liquefied gases at atmospheric pressure and the approximate application ranges of low-temperature steels.

B.2.3 Structural steels for full heat treatment

gases. These steels have the required weldability as long as a few rules are observed.

Steels with 0.25 to 0.50 % C: For drop-forged parts and machined components, EN 10083-2,-3 specifies a series of alloys with a variable hardness penetration (e. g. C45, 41Cr4, 42CrMo4 and 34CrNiMo6) or a variable hardening capacity (C22 - C60 or 25CrMo4 - 50CrMo4). This takes into account the wall thickness as well as water-hardening (% C $\downarrow$) or surface-hardening (% C $\uparrow$). DIN 17021 provides assistance in selecting steels. These steels are widely used in highly dynamically and cyclically loaded drive trains as well as in the wheel suspension and steering systems of commercial vehicles. Powder-forged automobile connecting rods made of CrNiMo-alloyed Sint-F30 (Chap. B.1, p. 142), quenched and tempered to $R_\mathrm{m} \approx 1000$ MPa, are one example of the use of sintered steel for highly loaded components. Forged drive axles for locomotives are made of e.g. 34CrNiMo6. Pre-forged and roll-formed solid wheels of $\approx$ C60 for railway wagons are sometimes given partial quenching and tempering of the wheel tread. In this case, the circumference of the austenitised wheel is sprayed while the wheel disk is cooled in air. After tempering, the tread has a QT structure, whereas the disk is normalised. Light wheels consist of a quenched and tempered wheel disk with a shrunk-on tyre. There are also plenty of applications in general mechanical and drilling engineering, e. g. as tool joints. These threaded sockets of 34CrNiMo6 are used to join drill strings for oil wells. Mining applications include castings of G42CrMo4 (specified by EN 10293) as tool holders in coal ploughs or as machine elements in chain conveyors.

DIN 267 divides nuts and bolts into various strength classes. Classes 8.8 to 14.9 ($R_\mathrm{m} = 800$ to 1400 MPa) require quenching and tempering. Parts with a diameter of up to ≈ 25 mm are shaped by cold deformation of wire, which requires good cold-heading properties. Besides a low sulphur content, spheroidising annealing (Chap. A.3, p. 59) is advisable. Accelerated cooling from the hot-rolling temperature is used to produce a bainitic/martensitic microstructure in the wire to facilitate the formation of spheroidal cementite during soft annealing. As the carbon and alloying contents increase, the flow curves are shifted to higher stresses (Figure B.2.18), thus hampering cold-working. Most of the QT steels specified in EN 10263-4 for cold-heading and cold extrusion correspond to those in EN 10083. Small amounts of boron (0.0008 to 0.005 %) have a favourable effect (p. 192). This leads to a significant and cost-effective increase in the hardenability without an increase in the annealing resistance, which would be the case for e. g. chromium and molybdenum (Tab. B.2.5).

In addition to final QT treatment of shaped fastening elements, cold-deformation of QT wire is also used. The greater strength increases the load on the tool. However, it increases the useful strength from cold-working as well as residual compressive stresses e. g. in the rolled root of the threads. For stamping applications, EN 10132-3 regulates cold strip, annealed and with an optional skin-pass. It is intended to be quenched and tempered after

B.2 High-strength materials

Table B.2.5 Influence of boron in QT steels for cold heading and cold extrusion. In the soft-annealed state, the tensile strength of boron-alloyed steels is less than those alloyed with CrMo or CrNiMo with a comparable hardenability (cf. steels 2 and 3 or 5 and 6), which require a greater force to work the metal. These examples have been taken from EN 10263-4.

Steel Nr.	Name	R_m (AC)[1] [MPa]	Diameter[2] [mm]	Core hardness[3] [HRC]
1.	23B2	490	9	40
2.	23MnB4	520	14	40
3.	25CrMo4	580	13	41
4.	33B2	550	11	45
5.	32CrB4	550	30	46
6.	34CrNiMo6	720	31	46

[1] spheroidised (AC), maximum value
[2] d_{max} for $\geq 90\%$ martensite in the core (Q)
[3] core hardness of d_{max}

cold-shaping. EN 10277-5 regulates unalloyed and alloyed QT steels as well as quenched and tempered bright steel that has been drawn, peeled or ground.

For the manufacture of heavy-duty machines, large forgings made of highly hardenable steels are produced according to Stahl-Eisen Materials Guidelines

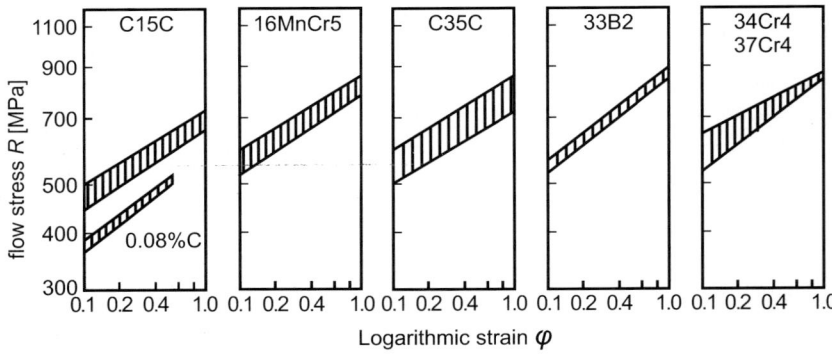

Fig. B.2.18 Scatter bands of flow curves: Unalloyed and low-alloy cold extrusion steels in the soft-annealed state, evaluated using the equation $R = c \cdot \varphi^{n'}$, which is simplified with respect to that given in Figure A.4.9 c, on p. 90 (from G. Robiller, W. Schmidt, C. Straßburger).

B.2.3 Structural steels for full heat treatment

550 and 555. A typical example of this is a power generator shaft with a diameter of more than 1 m. The large ingots required for this tend to contain macrosegregations (Figure A.2.8, p. 27) and their localised phenomena, in particular. Post-treatment of the steel (including purging and degassing of the steel melt in the ladle) results in lower levels of oxide and sulphide impurities, influences the sulphide shape, lowers the hydrogen content to avoid flaking (Chap. A.3, p. 58) and produces the lowest phosphorus contents. The latter reduces temper embrittlement, which starts during slow transition through the embrittling temperature range (about 500°C)) during cooling from the tempering temperature at $>600°C$, (Figure B.2.19). Mo dissolved in the ferrite slows down P segregation at the grain boundaries and thus embrittlement. Low levels of Sn, As and Sb also reduce temper embrittlement. Such levels can be achieved by the use of the correspondingly pure pig iron or scrap. This results in very pure steels, known as 'clean steels'. They have the lowest contents of tramp elements, namely, O, S, Mn, Si, H, P, Sn, Sb, As, whose

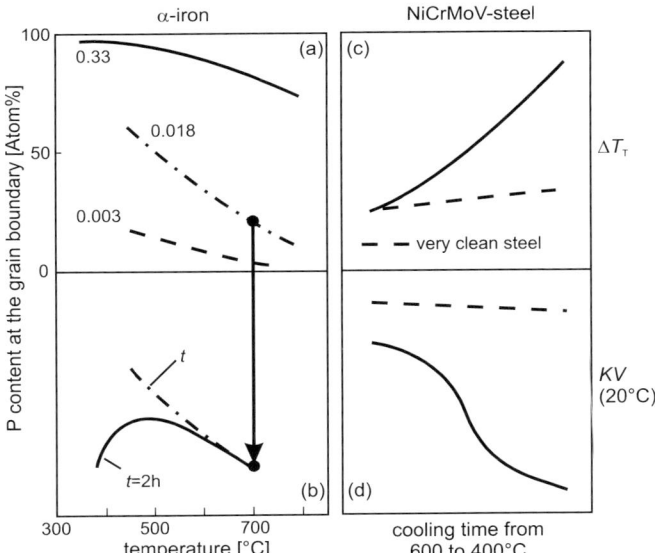

Fig. B.2.19 Temper embrittlement: (a) Grain-boundary segregation of phosphorus in iron up to the equilibrium. (b) The holding time t required for this increases as the temperature decreases (from H.J. Grabke). The diagram shows that tempering at 700°C and holding for 2h at a lower temperature produces a maximum P content at the grain boundary. (c) Slow cooling from the tempering temperature causes a similar enrichment of phosphorus. This weakens the grain boundaries, and the transition temperature increases by ΔT_T. (d) The notched-bar impact bending energy KV at room temperature approaches the lower shelf region as T_T increases. Temper embrittlement is low in bainitic steels from a very clean melt.

enrichment in local segregations is correspondingly lower. High alloy contents, as in e. g. steel 26NiCrMoV14-5, and water-spray hardening of a slowly rotating heavy shaft ensure that the workpiece is essentially through-hardened. Unlike a quenching bath, the spray intensity can be controlled so that hardening cracks due to internal stresses can be avoided.

Steels with 0.5 to 0.6 % C: A high hardening capacity due to carbon in combination with low tempering temperatures between those causing blue brittleness and temper embrittlement (Figure A.2.25 a, p. 49) affords strengths from 1300 to more than 2000 MPa. A degree of hardening R_H close to unity produces a yield ratio $R_{p0.2}/R_m > 0.9$. This means that steels such as 51CrV4, 55Cr3 and 61SiCr7 are suitable for leaf, coil, torsion-bar and cup springs (EN 10089). In commercial vehicles, the changeover from a laminated trapezoid spring to the parabolic spring has led to thicker individual leaves. The alloy content must be increased to ensure through-hardening: 52CrMoV4. Steel 56NiCrMoV7 can be used for heavy-duty coil springs with a rod diameter of up to 80 mm e. g. for damping vibrations in foundations or from explosions. Thermomechanical rolling of parabolic springs can reduce the embrittling effect of phosphorus, particularly at high strengths (> 2000 MPa), and the correspondingly lower tempering temperature (< 300°C) improves the fatigue strength. The surface of the springs is usually left untreated after hot-working and subsequent QT treatment for reasons of economy. The rougher and slightly decarburised surface layer promotes the formation of fatigue cracks. Shot peening is thus used to introduce residual compressive stresses in the surface layer to a depth of a few tenths of a millimetre, thus delaying crack initiation (Figure B.2.20). The decarburisation depth is determined according to EN ISO 3887.

Smaller spring elements can be made from cold strip of unalloyed and alloyed spring steels (regulated by EN 10132-4). They are up to 6 mm thick, soft-annealed with an optional skin-pass or up to 3 mm thick and already quenched and tempered. Spring steels are also suitable for parts subject to wear, such as excavator bucket teeth, mixer blades, shredder hammers and demolition chisels. Some of these applications require harder cutting edges on a tougher body, and this is achieved by partial hardening or tempering (Figure B.2.21).

B.2.3.2 Ultrahigh-strength steels

On p. 140 and p. 184, we discussed patented drawn wires with strengths of 2500 to 4000 MPa, but only those with thin cross-sections. Steels that can be used for components with larger cross-sections are listed in Tab. B.2.6. They have an elongation at fracture of approximately 3 to 6 %. As the cross-section becomes larger, the alloying costs increase along with the tendency to a segregation-related reduction in the toughness.

B.2.3 Structural steels for full heat treatment

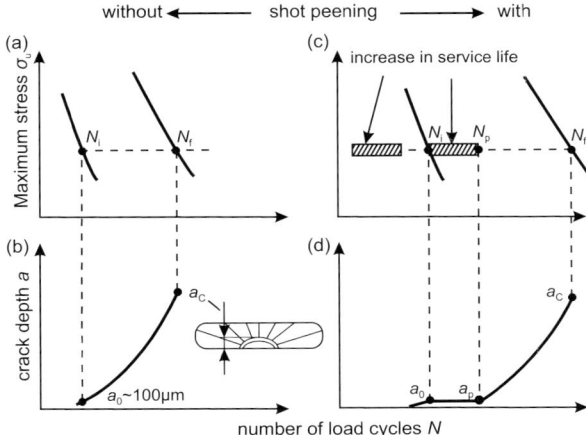

Fig. B.2.20 Influence of shot peening on the fatigue strength: Sections (100·15·400 mm) of 51CrV4 in the 4-point bending fatigue test. (a) Fatigue strength diagram. (b) Number of cycles N_i to initiation of an incipient crack $a_0 \approx 0.1$ mm measured ultrasonically; stable crack propagation to the critical depth a_c at N_f (number of cycles to failure); instant fracture of the remaining cross-section. (c) Delayed crack initiation due to residual compressive stresses in the surface produced by shot peening; crack arrest between N_i and N_p owing to a maximum residual compressive stress at 0.2 to 0.3 mm beneath the surface. (d) Stable propagation from N_p after dissipation of the maximum residual compressive stress by cyclic microplastic deformation; instant fracture of the remaining cross-section at a_c (from L. Weber).

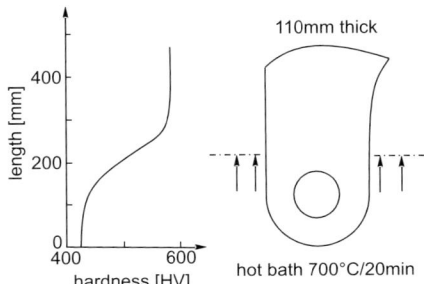

Fig. B.2.21 Partial tempering: Shredder hammer of G55NiSiCrMoV6 used to cut up scrap; hardened at 890°C/oil and tempered at 300°C/air. Toughness of the eye end was increased by tempering to a lower hardness in a hot bath. A silicon content of 1.5 % was selected to shift the region of blue brittleness (Figure B.2.22).

B.2 High-strength materials

Table B.2.6 Examples of ultrahigh-strength steels: d = diameter at which the tensile strength R_m is reached; C and D with degree of hardening $R_H \to 1$; E also for larger diameters; however, with a progressive decrease in the toughness; K = relative costs of billets (alloy contents in mass-%)

	Type of steel		d [mm]	R_m [MPa]	K
			\multicolumn{3}{l}{Approximate values of}
A.	Steel with 0.7 to 0.8 C (optionally microalloyed), isothermally transformed to lower pearlite, cold-drawn (patented drawn wire)				
		tyre cord	<0.25	>3000	1
		spring wire	1.0	2500	1
B.	71Si7 0.7 C; 1.7 Si; 0.7 Mn, isothermally transformed to lower bainite at 270°C		<10	2100	1
C.	41SiNiCrMoV7-6 0.4 C; 1.7 Si; 0.6 Mn; 0.8 Cr; 0.4 Mo; 1.5 Ni; 0.1 V, hardened from 900°C and tempered at ≈ 300°C		<50	2000	4
E.	X41CrMoV5-1 0.4 C; 1 Si; 5 Cr; 1.3 Mo; 0.5 V, hardened from 1020°C and secondary tempered at ≈ 550°C		<100	2000	6
D.	X2NiCoMoTi 18-9-5 18 Ni; 9 Co; 5 Mo; 0.7 Ti; 0.1 Al solution-annealed at 820°C and artificially aged at ≈ 500°C s. Figure B.2.23		<200	2100	20

(a) Properties

Isothermal transformation of carbon-rich steels to lower bainite without a tempering treatment leads to finely dispersed retained austenite and thus to good ductility and a high strength. To ensure that bainite formation is complete within a reasonable time, this treatment is limited to low-alloy steels such as 71Si7 (compare ADI, p. 208). For a higher alloy content see Tab. A.3.2, p. 71. Spring steels tempered at a low temperature of e.g. 350°C reach tensile strengths of ≈ 2100 MPa. Further lowering of the tempering temperature leads to blue brittleness (Figure A.2.25 a). Silicon shifts this embrittlement range to higher temperatures so that the steels can be tempered below 350°C (Figure B.2.22)). The reason for this lies in the inhibition of embrittling cementite precipitates within the retained austenite, presumably due to the poor solubility of silicon in cementite. Even grades with carbon contents

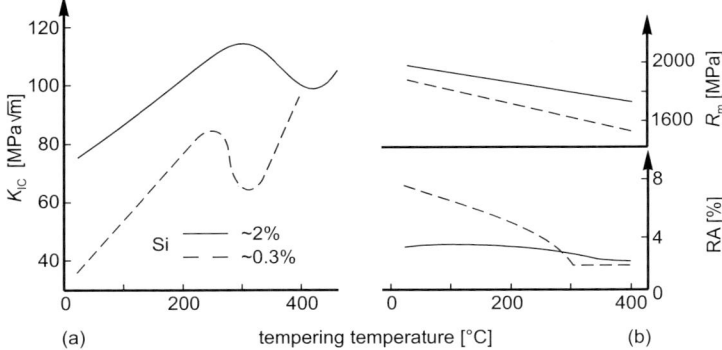

Fig. B.2.22 Influence of silicon on the tempering behaviour: CrNiMo-alloyed QT steel with $\approx 0.3\,\%$ C (from E. R. Parker). (a) Addition of silicon shifts the range of blue brittleness to higher temperatures. Silicon refines the temper carbides and increases the fracture toughness K_{Ic}. (b) Owing to the more finely dispersed carbides, silicon improves the temper-resistance along with the strength R_{m}. The precipitation of cementite bands from the retained austenite (RA) contributes to blue brittleness in low-silicon steel.

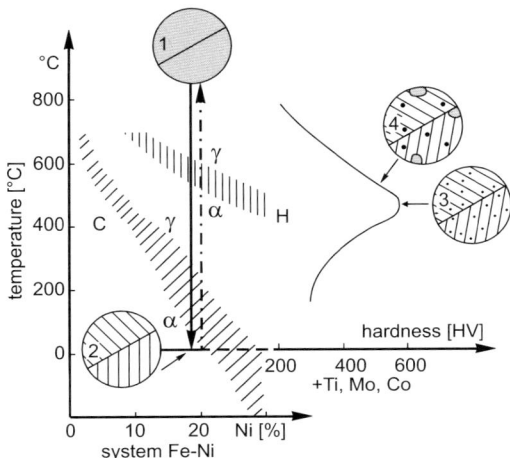

Fig. B.2.23 Precipitation-hardening of martensite (maraging) in nickel steel (schematic): (1) Solution-annealed austenite, (2) soft nickel martensite, (3) precipitation-hardening of martensite (maraging) by small intermetallic precipitates during artificial ageing, (4) overageing and reverse transformation to austenite. Transformation scatter bands: C for cooling, H for heating.

below those of spring steels can thus attain a strength of ≈ 2000 MPa, e.g. 41SiNiCrMoV7-6. The toughness is increased by influencing the sulphide shape and by thermomechanical working. Secondary hardening by precipitation of alloy carbides such as VC and Mo_2C during tempering allows the tempering temperature to be increased by ≈ 250°C without affecting the strength. Example: X40CrMoV5-1. (Figure A.2.25, p. 49). This relieves residual stresses and improves the toughness. At the same time, these materials can be used for applications at elevated temperatures during which softening is expected only after an extended period. According to Eq. A.3.3, p. 69, the tempering effect of a service temperature of e.g. 450°C for ≈ 2000 h corresponds to a tempering treatment at 550°C for 3 h. If precipitation-hardening of martensite is not due to alloy carbides but to intermetallic phases instead, they are referred to as maraging steels (Figure B.2.23).

In the Fe-Ni alloy system there is a transformation/temperature hysteresis between cooling (C) and reheating (H). For 18 % Ni and a solution-annealing temperature of 820°C, austenite (1) is present; its transformation to soft nickel martensite (2) is complete just above room temperature. Back-transformation of nickel martensite to austenite, however, starts at > 500°C thus allowing precipitation-hardening of the martensite (maraging) by intermetallic phases, e.g. on ageing at 480°C, for 4 h. This presupposes the appropriate alloying additives, such as Ti and Mo, in order to achieve finely dispersed precipitates of Ni_3Ti, Fe_2Mo, etc. that increase the hardness HV (3) (Tab. A.2.1, p. 52 and Figure A.2.25 c, p. 49). Cobalt promotes precipitation-hardening indirectly. Above 500°C the hardness drops again due to coarsening (overageing) of the precipitates and their dissolution in the regenerated austenite (4). In components made of maraging steel X2NiCoMoTi18-9-5 with 0.7 % Ti and 0.1 % Al that have been solution-annealed and machined, the tensile strength increases during low-distortion artificial ageing from approx.1000 to 2100 MPa.

(b) Applications

Lightweight engineering requires a high ratio between the yield point and the density. Forgings and turned parts made of ultrahigh-strength steels are used to join aircraft engines to the wings and the wings to the fuselage. Some steels are specified as aircraft materials, e.g. in the German Aviation Materials Handbook. The creep resistance of secondary hardening steels is advantageous for aircraft engines and for creep-resistant springs (see DIN 17225). Grain refinement during thermomechanical working has a favourable effect on the product of $R_{p0.2} \cdot A$ or $R_{p0.2} \cdot K_{Ic}$. Steel X41CrMoV5-1 is suitable for low-temperature thermomechanical treatment (ausforming) owing to the sluggish transformation field between the pearlite and bainite noses (Figure B.2.24). One example of this is in the manufacture of roll-formed tubular bodies with a tensile strength of up to 2800 MPa.

Maraging nickel steels are important in aerospace engineering for manufacturing rocket shells because they are suitable for welding and allow

B.2.3 Structural steels for full heat treatment

low-distortion formation of martensite. Slow cooling is sufficient after solution annealing at ≈ 820°C. The low degree of cold-work hardening allows high levels of cold-working in this condition, which is followed by low-distortion artificial ageing. This is how roll-forming is used to produce highly precise tubes with a diameter of e. g. 150 mm and a wall thickness of < 0.4 mm. Tubes with electron-beam welded caps are used in ultracentrifuges rotating at > 50.000 rpm during enrichment of uranium isotope 235 from gaseous UF_6. Cold-working and artificial ageing of grades with a titanium content of ≈ 1.5 % can produce tensile strengths of > 3000 MPa. The high strength of the steels can only be exploited if precautions are taken with respect to corrosion and fatigue, because they are susceptible to hydrogen embrittlement and stress corrosion cracking. This can be counteracted by baking and surface protection to prevent attack by hydrogen generated by corrosion during service.

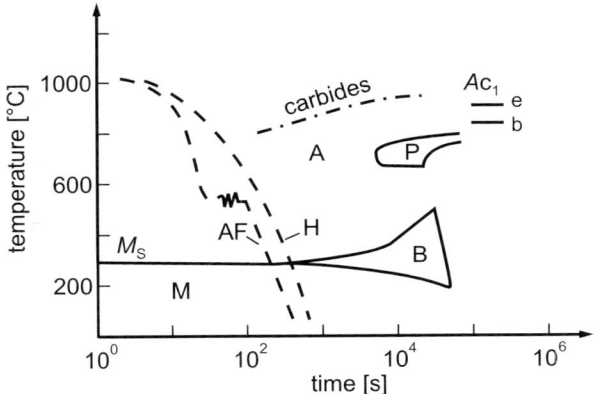

Fig. B.2.24 Ausforming: Steel X41CrMoV51 austenitised at 1020°C, H = oil-hardening, AF = ausforming, i. e. working after cooling in a hot bath to ≈ 500°C, followed by air cooling. Ausforming refines the grains and precipitates during tempering.

B.2.3.3 Hard steels

The hardness of ultrahigh-strength steels generally lies below 58 HRC and is associated with only a few percent elongation at fracture. Hard steels can be hardened to > 60 HRC by dissolution of 0.7 % C during austenitising. As shown by Figure A.4.14, p. 97, the transition temperature T_T at this hardness appears to favour service under compressive stress. Figure A.4.15 indicates that small microstructural defects, e. g. non-metallic inclusions, lower the permissible stress. Harder steels are thus used for machine elements subjected to

Hertzian contact stress that have to transfer high compressive forces. Outstanding examples are the low-alloy, through-hardenable roller-bearing steels with 1 % C, which are regulated by EN ISO 683-17.

(a) Properties

The hardenability of the standard steel 100Cr6 can be tailored to thicker cross-sections by changing the alloy composition, e.g. 100CrMnSi6-6 or 100CrMo7-3. Undissolved carbides remain after austenitisation slightly above Ac_{1e} (Figure A.2.9 p. 29). Only a little retained austenite is thus formed during hardening and its proportion drops even further on sub-zero cooling. Tempering at 150 to 180°C relieves residual microstresses and improves the dimensional stability (Chap A.3, p. 73). Isothermal transformation to lower bainite can also be carried out to reduce hardening stresses and distortion (Chap. A.3 p. 71). A low content of retained austenite enhances the size stability during service. Favourable residual compressive stresses develop in the surface layer during quenching in a salt bath to the bainite temperature. This may be due to plastic stretching in the more rapidly cooled surface layer. As cooling proceeds, the core contracts and the initial tensile stresses in the surface layer are converted into compressive stresses, which are retained during the subsequent transformation. In contrast to this isothermal bainite transformation, continuous cooling with through-hardening to martensite leads to residual tensile stresses in the surface layer (Chapt. A.3, p. 74). Contact stress may be as much as 3000 MPa in heavy-duty bearings. The maximum load induced by rolling contact acts just beneath the surface of the material, where it initiates plastic deformation, which in turn leads to residual stresses in this zone that increase with the number of rolling contacts: the martensite softens, carbides are rearranged and the material cracks. These cracks propagate to the surface and lead to small break-outs (pit formation, Figure A.4.17 c, p. 106). This contact fatigue (Figure A.4.18, p. 108) limits the service life or causes the component to fracture. Brittle hard phases, such as non-metallic inclusions (particularly oxides), coarse segregation carbides and titanium carbonitrides represent internal notches that initiate cracking (Figure A.4.15, p. 97). As their size and number increase, they lower the number of rolling contacts before cracks are initiated and thus, in general, the service life (Figure B.2.25). Their influence on the minimum fatigue strength and endurance limit increases even more with increasing matrix hardness. A good cleanness is thus particularly important for rolling bearing and other hard steels as well as surface layers. Titanium can e.g. be introduced as a contaminant in ferrochrome, whose costs increase with its cleanness. It forms angular Ti(C,N) particles that act as acute internal notches.

In addition to an analysis of the microscopic cleanness, the stepped turning test (specified in Stahl-Eisen Testing Guideline SEP 1580) and the blue fracture test (SEP 1584) are used to assess the macroscopic cleanness, i.e. detecting coarse bands of inclusions. Advanced analytical techniques provide a more cost-effective method of determining oxygen, and the results correlate with the microscopic cleanness. Proeutectoid-precipitated grain-boundary carbides (Figure A.2.10, p. 30) have a negative effect on the service life. They can be avoided e.g. by a low final rolling temperature with accelerated cooling. The almost 1 µm-sized spherical carbides remaining in steel 100Cr6 after hardening are assumed to reduce the metal contact between the rolling elements, which is advantageous if slip occurs between these elements during acceleration. However, from the fatigue standpoint, they also represent small internal notches. As a result, it has been repeatedly suggested that hypereutectoid roller bearing steels should be replaced by near-eutectoid grades. The composition of such a steel is (%) 0.68 C, 1.5 Si, 1.5 Mn, 1.1 Cr and 0.25 Mo. Overheating during austenitising e.g. during surface-hardening, does not lead to a drop in hardness due to a greater proportion of retained austenite. Low-stress, low-distortion air-cooling is sufficient for most wall thicknesses. The more highly alloyed retained austenite has a greater thermal stability, so that even a low tempering temperature ensures good dimensional stability.

(b) Applications

Because core segregations that form during solidification of ingots or continuously cast billets protrude from the rolling surface of balls, the cleanness and homogeneity of this zone affect the service life. Owing to the more sharply defined core segregation in continuous castings, they were thus initially used for rings, cylinders, barrels and cones. Balls with diameters between 0.05

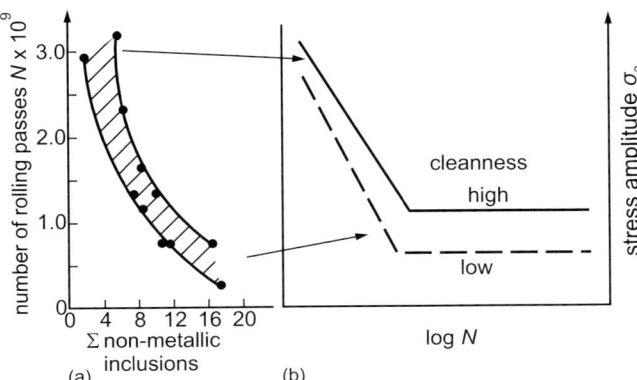

Fig. B.2.25 Cleanness and service life of rolling bearings: (a) Number of rolling contacts (N) until failure of angular ball bearings subjected to a Hertzian contact stress of $p_0 \approx 2600$ MPa. Eleven 100Cr6 melts were evaluated. Their non-metallic inclusions were weighted according to SEP 1570 and are given as a total. (b) Schematic fatigue curve for conventional and clean steels (from H. Schlicht, E. Schreiber and O. Zwirnlein).

and 35 mm made of through-hardenable steels, such as 100Cr6, are formed from wire sections by cold-heading. After spheroidising (AC according to Chapt. A.3, p. 59), the rolled wire is drawn to a narrow dimensional tolerance to achieve a precise filling of the tool with only small burrs. The races of deep-groove ball bearings are produced by hot-pressing a piece of 100Cr6 rod to a preform, from which the inner and outer races can be machined with little waste. Spheroidising is also suitable for components that are to be machined. In this case, the resulting carbides should not be too small (annealed hardness ↑) and not too large (hardenability, ductility ↓). During hardening and tempering to 59 - 63 HRC, the races are usually left somewhat softer than the rolling elements. Rolling bearings made of through-hardenable steels are virtually ubiquitous in all fields of engineering. Nowadays, as a consequence of the continuously improving cleanness, damage to surfaces under cyclic rolling contact is often caused by entrained particles of hard dust that leave small indentations. A higher fraction of retained austenite is locally transformed to martensite, which introduces residual compressive stresses around the indentations thus inhibiting crack initiation. However, because the dimensional stability suffers under an increasing proportion of retained austenite, carbonitriding can be used to achieve e.g. 20 % retained austenite in only a thin layer (Chap. B.3, p. 238). In larger conical bearings, the tangential tensile stresses are absorbed by ring-rolled, case-hardened races. Case hardening steel, namely 18NiCrMo14-6, is also suitable for large self-aligning roller bearings with a diameter of e.g. 2.5 m. Owing to the simple cylindrical shape, hard machining of the machining allowance for distortion is less complicated than for large gearwheels. For large bearings with a diameter of up to ≈ 7 m, such as those used in excavators, tunnel boring machines and off-shore crude-oil transfer towers, ring-rolled races of QT steels, e.g. 48CrMo4 or 45CrNiMoV5, are subjected to surface-hardening. Progressive hardening of the contact surfaces of the race ends just before it reaches the starting point in order to avoid cracks caused by overlapping of hardened areas. The remaining narrow soft zone can be repositioned from a parallel alignment with the bearing axis to an angle of 45°C by inclining the torch. This reduces the contact between the soft zone in the race and the rolling element from a line to a point. A little local grinding relieves the soft zone even further. Multi-part races made of surface-hardened steels are used in even larger bearings with a diameter of e.g. 15 m, such as those used in the swivelling frame of large excavators in open-cast mining of lignite coal.

Rolling bearings used for pumps in the chemical industry or for equipment influenced by a maritime atmosphere (shipping, off-shore engineering, transatlantic aircraft) require stainless martensitic steels with a high chromium content (Chap. B.6, p. 325). Many results indicate advantages of steel X30CrMoN15-1 with 0.35 % N, produced by pressure metallurgy, compared to conventional steels X65Cr14 and AISI 440C ($\approx$ X108CrMo17). We are already familiar with the advantage of the greater temper-resistance of high-strength secondary hardening steels for service at elevated temperatures.

For example, if a roller bearing of 100Cr6 is operated for a lengthy period in contact with hot hydraulic fluid or motor oil below the tempering temperature, this has an adverse effect on the dimensional stability. Bearings in jet turbines must withstand temperatures of up to $\approx 400°$C. To prevent them softening, WMoV-rich steels, such as X82WMoCrV6-5-4 ($\approx$ HS6-5-2, Tab. B.5.5, p. 300) are used because they exhibit significant secondary hardening and have tempering temperatures $> 550°$C (Figure A.2.25 b, p. 49). At very high engine speeds, the tangential tensile stresses due to the centrifugal force in the races are so high that they have to be compensated by the residual compressive stresses resulting from case hardening. This is achieved by lowering the carbon content of the WMoV steels to e.g. 0.12 % and adding 4 % Ni to prevent the formation of δ-ferrite. There are no specific steels for low-temperature applications, such as refrigeration engineering and aerospace, just those already mentioned.

Another field of application of hard steels concerns compressive contact between a camshaft and the valve train. To reduce weight, the assembled shaft consists of cam rings of 100Cr6 pressed onto a tube of S355J2G3. The grooved toothing inside the ring engages in the locally rolled surface of the tube. If the bucket tappet is replaced by a cam follower with roller actuation, this results in lower friction losses and repeated rolling passes with linear contact between the rollers and cam lobes.

B.2.4 Cast iron for full heat treatment

As part of the near-net-shape manufacturing process, components of grey iron are cast and their mechanical properties are then tailored by subsequent quenching and tempering or transformation in the bainite range.

B.2.4.1 Quenching and tempering

Hardening with through-austenitising and rapid cooling is only commonly used for cast iron with spheroidal graphite because other graphite shapes act as internal notches and cause hardening cracks during quenching. Heat-treatable cast irons are alloyed because a pearlite-free martensitic microstructure is only possible in unalloyed cast iron if the cooling time is extremely short ($t_{8/5} \leq 6$ s; Figure A.2.18, p. 40). Ni (<3 %), Cu (<1.5 %) and molybdenum (<1 %) significantly lengthen the critical $t_{8/5}$ cooling time, whereby 0.4 % Mo produces a deeper hardness penetration than 2 % Ni (Figure B.2.26).

If an alloyed cast iron is used, account must be taken of the effect of these elements on the microstructural development during solidification. Ni and Cu promote grey solidification, whereas molybdenum has a weakly carbide-stabilising effect. In addition, nickel inhibits embrittling copper precipitates by significantly increasing the copper solubility, which is lowered by magnesium, in Fe-C-Si melts. A combination of nickel and molybdenum has proved beneficial because nickel compensates the carbide-stabilising effect of molybdenum.

B.2 High-strength materials

Molybdenum also stabilises the ferrite and is responsible for the formation of ferrite shells around the graphite spherulites in martensitic / bainitic matrices.

Martensitic cast irons with spheroidal graphite hardened at 900 to 950°C achieve an as-quenched hardness of up to 60 HRC, which progressively decreases during tempering as a result of carbide precipitation from the martensite. The higher phosphorus and antimony contents compared to steel promote temper embrittlement, which can be effectively compensated with as little as 0.15 % Mo. Owing to its carbide-stabilising effect, molybdenum also delays the start of graphite precipitation in unalloyed cast iron, which commences at about 450°C (Fig. A.2.25 d, p. 49). This enables the hardened cast iron to be tempered between 450°C and 550°C to achieve a maximum tensile strength of 1000 MPa with an elongation at fracture of 1 % (Figure B.2.27 a).

B.2.4.2 Transformation in the bainite range / ADI

Transformation of cast iron in the bainite range after austenitising produces an austenitic-ferritic matrix, also known as ausferrite (Chap. A.3, p. 70). This treatment defines bainitic spheroidal graphite cast iron alloys as an independent group of materials that are regulated by EN 1564. They are also referred to as austempered ductile iron (ADI). These alloys are characterised by a very high strength in combination with a good ductility (Figure B.2.27 b). Because austenite is able to undergo a deformation-induced transformation with a TRIP effect (Chap. A.2, p. 41), these alloys have certain parallels to TRIP steels (Chap. B.2, p. 170). For example, an ADI with a tensile strength of 1000 MPa has an elongation at fracture of at least 5 % and is thus in competition with quenched and tempered cast steel. ADI is thus becoming increasingly common: its annual global tonnage doubled in only 5 years.

Unalloyed austempered ductile iron contains approximately 3.2 to 3.8 % C and 2 to 2.5 % Si. Its properties are obtained by austenitising, quenching

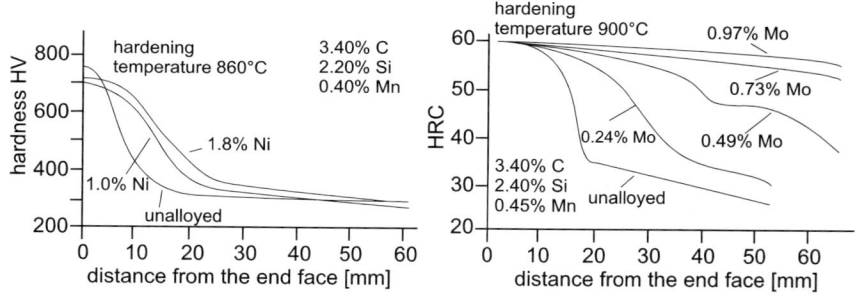

Fig. B.2.26 Hardenability of alloyed cast iron: Influence of nickel and molybdenum on the hardenability of GJS grades with approximately the same composition that were determined by a Jominy test (from S. Hasse)

B.2.4 Cast iron for full heat treatment

and isothermal holding in the bainite range (Chap. A.3, p. 70). At austenitising temperatures between 900 and 1000°C the austenite is enriched with carbon originating not only from the dissolution of carbides but also from primary graphite spherulites. Rapid dissolution of carbide in the initial pearlitic microstructure coupled with short diffusion paths and fine spherulites allows holding times well below 60 minutes so that the matrix grain size remains small. Cooling must be fast enough to suppress the formation of pearlite, which is promoted by the typical alloying elements Ni, Cu and Mo in the concentrations given above. Addition of molybdenum on its own is unfavourable because segregation can lead to poorly soluble carbides on the grain boundaries. Liquid quenching media (salt bath) have proven suitable because they provide a homogeneous heat transfer and the temperature required during isothermal treatment can be kept constant within narrow limits.

At transformation temperatures above 350°C, ferrite laths, which are initially low in carbon, start to grow into the austenite grains from the austenite grain boundaries. Silicon suppresses carbide precipitation so that the austenite is progressively stabilised during ageing by the increasing carbon content. At this point, the microstructure consists of graphite spherulites, ferrite and 20% to 40% retained austenite. It is therefore known as ausferrite (Figure B.2.28).

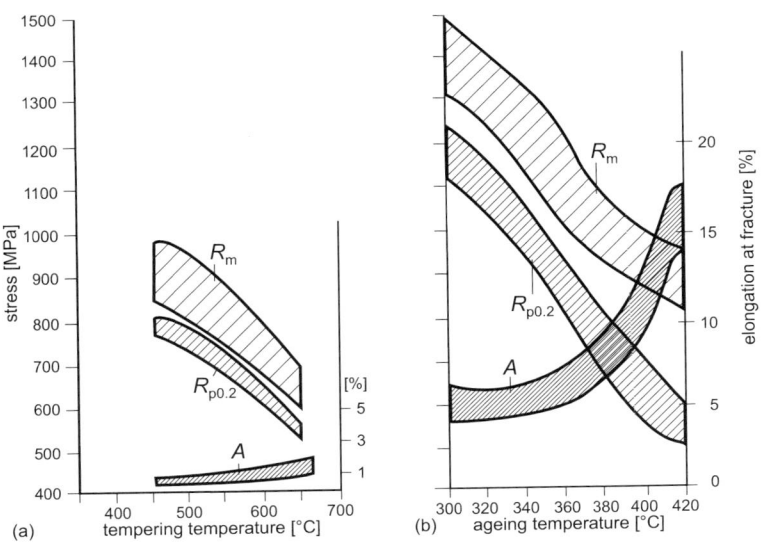

Fig. B.2.27 Mechanical properties of cast iron with spheroidal graphite: (a) After hardening from 900°C as a function of the tempering temperature (from S. Hasse) (b) After austenitising as a function of the isothermal holding temperature for the transformation in the bainite range (from E. Dorazil et al.)

Table B.2.7 Mechanical properties of grey cast iron after transformation into the bainite range (EN 1564)

	R_m [MPa]	$R_{p0.2}$ [MPa]	A [%]	Hardness [HB]	Impact energy at RT [J][3]	K_{Ic} [MPa$\sqrt{m}$][3]
GJS 800	800	500	8	260-320	100	62
GJS 1000	1000	700	5	300-360	80	58
GJS 1200	1200	850	2	340-440	60	54
GJS 1400	1400	1100	1	380-480	35	50
GJS 1600[1]	1600	1300	1	400-510	20	40
GJL 350[2] (AGI)	400	–	0.5	350-370	6	–

[1] according to ASTM 897, [2] no standard for this material, [3] literature data

Apart from these graphite spherulites, this microstructure is similar to that of TRIP steels (Chap. A.3, p. 70 and A.2, p. 41). In this condition, the high-strength ADI has an excellent fracture toughness K_{Ic} (Tab. B.2.7), which can be explained by stress-induced martensitic transformation in front of the crack tip. The associated increase in volume introduces residual compressive stresses, which are superimposed by tensile load stresses. This effect is also responsible for the high bending fatigue strength and rolling contact fatigue strength of this group of materials. As the ageing time progresses, silicon stops suppressing the formation of carbides and they start to precipitate, which increases the hardness and decreases the ductility. Below 350°C, some of the retained austenite transforms into martensite after a short holding time during cooling, and although this increases the hardness, it also has a strong embrittling effect. The influence of the austenitising temperature, the

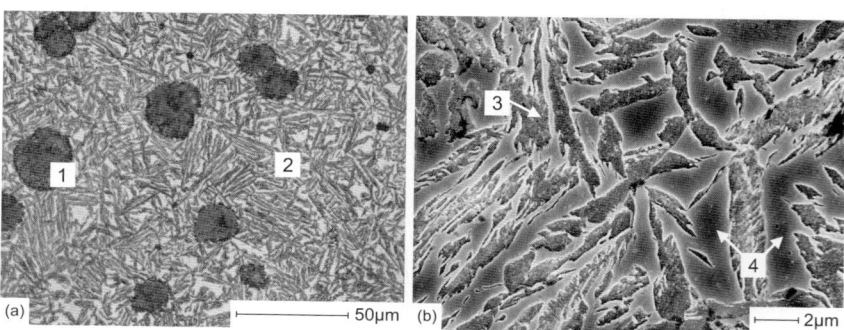

Fig. B.2.28 Microstructure of austempered ductile iron (ADI): (a) Graphite spherulite (1) in a metallic matrix (2) of austenite and ferrite. (b) Metallic matrix: ferrite (3), austenite (4)

transformation temperature and the holding time on the mechanical properties is shown in Figure B.2.29. The best toughness values are obtained by holding at 400°C for 15 to 60 minutes as a consequence of the high proportion of retained austenite (Figure B.2.29 a, bottom), for which a moderate Si content of 2.6 % is favourable. Even with its high strength, ADI exhibits a remarkable resistance to crack propagation (Tab. B.2.7). For example, GJS 1400, with a 0.2 % proof strength of 1100 MPa, exhibits K_{Ic} values of 50 MPa$\sqrt{m}$, which are higher by a factor of 3 compared to those of pearlitic spheroidal graphite cast iron with a yield point of only 500 MPa. As a consequence of the high proportion of austenite, there is no obvious transition temperature with respect to the impact energy, which is about 60 J for GJS 800 and still 25 J for GJS 1200 at -60°C. The thermal and mechanical stability of the austenite is important here. Whereas holding at -100°C leads to only a slight reduction in the proportion of retained austenite, the austenite that was insufficiently saturated with carbon already transformed under slight stress to α'-martensite at room temperature. This transformation can be induced in large castings by

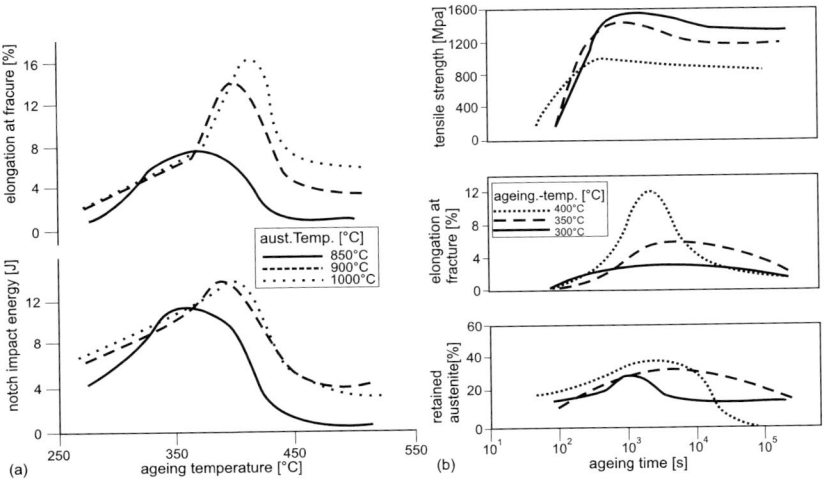

Fig. B.2.29 Heat treatment and mechanical properties of ADI (from E. Dorazil): (a) Austenitising and ageing temperature: a carbon content slightly above 1 % dissolves during austenitising between 850°C and 1000°C (30 min). The C content in the austenite increases to approx. 2 % during precipitation of ferrite on ageing at 400°C. This leaves retained austenite, thus ensuring the maximum ductility and toughness. (b) Ageing temperature and time: the processing window to manufacture ADI is restricted by the ageing time. If the holding time is too short, the ADI becomes brittle due to the growth of martensite. Carbides precipitate if it is too long. Silicon shifts this window to longer times.

their dead weight or by redistribution of residual stresses during machining. The resulting increase in volume may lead to an undesirable size change.

The high product of the yield point multiplied by the elongation at fracture gives ADI a fatigue strength that is double that of ferritic-pearlitic cast iron. As a direct comparison, GJS 1000 has a rolling contact fatigue strength which is four times higher than that of QT steel 42CrMo4 and which is thus close to that of surface-hardened steels. Further increases are possible by surface-strengthening measures such as shot peening and surface rolling. Cold-working of austenite and stress-induced transformation of martensite produce residual compressive stresses in the surface layer.

Cold-work hardening and stress-induced transformation of martensite are also the reasons for a high resistance to sliding and grooving abrasion. For example, the initial hardness of 400 HV in alloy GJS 1400 can be increased by wear loading to 600 HV in the sub-surface region.

The improved strength-to-toughness ratio of ADI compared to ferritic-pearlitic cast iron with spheroidal graphite is attributed to the special microstructure of the matrix. Nevertheless, ADI also has advantages over cast steel. The good fluidity of ADI melts during casting as well as the short ageing time during heat treatment keep the manufacturing costs low. The lower specific weight allows lighter designs at the same strength, but with a higher fatigue strength and better damping properties. ADI is thus particularly suitable for
applications in the automobile industry where it is used for crankshafts and wheel hubs in cars as well as spring seats, differential housings and timing gear wheels in commercial vehicles. In large planetary gear units, this material closes the gap between gear wheels made of QT steel and case hardened steels. After ADI had proven successful for trailing wheels of rail-mounted cranes, some railway companies switched to the more cost-effective ADI for wheels of locomotives and passenger cars. The good wear properties have encouraged applications in which there is abrasive loading. These include sprocket wheels and chains in conveying systems, impellers and housings of slurry pumps as well as teeth and cutting edges for excavators and agricultural machinery.

The resistance of ADI to wear by minerals can be increased by intentional precipitation of carbides. A carbidic ADI, marketed as CADI, is alloyed with chromium and molybdenum so that approximately half of the carbon is bound as eutectic carbides. Although some of these carbides (Figure B.2.30) decompose during subsequent heat treatment, the wear resistance of the materials lies between that of white cast iron and conventional ADI. If a high damping capacity is required, but the attainable strength of ferritic/pearlitic GJL materials is insufficient, grey cast iron with flake graphite can also be transformed in the bainite range. The resulting austempered grey iron (AGI) is used by American manufacturers for cylinder liners in diesel engines not only because of its good wear resistance.

B.2.4 Cast iron for full heat treatment

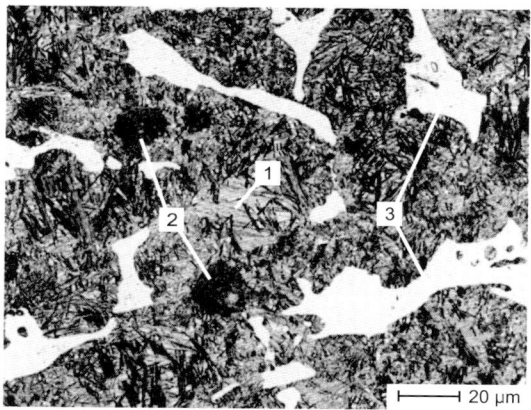

Fig. B.2.30 Light microscope image of carbidic ADI: 1 = ausferrite; 2 = graphite; 3 = Fe_3C

Inductive surface hardening of a cold roll
(Deutsche Edelstahlwerke GmbH, Witten, Germany)

B.3 Materials for surface layer treatments

Surface engineering employs a number of processes to coat or treat the surface (Chapt. A.3, p. 71), and optimised materials are available for some of them. We will now discuss one thermal and two thermochemical heat treatments that affect the surface layer and also present the appropriate materials. All three processes have a common objective: above a ductile core, the hardness is to be increased and residual compressive stresses introduced into the surface layer in order to improve the wear resistance and the fatigue strength.

The hardness gradually decreases from the near-surface region to the core, which means that a minimum hardness has to be specified for a certain depth below the surface. For surface-hardening, EN 10328 defines this specific minimum hardness as follows: hardness limit = $0.8 \cdot$ minimum required surface hardness, measured in HV1. The effective depth of hardening DS is given in mm. DIN 50190-3 refers to the effective depth of hardening after nitriding Nht in mm and the hardness limit = measured core hardness + 50 (HV0.5). For case hardening, the hardness limit is 550 HV1 and the effective depth of case hardening, abbreviated to CHD in EN ISO 2639, is given in mm (previously DC in EN 105). Special cases must be agreed upon in accordance with the appropriate standard. In reality, the effect of heating or chemical modification penetrates more deeply than the surface layer thickness specified by the hardness limit.

B.3.1 Materials for surface-hardening

B.3.1.1 Process engineering aspects of surface-hardening

In this thermal process, the surface layer of the workpiece is heated – and thus austenitised – by a flame, electromagnetic induction, a plasma, a laser beam or an electron beam at a very high rate so that the core is not heated to an appreciable extent. In thin layers or in high-alloy grades, self-quenching by the cold core is sufficient to harden the surface layer. However, a quenching medium is generally used to dissipate the heat outwards. The depth of hardening ranges from e.g. < 0.1 mm (laser beam with self-quenching) up to 50 mm (induction and cooling with a water spray). We differentiate between progressive hardening and hardening of the entire surface (Figure B.3.1). Overlaps in progressively hardened tracks are susceptible to cracking. Thus small gaps are usually left untreated in which no hardening takes place. This problem does not affect hardening of the entire surface so that not only water-quenching, but also the milder oil-quenching can be used. The latter is out of the question for progressive hardening owing to its flammability.

The heating rate ranges from e.g. 10^{1}°C/s for hardening of the entire surface with an oxyacetylene flame, up to 10^{4}°C/s for progressive hardening with a laser or an electron beam. In contrast to near-equilibrium austenitising in a furnace, rapid heating of the surface layer means that there is not enough diffusion time for the ferrite/austenite phase transformation and dissolution of the carbides. Austenitisation is shifted to a higher temperature, as shown by the time-temperature-austenitising diagram (TTA) for steel 42CrMo4 in Figure A.3.3, p. 65. The temperatures Ac_1 and Ac_3 increase with the heating rate. Although the carbides are dissolved in the austenite above Ac_3, the alloying elements enriched in them are still inhomogeneously distributed because they diffuse much more slowly than carbon. As the temperature increases, the austenite becomes more homogeneous and the hardness increases. Unfortunately, the grain size increases as well. In practice, raising the heating rate increases the surface-hardening temperature by about 50 to 250°C above that used for heating in a furnace. The higher the energy density of the heat source is, the steeper is the temperature gradient within the heated surface layer. Therefore, as the thickness of the austenitised layer increases, so does the risk of overheating on the surface, or it may even start to melt. Depending on the type of heat source, different strategies are used to increase the depth of hardening without damaging the material. For flame and induction hardening, the entire workpiece can be preheated to e.g. 500°C. During induction hardening, the depth of heat penetration increases as the frequency decreases. The simultaneous dual-frequency process produces a more uniform depth of hardening in components with a highly contoured shape. For example, toothed wheels can be given a near-contour surface layer during hardening of the entire surface by using two frequencies simultaneously: a medium frequency to heat

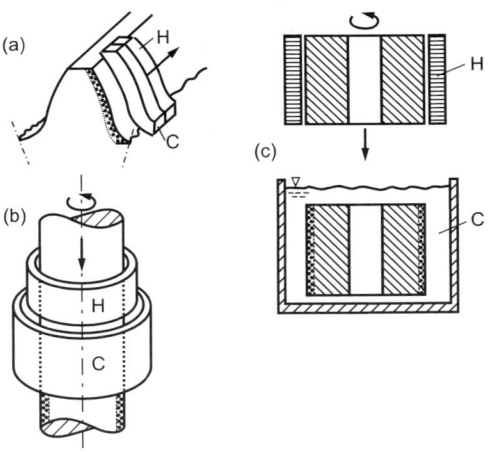

Fig. B.3.1 Surface-hardening: (a) Progressive hardening of a tooth flank; heating H with a gas torch, cooling C with a water spray. (b) Progressive spin hardening of a shaft or roll, heating by an induction coil or gas burner, cooling with a water spray. (c) Hardening of the entire surface by heating as for (b), cooling by immersion in e.g. an oil bath.

the roots of the teeth and a high frequency for the tips. Because the power ratio of both oscillating circuits (e.g. 10 to 25 kHz and 150 to 350 kHz) can be regulated, together with the selectable frequency ratio, it offers a number of possibilities to control the local input of energy, indeed, with a significantly lower energy consumption compared to case hardening. During laser hardening, the beam is pulsed to avoid overheating. One pulse cycle equals the pulse duration plus the pause. The ratio of the pulse duration to length of the pulse cycle is known as the duty cycle, and the number of cycles per second is the pulse frequency. For a pulse frequency of 50 Hz and a duty cycle of 20 %, the pulse duration is 4 ms. During stationary hardening, the required average surface temperature is reached by varying the two parameters so that the material does not start to melt. Pulsed operation can also be achieved by deflecting the laser beam (scanning) in the cw mode (continuous wave). For progressive spin hardening in the cw mode, the pulse frequency increases with the number of rotations per second. Steel 42CrMo4 exhibits a significantly greater hardening depth after e.g. ten repetitions of the following pulse cycle: heating at a rate of 1000°C/s to 1150°C and quenching to a temperature just above M_s. Independently of the heating method, carbides with a small size and a homogeneous distribution in the initial state accelerate austenitising of the surface layer and also increase the depth of hardening. The Ac_3 and Ac_c temperatures are increased to a lesser extent by rapid heating if the initial microstructure is changed from spheroidised to ferritic-pearlitic to QT (Figure B.3.2).

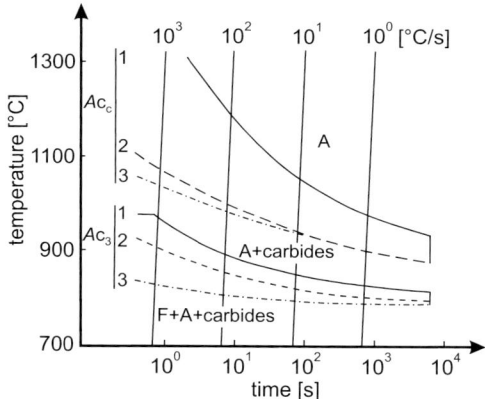

Fig. B.3.2 Influence of the initial microstructure on surface hardening: Steel 50CrMo4, condition 1 = spheroidal cementite after cycle annealing at 745 ± 20°C, 7 h; condition 2 = upper pearlite after austenitising at 850°C and isothermal transformation at 650°C, condition 3 = QT microstructure after hardening from 850°C and tempering at 500°CC (TTA diagram from J. Orlich, A. Rose, P. Wiest).

B.3.1.2 Materials and the surface layer

As a consequence of advances in steel cleanness and in process engineering of the surface-hardening process, DIN 17212 (steels for flame and induction hardening) has become obsolete and is now incorporated in EN 10083 (steels for quenching and tempering). Part 2 regulates unalloyed steels (C35E to C55E) both in the normalised as well as in the quenched and tempered state. Part 3 regulates alloyed QT steels (37Cr4 to 50CrMo4). The surface-hardness of grades that are intended for surface-hardening (Tab. B.3.1) increases with the C content. After stress-relief tempering at 150 to 180°C, they exhibit values of 48 to 58 HRC (Figure B.3.3).

This hardness range also includes precipitation-hardened ferritic-pearlitic (FP-PH) steels 38MnVS6 and 46MnVS6 (Tab. B.2.3, p. 189), which are

Table B.3.1 Steels for surface-hardening (I) and nitriding (II): N = normalised, QT = quenched and tempered, P = precipitation-hardened (precipitation-hardened ferritic-pearlitic steel, FP-PH)

	Steel	EN	core [1]			surface [2]	
			$R_{p0.2}$	R_m	A	hardness	
			[MPa]	[MPa]	[%]	[HV]	[HRC]
	C45E+N	10083-2	305	580	16	(596)	55
	C55E+N		330	640	12	(655)	58
I	37Cr4+QT	10083-3	510	750	14	(527)	51
	42CrMo4+QT		650	900	12	(560)	53
	50CrMo4+QT		700	900	12	(655)	58
	46MnVS6+P	10267	580	900	10	(578)	54
	31CrMo12+QT	10085	785	980	11	800	(64)
II	33CrMoV12-9+QT		850	1050	12	900	(67)
	34CrAlNi7-10+QT		650	850	12	950	(68)
	41CrAlMo7-10+QT		720	900	13	950	(68)

[1] minimum values for a rod diameter of 40 to 100 mm

[2] approximate values for the surface: I) after surface-hardening and tempering at 150 to 180°C, II) after nitriding; reference values in accordance with EN ISO 18265 are given in brackets

regulated by EN 10267. In addition to this hardening capacity, the hardness penetration and thus the alloy content become more important as the thickness of the surface layer increases. In alloyed QT steels, the alloying levels are selected so that quenching and tempering produces a deep hardness penetration or even through-hardening (Figure A.3.5, p. 66). This generally means a sufficient level of hardness penetration for surface-hardening, where internal self-quenching is supported by external water-spraying. If milder cooling e. g. with oil is to be used to avoid hardening cracks, the selected steel must be sufficiently hardenable.

Cracks may arise because the treated materials contain a high carbon content, experience a steep temperature gradient and may have notched contours (Chapt. A.3, p. 73). The higher the QT strength is in the initial state, all the more noticeable is the drop in hardness at the transition between the surface layer and the core because it is at this point that the tempering temperature of the previous QT treatment is exceeded by the surface-hardening temperature. The higher volume of the growing martensite introduces residual compressive stresses in the surface layer. These stresses in conjunction with load-induced stresses lower the crack-initiating amplitude of the tensile stress in the surface layer, especially under cyclic bending loads, and thus increase the fatigue strength. For a large effective depth of hardening and a low degree of self-quenching (core cross-section too small, preheating), undesirable residual tensile stresses may develop in the outer region of the surface layer due

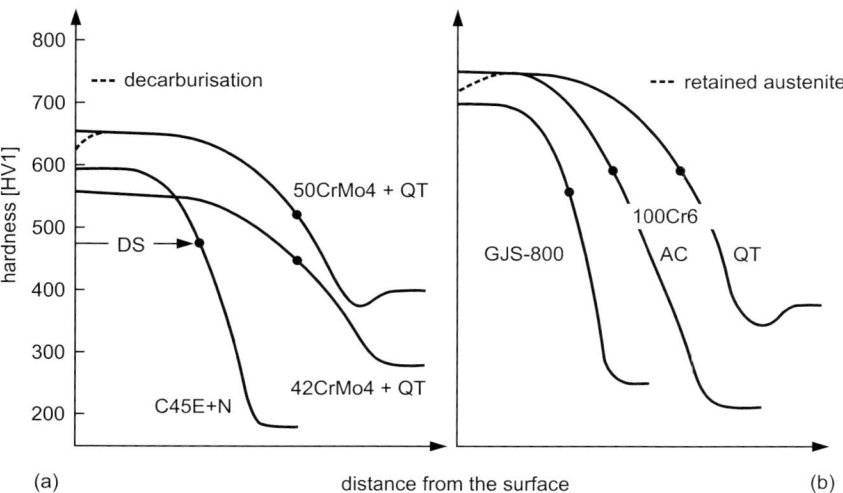

Fig. B.3.3 Hardness curves after induction hardening and tempering at 160°C: (a) Influence of carbon content, alloy content and heat treatment condition of hypoeutectoid steels, (b) hypereutectoid steel, soft-annealed or QT-treated, and pearlitic cast iron with spheroidal graphite.

to the progressive growth of martensite from the surface towards the core. As mentioned above, components are tempered after surface-hardening to relieve residual microstresses in the microstructure without significantly decreasing the residual macrostresses in the component and the hardness of the surface layer. After inductive hardening, the material can also be immediately inductively stress-relieved, which does not interrupt the production line with a tempering step in a furnace.

In addition to the previously discussed hypoeutectoid structural steels, near-eutectoid steels, such as C70 or 85CrMo7, are also surface-hardened. The matrix of cast iron is hypo- or near-eutectoid. In the example shown in Figure A.2.11, p. 31, the eutectoid carbon content is approx. 0.6 %. Only the C fraction bound in the pearlite dissolves on rapid heating. There is not enough time for a significant amount of graphite to dissolve in the austenite. The C content in the austenite that is responsible for increasing the hardness thus rises with the pearlite/ferrite ratio in the initial microstructure. Because the strength of cast iron also increases with this ratio, the higher strength grades, such as GJL-350, GJV-500 and GJS-800 (Tab. B.1.2, p. 146) are expected to exhibit a higher surface layer hardness. The softer low-pearlite grades and white heart malleable cast iron are not suitable for surface-hardening; however, they can be made suitable by pearlitising annealing (Chapt. A.3, p. 63). The resulting surface hardness of cast iron ranges from approx. 45 to 60 HRC. These materials are usually stress-relieved at 160 to 200°C. The hardness penetration depends on the alloy content, but this is not indicated by its designation. The elements in question are Mn, Si, Cu and, in some cases, low amounts of Cr and Mo. In GJL with an elevated P content, account must be taken of the fact that the phosphide eutectic (steadite) melts at around 950°C, and consequently, this temperature should not be reached to avoid damage.

The internal stresses introduced by the temperature gradients during heating and cooling are concentrated around the graphite precipitates whose stress concentration effect decreases from flake to vermicular to spheroidal graphite. Therefore a decreasing susceptibility to cracking can be expected in this order in surface-hardened materials. Inductive heating is very popular because it generates heat directly within the surface layer and is successful with gentler heating gradients than if the surface is heated by a radiant source. A preheated water/polymer emulsion is recommended as the quenching medium.

Just as the graphite does not dissolve during surface-hardening of cast iron, the eutectic carbides in cold-work and high-speed steels, such as X153CrMoV12 and HS6-5-2, do not contribute to austenitisation. Only the secondary carbides in the spheroidised matrix dissolve in the austenite. In contrast to hypoeutectic steels, however, there is a risk of too much carbon and alloying elements dissolving in the austenite so that the M_f temperature drops below room temperature, a high proportion of retained austenite remains and the required hardness is not achieved. If the temperature on the surface is too high, a hardness maximum develops below the surface within the hardened layer. This can also occur in stainless steels such as X46Cr13 or X70CrMo15. There may be differences in the corrosion resistance between hardened and non-hardened regions. In such cases, a higher Cr/C ratio, as in e.g. X39CrMo17-1, has a favourable effect. The above-mentioned high-alloy steels harden in air. One reason for a decreasing hardness towards the surface, which applies to all materials, is decarburisation. Carbon depletion of the surface layer resulting from hot working should be countered by premachining. As the hardening depth increases, the residence time at the high temperature increases and a soft skin can develop during surface-hardening. This skin can only be removed by final machining.

B.3.1.3 Applications

Inductive surface-hardening is used to increase the wear resistance of crankshaft bearing faces. If the transition to the web is included, this also increases the fatigue strength. Partial hardening of limited surface areas of control cams and shift gates has proven beneficial. Sprockets for combustion engines are given a homogeneous depth of hardening using the dual-frequency process. The flanks of the teeth on large gear wheels are individually progressively hardened by induction or with a flame. In slender components, it is important to limit shape changes to minimise straightening costs and rejects. For example, sliding strips for machine tools are either pre-bent or surface-hardened on both sides. In ball screw spindles, longitudinal expansion causes an error in the pitch. If this is taken into account as a short length, the longitudinal expansion can be measured during inductive progressive spin hardening and the depth of hardening then regulated so that the pitch corresponds to the specification. In toothed racks for car power steering systems, the greatest deflection can be expected for hardening of the tooth flanks and roots, which is the reason why inductive progressive spin hardening is generally used because it also hardens the surface opposite the row of teeth. This not only reduces the shape change, but it also increases the strength of the component, thus protecting it from permanent deformation if the driver hits the kerb. For forged or cast crane running wheels or caterpillar track rollers, flame or induction hardening improves the rolling-contact fatigue strength and wear resistance. Heavy cold rolls are e.g. given a preliminary QT treatment followed by preheating via inductive progressive spin hardening (Figure B.3.1 b) to surface-harden them to a depth of 50 mm. Non-circular components, such as heavy eccentrics and control cams, can be subjected to surface-hardening of the entire circumference by gas burners distributed around the periphery that are kept at a constant radial distance by another control cam with an identical contour. The ends of the rails of a jointed track are given a pearlitic microstructure with very fine lamellae by inductive heating with self-quenching. This prevents deformation and break-outs.

In addition to medium-frequency hardening of heavy components, high-frequency hardening is commonly used for small parts, such as those used in precision engineering where hardening depths of tenths of millimetres are required. Laser hardening can also be used in this case. Laser-hardening of the insides of large-diameter cylinders takes advantage of the simple handling of the laser beam using optical mirrors and lenses. An axial beam is radially deflected and focussed. By rotating and/or axial movement, hardened longitudinal or spiral tracks with a depth of e.g. 1 mm are produced on the inside surface. The development of powerful diode lasers (e.g. 5 kW) has increased the popularity of laser hardening. Other advantages include the compact design, the rectangular intensity distribution at the focal spot and the short wavelength, which leads to the absorption of a higher amount of beam energy compared to Nd-YAG and CO_2 lasers. Previously, the cutting edges of cast tools for large stamping machines used in the automotive industry had to be progressively flame-hardened by hand. Nowadays, this is achieved by a single laser mounted on a multi-axial robot arm. Some drawing dies for car roofs are made of cast iron, and the drawing edges are laser-hardened to protect them against wear. In large injection moulds for plastics, progressive laser hardening is used to protect the closing edges - which are often several metres long - against wear and permanent deformation.

The development of the high-speed scanning mode for electron beams (EB) has made this tool useful for surface-hardening. A high surface hardness can be obtained for a layer thickness of < 1 mm, e.g. just below 1000 HV 0.1 for 50CrV4 QT – because stress-relief annealing is not necessary. Attractive, but cost-intensive, is its combination with other surface-hardening methods: (a) EB-hardening produces a higher level of dissolution in the surface layer of ledeburitic Cr cold-work tool steels so that subsequent nitriding produces a greater degree of hardening. (b) EB-hardening below a borided layer can improve the supporting effect of the underlying material. The thin boride layer is virtually unaffected. (c) In nitrocarborised layers, EB-hardening dissolves C and N and increases the hardness.

B.3.2 Nitriding steels

B.3.2.1 Process engineering aspects of nitriding

Nitriding is a thermochemical process. As the workpiece is completely heated to the nitriding temperature $T_N < 590°C$, its surface layer absorbs nitrogen from the surrounding gas atmosphere. Because the temperature corresponding to A_1 is 590°C in the Fe-N system (Fig. B.3.9 a), the steel is in the bcc phase, which only dissolves $< 0.1\%$ N. The resulting nitrides coalesce to create a continuous, hard compound layer on the surface with an underlying precipitation or diffusion layer that is precipitation-hardened. Because the N_2 molecule contains a stable triple bond, there is virtually zero thermal dissociation to nascent nitrogen within the temperature range used for nitriding. Nitrogen gas is thus only suitable as an N donor after it has been dissociated in a plasma. In contrast, ammonia NH_3 exhibits a suitable dissociation rate in contact with the steel surface, and it is often used as a process gas. The properties of the surface layer can be modified further by adding water vapour (oxynitriding) or CO (nitrocarburising). Reaction potentials can be calculated for the various equations from the partial pressures p of the participating gas components according to the Law of Mass Action. These values describe the chemical potential or the nitriding activity of the process gas.

$$\text{NH}_3 \rightarrow [\text{N}] + \frac{3}{2}\text{H}_2 \qquad K_\text{N} = p(\text{NH}_3)/p(\text{H}_2)^{3/2} \qquad \text{(B.3.1)}$$

$$\text{H}_2\text{O} \rightarrow [\text{O}] + \text{H}_2 \qquad K_\text{O} = p(\text{H}_2\text{O})/p(\text{H}_2) \qquad \text{(B.3.2)}$$

$$\text{CO} + \text{H}_2 \rightarrow [\text{C}] + \text{H}_2\text{O} \qquad K_\text{C} = p(\text{CO}) \cdot p(\text{H}_2)/p(\text{H}_2\text{O}) \qquad \text{(B.3.3)}$$

A further carburisation reaction is based on the CO/CO_2 Boudouard equilibrium. The amount of methane produced in technical reactors is usually negligible. The nitriding potential K_N, the oxidising potential K_O as well as the carburising potential K_C of the heterogeneous water-gas shift reaction are interrelated via H_2 and CO. These combined potentials determine the composition of the surface layer, where there is a local equilibrium between the activity of N, O and C in the gas and in the steel. The concentrations of these elements decrease within the surface layer until they equal the core composition (Figure B.3.4). For nitriding in NH_3, the composition of the iron surface is shown in the Lehrer diagram. The composition is dependent on K_N and T_N (Figure B.3.5 a). The precipitating phases are cubic γ'-nitride Fe_4N or the hexagonal ε-nitride Fe_{2-3}N (or $\text{Fe}_2\text{N}_{1-x}$). K_N can be calculated from $p(\text{H}_2)$ in the furnace, measured with a hydrogen sensor, and the amount of added H_2 diluent, measured with a flowmeter. In the case of nitrocarburisation, the flow rates of the other additives are also measured and $p(\text{H}_2\text{O})/p(\text{H}_2)$ is measured with a probe. Figure B.3.5 b shows that the addition of carbon stabilises the ε-carbonitride with respect to the γ' phase. Oxynitriding is used to destroy passivating oxide layers that hinder the absorption of nitrogen. If K_O is adjusted by means of an oxygen sensor so that iron is oxidised but not iron nitride, a uniform surface layer can be obtained, even at a comparatively low nitriding temperature, because the oxides are converted to nitrides and the layer grows more quickly. As post-oxidation, oxygen is introduced towards the end of the nitrocarburisation process to achieve a high corrosion resistance. The potentials defined in Equations B.3.1 to B.3.3 are regarded as being approximately valid for low-alloy nitriding steels as well because the proportion of nitrogen bound to the alloying elements remains relatively small with respect to that bound to iron. By the use of gas sensors, computer-aided calculation of the potentials and controlled gas flow rates, gas nitriding (nitrocarburising, oxynitriding) has developed into a controllable surface-hardening process. Typical nitriding depths of 0.1 to 0.8 mm are obtained for a nitriding time of e.g. 2 to 50 h between 500 and 580°C. The brittle compound layer can be made thinner if an appropriate control system is used; however, it cannot

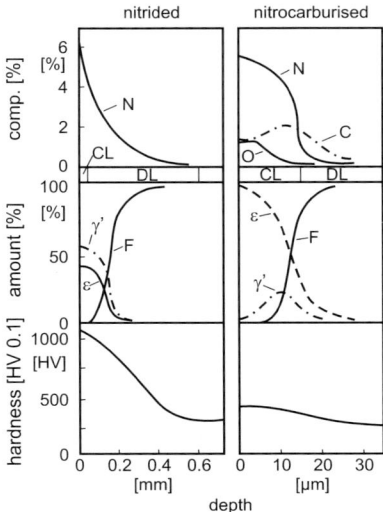

Fig. B.3.4 Nitrided layer: Approximate values for the chemical composition, amounts of different phases and microhardness curves after nitriding a steel containing $\approx 1\,\%$ Al in ammonia or after short-term nitrocarburisation of an unalloyed steel with $0.15\,\%$ C in a salt bath. (CL = compound layer; DL = diffusion layer; N, C, O = nitrogen, carbon, oxygen; γ', ε = iron nitrides; F = ferrite).

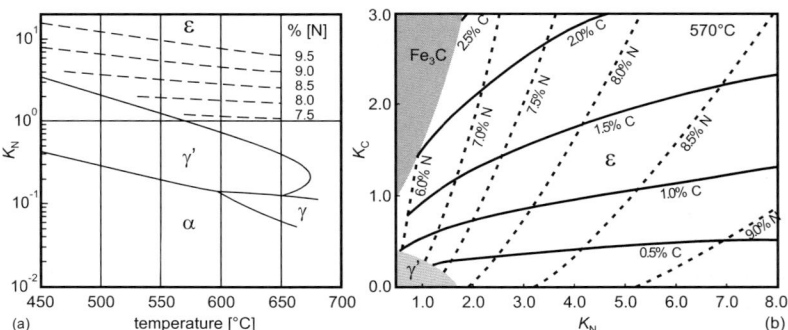

Fig. B.3.5 Influence of the nitriding potential K_N and the carburising potential K_C on the surface of iron, from H.-J. Spies et al.: (a) Phase diagram for nitriding in NH_3/H_2 mixtures, from E. Lehrer. Nitrogen contents calculated according to J. Kunze, (b) Nitrogen and carbon contents for nitrocarburisation, calculated according to J. Kunze.

be completely avoided. Surface areas that are not to be nitrided must be covered with a layer of tin.

During plasma nitriding in $N_2(+H_2)$, the glow-discharge conditions can be adjusted so that material is ablated from the surface by sputtering. This counteracts the mass increase caused by the absorption of nitrogen, and formation of the compound layer can be essentially avoided. Partial nitriding can be achieved more easily by shielding the remaining areas. Complete heating of the workpieces requires a high plasma power, which results in temperature differences, particularly for contoured designs. This drawback can be counteracted by heating the walls of the furnace and/or pulsing the plasma power (pulsed plasma nitriding). Initial sputtering removes any passivating layers, even on stainless steels, thus resulting in a uniform layer. The installation of a negatively biased screen close to the wall between the positively poled furnace wall and the negatively poled workpiece holder produces glow discharge from this active screen. The applied bias voltage causes a reactive mixture of charged particles to bombard the workpiece, resulting in uniform nitriding, even in high-alloy steels, because the passivating layers can be removed without sputtering. It is assumed that these surprisingly long-lived particles are e.g. FeN^+ or NH_x^+, and that nitrogen diffuses into the surface from these chemisorbed particles in a series of steps, namely $FeN \rightarrow Fe_{2-3}N \rightarrow Fe_4N$. This plasma nitriding technique is known as ASPN (active screen plasma nitriding).

In the salt bath nitrocarburising process, neutral carbonate-containing salts are mixed with potassium or sodium cyanate so that they are present as a melt at the process temperature of e.g. 560 to 585°C. During typical processing in a heated crucible in the air, the cyanate decomposition reaction

$$3CNO^- \rightarrow CO_3^{--} + CN^- + C + 2N \qquad (B.3.4)$$

is superposed by a cyanate/cyanide equilibrium that shifts towards the non-toxic cyanate in the temperature range used for nitrocarburisation.

$$CN^- + \frac{1}{2}O_2 \rightarrow CNO^- \qquad (B.3.5)$$

Oxidation of the cyanate liberates nitrogen but not carbon at the surface of the material. On the other hand, the solubility of oxygen in the melt depends on the cation composition (K^+, Na^+, Li^+) in the molten mixture, which thus influences the N/C ratio in the layer. The salt-bath process can be used very flexibly and is thus particularly suitable for treating different workpieces that frequently change. However, the subsequent washing step involves additional work.

B.3.2.2 Materials and the surface layer

Nitriding steels, as defined in EN 10085, are alloyed steels for quenching and tempering (Tab. B.3.1). We differentiate between grades alloyed with Cr and those with Cr and Al. Alloying of nitriding steel C10 with both elements can increase the surface hardness from e.g. 300 HV0.1 to values exceeding 1000 HV0.1 (Figure B.3.6). In addition, these elements improve the depth of hardening so that holes drilled through the core have a quenched and tempered microstructure that exhibits more homogeneous layer properties than e.g. a ferrite/pearlite microstructure. Alloying with Mo and V leads to the same type of effect as that obtained with Cr; however, it also improves the tempering properties. This is of importance for workpieces with a high core strength. For example, if steel 42CrMo4 is to have a tempered strength of 1100 MPa (≈ 35 HRC), Figure A.3.6, p. 68 shows that the tempering temperature at $R_H=0.85$ should be $\approx 530°C$. The core strength would have decreased on nitriding at 570°C. If steels with a secondary hardness, such as 33CrMoV12-9, are used, the tempering temperature can lie above the nitriding temperature so that the core strength is retained on nitriding. This also has a beneficial effect on the high-temperature strength, which barely decreases over longer operating periods if the tempering temperature was e.g. 50°C above the operating temperature. Owing to the good hot hardness of the nitrided layer, this

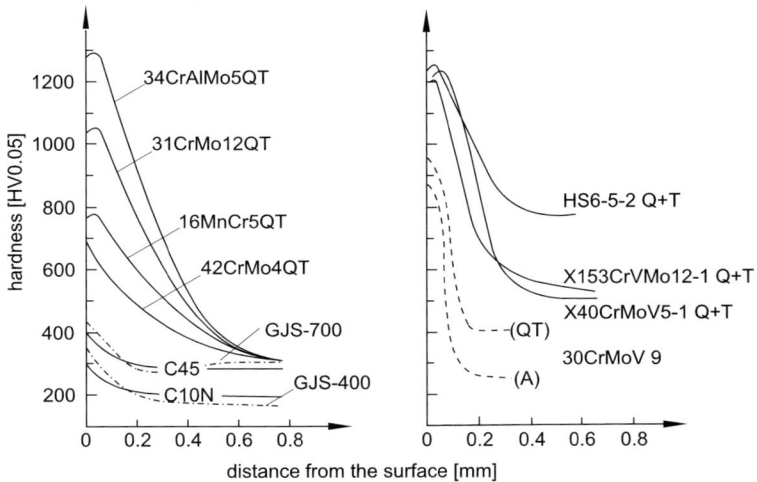

Fig. B.3.6 Microhardness curves after nitriding: Initial state A = soft-annealed, N = normalised, QT = quenched and tempered, Q+T = hardened and secondary tempering; solid curves: gas nitrided in NH_3, 520°C, 50 h; dotted curves: nitrocarburised in a salt bath, 570°C, 2 h; dash/dot curves: as above, 580°C, 3 h, + oxidation, 370°C, 10 min.

steel can be used at temperatures up to 500°C. The Cr, Mo and V fractions, enriched in the carbides after tempering, are no longer available in the surface layer for precipitation-hardening of the matrix. Aluminium is not affected because it avoids forming carbides. Although a low C content would be more favourable for a high case hardness, a high content increases the core strength. However, the diffusing nitrogen leads to carbon enrichment underneath the compound layer, which results in the precipitation of brittle cementite on the grain boundaries parallel to the surface.

Continuous casting with a tundish and a submerged nozzle excludes atmospheric oxygen and thus increases the cleanness of Al-alloyed grades, which have a particular affinity for oxygen. The production quantities are usually insufficient for this production method, and the cleanness of Al-free steels can be controlled more effectively in ingot casting. Polished surfaces, in particular, exhibit so-called comet tails that emanate from non-metallic inclusions. These comet tails are much more conspicuous than the inclusions themselves. Nitriding generally improves the polished surface because the difference in hardness between the oxide inclusions and the surrounding surface is lower. It is often expedient to remove the compound layer by polishing or by a combination of shot peening and polishing because, if the N (+C) content is high, this layer may contain incipient cracks parallel to the surface due to residual stress. If these cracks propagate during service, they lead to flaking and consequently to a rough surface. Owing to its high hardness, this surface rapidly abrades its counterbody in the friction pairs used in machine elements. The compound zone can also be weakened by a porous zone. Nascent hydrogen liberated by the decomposition of ammonia appears to perform a certain amount of preparatory work: because it is the smallest atom in steel and thus very mobile, it recombines to H_2 in the smallest voids and enlarges incipient pores so that nitrogen atoms, which diffuse more slowly, can recombine to N_2 on the pore walls. The associated increase in pressure then enlarges incipient pores to micropores.

As a rule, the workpieces are slowly cooled after nitriding to minimise distortion. However, slow cooling can also lead to changes in the microstructure of the surface layer and to residual stresses. An established method for unalloyed steels is quenching in water or oil to retain a higher concentration of dissolved nitrogen and thus improve the fatigue strength. The development of residual stresses in the surface layer of an alloyed steel during nitriding and slow cooling is shown in Figure B.3.7. The nitrogen diffusing into the surface increases the volume as well as the residual compressive stresses. These stresses subsequently decrease when nitrides are precipitated in the diffusion zone. Differences in the coefficients of thermal expansion between the individual layers and the core lead to further changes in the overall residual stress during cooling. The maximum residual compressive stress generally lies below the surface and decreases with increasing nitriding temperature (Figure B.3.8). This lowers the creep limit and increases the relaxation of the stresses in the surface layer caused by nitriding. Thus the maximum stress decreases as the local hot hardness decreases (Figure B.3.8 a). However, it decreases less with increasing content of alloying elements that improve the high-temperature strength.

In contrast to the previously discussed process carried out at a temperature of $T_N < A_1$, nitriding is also carried out to a limited extent at $T_N > A_1$. The abscissa in the Fe-N phase diagram gives the nitrogen content of the surface layer, which increases from the core to the surface. Austenitisation starts at $A_1 = 590°C$ (Figure B.3.9 a). At the selected temperature of 620°C, the austenite grains contain approx. 2 % N in the two-phase $\alpha + \gamma$ region and approx. 2.5 % N in the $\gamma + \gamma'$ region. After quenching unalloyed steels, in addition to martensite, they can be expected to contain predominantly retained austenite that has a high degree of solid-solution hardening due to dissolved N atoms (Figure B.3.9 b). Tempering or an oxidative secondary treatment at e.g. 380°C converts the retained austenite to high-strength bainite that supports the outer nitride layer. Although the thickness of the support layer increases with T_N, it is usually thinner than the compound layer. Nevertheless, in spite of this low thickness, it provides a gradual transition of the hardness from the hard nitride layer to the softer core.

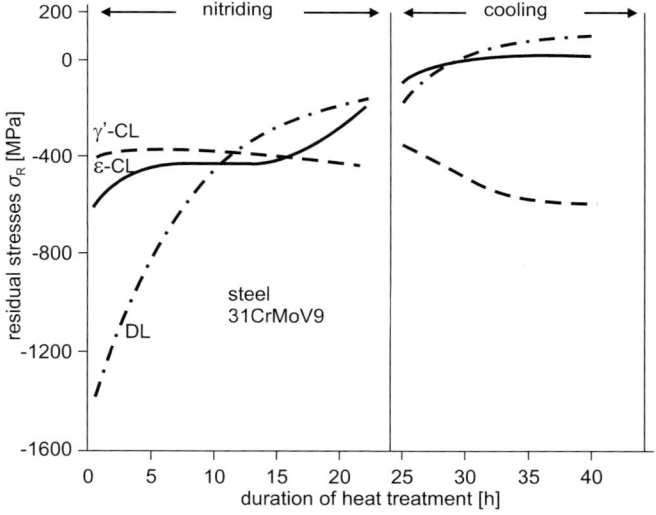

Fig. B.3.7 Development of residual macro-stresses, in the compound layer CL and the outer diffusion layer DL after heating to $T_N = 520°C$ during nitriding (24 h, $K_N = 1.7$) and slow cooling. Measured using the interference lines {200} γ' - Fe$_4$N, {102} ε - Fe$_{2-3}$N for CL and {211} α - Fe for DL (acc. to D. Günther et al.).

B.3.2 Nitriding steels

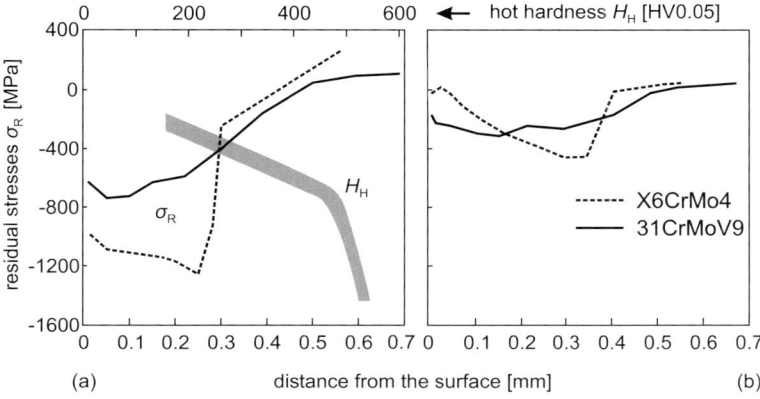

Fig. B.3.8 Residual stress curves in the nitrided layer: (a) nitrided at 500°C, 96 h, $K_N = 9$, (b) nitrided at 570°C, 32 h, $K_N = 3$. Diagram (a) also shows the relationship between the maximum residual compressive stress of differently alloyed steels and the hot hardness measured at the same distance from the surface for a nitriding temperature of 500 or 570°C (from H. J. Spies et al.).

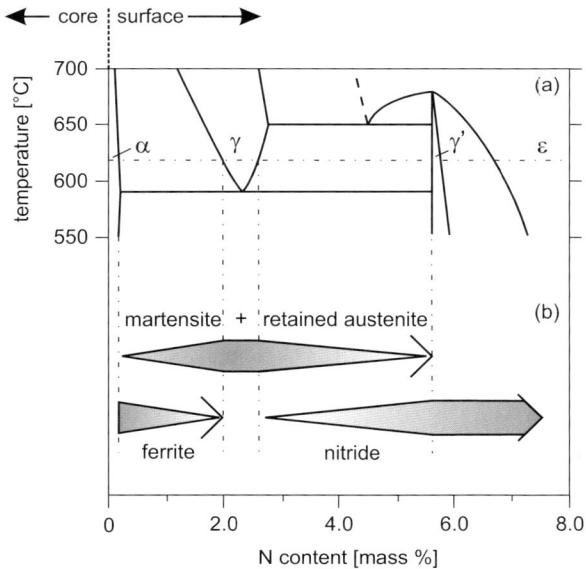

Fig. B.3.9 Austenitising nitriding of iron above 590°C: (a) Phase diagram of Fe-N, (b) Change in the amounts of the different phases (schematic) after quenching from 620°C.

B.3.2.3 Applications

Whereas the size change during surface-hardening discussed in Section B.3.1.1 is kept low by limiting the heated volume of the workpiece, the entire volume is heated during nitriding. However, slow heating and cooling without a phase transformation in the core means that the size change remains small. The actual change is limited to an increase in layer thickness, which in unalloyed steels can be as much as $\frac{1}{5}$ of the thickness of the ε compound layer and can even increase to $\frac{1}{2}$ for a pore volume of 30 %. This ratio may even increase to > 1 in alloyed steels. Nevertheless, the dimensional change is low with respect to the thickness of the component. The change in shape can be limited by stress-relief annealing prior to nitriding followed by mild cooling. The advantages of nitriding in practical applications are the high hardness and the low distortion. This does not apply if the surface of the workpiece is large with respect to its volume, as in e.g. sheet components or perforated plates. Quenching of unalloyed steels also promotes distortion.

Workpieces that are particularly susceptible to distortion include slender components, such as cylinders and screws for plastic extruders, calendering rolls used in the manufacture of films, paper, and non-wovens, and also threaded spindles or even heavy crankshafts for commercial vehicles and diesel locomotives. On account of its high cleanness, the Al-free steel 31CrMo12 can be used for these applications; it is suspended in the gas during nitriding. Formerly, nitrided automobile crankshafts were commonly used; however, the cost advantage of induction-hardened precipitation-hardened ferritic-pearlitic steels (Tab. B.2.3, p. 189) has not been ignored. The required hardness penetration of large crankshafts calls for a higher alloy content of the quenched and tempered steels. However, their lower carbon content means that the increase in hardness on induction hardening is insufficient and they must therefore be nitrided to increase their wear resistance. The nitrided layer, with its advantageous compressive stress, is too thin with respect to the component cross-section to provide a significant increase in the fatigue strength, i.e. the bending and torsional stresses underneath the layer have not sufficiently declined. The alloying costs and availability of steel grade 42CrMo4 make it suitable for heavy calendering rolls with body dimensions of e.g. $\varnothing\, 500 \cdot 4000$ mm. The concentricity precision of e.g. < 10 μm is achieved after nitriding by hot grinding at the service temperature of e.g. 220°C. One of the steels used for pressure spindles in friction spindle presses is 34CrAlNi7-10. In this case, final machining must ensure that the compound layer does not flake away because roughening of the spindle surface during operation would destroy the gunmetal nut. As a consequence of the greater slenderness of extruder screws and the increasing viscosity of plastic melts, the required torque also increases so that it is advisable to use secondary hardening steels with a high core strength, e.g. 33CrMoV12-9 or X40CrMoV5-1. Distortion problems usually limit the maximum production diameter of carburised gear wheels to 2.5 m.

In contrast, gas-nitrided gear wheels have been made with a diameter of up to about 4 m.

In addition to standardised nitriding steels, large quantities of small stamped and turned parts made of unalloyed steels are nitrided or nitrocarburised (e.g. 580°C, 2 h / water) or oxynitrided to improve their wear and corrosion resistance as well as their fatigue strength. For components subjected to high Hertzian contact stresses, as found e.g. in the roller guides of automobile seats, the austenitising nitriding process at 620 to 680°C leads to a higher hardness penetration and greater support of the nitrided layer. The core strength of a seat frame made of unalloyed steel strip can also be increased by mild nitriding to the core. In parts made of sintered iron, the pore canals are generally closed if the porosity is < 8 volume-%, which prevents nitriding to the core. The nitriding behaviour of a cast iron matrix is similar to that of unalloyed steels (Figure B.3.6). However, a high Si content promotes passivation. Even low Cr and Mo contents increase the case hardness. Whereas precipitation-hardenable ferritic-pearlitic steels (Chap. B.2.3, p. 189) with a higher carbon content are designed for case hardening, the surface hardness of grades with a lower carbon content, such as 27MnSiVS6, can be increased to approx. 600 to 650 HV0.3 by nitriding or oxynitriding at 550°C, while the core hardness decreases only slightly.

Nitriding is also widely used to increase the wear resistance of tools, such as injection moulds for plastics made of quenched and tempered steel, pressure die-casting moulds, extrusion dies and forging dies made of hot-work tool steel, stamping tools made of cold-work tool steel and rams made of high-speed steel for semi-hot forging. This also takes advantage of the good high-temperature strength of the nitrided layer, and secondary hardening of the steel provides a high core hardness. The ceramic-like nitrided surface hinders undesirable bonding between the metallic workpieces and the tools. Nitrided ledeburitic cold-work tool steels, such as X153CrMoV12, are also used in precision-cast wear parts for thread guides and in grinding plates. Martensitic stainless steels, such as X20Cr13, can be given a high degree of nitride hardening on account of their high Cr content. However, the formation of chromium nitrides may have an adverse effect on the corrosion resistance.

Austenitic stainless steels, such as X5CrNi18-10, can be nitrided at low temperatures of approx. 400°C. The rate of ammonia decomposition is slower at this temperature. Therefore, these materials are usually nitrided in a plasma, sometimes with an active screen. Nitrogen dissolves in the surface layer and expands the austenite lattice (expanded austenite). The content at the surface may increase to approx. 10 % without nitrides being precipitated within the nitriding time of < 30 h. This supersaturated austenite is also known as the S phase. It can have a surface hardness of 1100 to 1500 HV0.1 without loss of corrosion resistance. Its resistance to pitting corrosion is usually even increased. Owing to the low nitriding temperature, the nitrided layer thickness is generally limited to < 25 μm. At $T_N > 450°C$ the formation of CrN is accelerated to such a degree that it can form during nitriding with an associated loss of the special properties of the S phase. A similar case hardened layer can also be formed by the absorption of approx. 6 % C or a combination of N and C.

B.3.3 Case hardening steels

B.3.3.1 Process engineering aspects of case hardening

In this thermochemical process, the workpieces of low-carbon steel are completely austenitised and the surface layer is enriched with carbon using a carburising agent. The resulting hardness decreases from the surface towards the core owing to the diffusion-dependent carbon concentration profile. The thickness of the surface layer or case is usually < 3 mm, but may be as much as 10 mm in some cases.

Originally, a solid carburising agent was used that consisted of a mixture of charcoal and an activator, such as barium carbonate. The workpieces were placed in boxes and then surrounded with this powder, which produced the carburising gas as a CO/CO_2 mixture when heated. Slow heating in the powder pack and lack of controllability of the carbon content in the surface layer are drawbacks of this process. Carburising in salt baths decisively shortened the heating time of preheated workpieces; however, the problem of direct control remained unsolved. This was eventually overcome with the development of carburising in a gas atmosphere as well as the corresponding instrumentation and control engineering.

(a) Gas carburising

Nowadays, widespread use is made of gas carburising in a carrier gas, obtained e.g. by catalytic cracking of natural gas and air at $\approx 1000°C$ to produce a mixture of N_2, CO and H_2 with a little CO_2 and H_2O (endogas, endothermic gas). The carbon activity a of the heterogeneous water gas shift reaction can be controlled by adding an enriching gas such as methane (CH_4) or propane (C_3H_8):

$$CO + H_2 \rightarrow [C] + H_2O \qquad a \sim p(CO) \cdot p(H_2)/p(H_2O) \qquad (B.3.6)$$

This is associated with a very much larger carbon mass-transfer coefficient for the transfer of C from the gas into the surface of the steel compared to methane decomposition $CH_4 \rightarrow 2H_2 + [C]$ or the Boudouard reaction $2CO \rightarrow CO_2 + [C]$. In practice, the ratio of the partial pressures $p(H_2)/p(CO)$ is thus usually close to unity. Control of the carbon content in the surface layer of the steel (C_S) is based on attaining an equal carbon activity in the gas and in the steel. According to Figure B.3.10 a, the carbon activity a in austenite increases with the content of dissolved carbon up to the graphite or carbide limit of 1. For the example point, a = 0.4 roughly corresponds to C_S = 0.7 %. The corresponding carburising gas is usually characterised not by its carbon activity but rather by its potential to regulate the carbon content in the surface layer, the so-called carbon potential, where $C_P = C_S$. Alloying elements such as Mn, Cr, Mo and V lower the activity of carbon in the austenite, but increase its solubility. Ni and Si have the opposite effect (Figure B.3.10 b). For a given gas activity a, a steel alloyed with Cr thus has a higher carbon content C_A in the surface layer than iron, whose carbon content in the surface layer corresponds to C_P. An alloying factor can be derived from

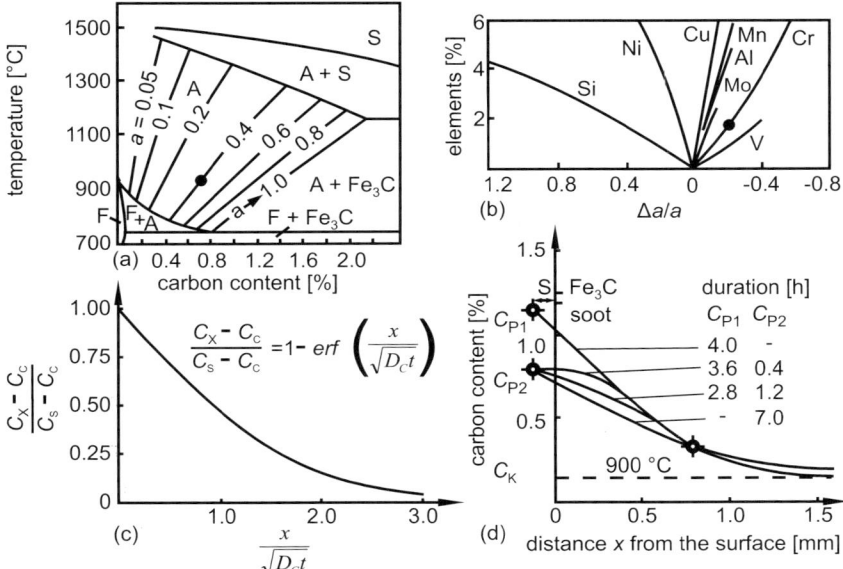

Fig. B.3.10 Carburising: (a) The required carbon content at the surface C_S at the carburisation temperature corresponds to a carbon activity 'a' in iron (example point: a = 0.4). (b) According to F. Neumann and B. Person, alloying elements modify a by Δa. (Example point: steel alloyed with chromium $\Delta a/a = -0.2$,, i.e. $\Delta a = -0.08$ and a in the steel $= 0.4 - 0.08 = 0.32$). Interactions must be taken into account if alloyed with more than one element. In order to reach the equilibrium concentration C_S att the surface, the carbon activity in the carburising gas must be equal to that in the steel. The gas has the required carbon potential $C_P = C_S$. (c) According to Fick's Second Law of Diffusion, the carbon content C_X decreases with the distance x from the surface to the core carbon content C_C. (d) If the rate of carbon-transfer from the gas into the surface of the steel does not keep up with diffusion towards the core, then C_P is only reached at a theoretical distance $S \approx D_C/\beta$ before the surface (D_C/β = diffusion coefficient/mass transfer coefficient for carbon). According to J. Wünning, overcarburising at C_{P1} just below the cementite and sooting limit with subsequent equilibration at C_{P2} gives time-savings of $\approx 40\%$ as well as a carbon profile that deviates from the Fick curve.

this: $k_A = C_A/C_P$. An expanded form of this equation describes the changes in the carbon content of the surface layer induced by alloying elements in case hardening steels with respect to iron:

$$\lg k_A = -0.055 \cdot \%Si - 0.011 \cdot \%Ni - 0.123 \cdot \%S$$
$$+0.012 \cdot \%Mn + 0.043 \cdot \%Cr + 0.009 \cdot \%Mo \qquad (B.3.7)$$

According to Eq. B.3.6, gas carburisation is associated with a change of oxygen partners, which means that this process can be indirectly monitored with an oxygen probe and regulated by the supply of enrichment gas. Initially, a certain time elapses until $C_S = C_P$. As the carburisation time increases, only the thickness of the diffusion zone increases, whereby C_S drops to the core carbon content C_C following Fick's Second Law of Diffusion (Figure B.3.10 c). Because the diffusion velocity increases with the concentration gradient, C_{P1} is initially increased to a value just below carbide formation in steel or soot formation in the gas. During subsequent holding at C_{P2}, overcarburisation is decreased to the required C_S value. The saved time is illustrated by an example in Figure B.3.10 d. The momentary carbon profile is continuously calculated and visualised. The largest time saving is achieved by increasing the carburising temperature; however, this is limited by grain coarsening in the steel. The temperature generally lies between 930 and 980°C.

By using a carrier gas, gas carburising is carried out at atmospheric pressure thus allowing the use of cheaper atmospheric furnaces, such as chamber, retort, bell or well-sealed continuous furnaces. The latter are operated with a slight overpressure to avoid uncontrolled ingress of air. Oil is generally used for quenching. Instead of an upstream endogas generator, a nitrogen-filled furnace can be injected with methanol (CH_3OH) to produce a carrier gas, although more hydrogen dissolves in the steel. Starting from e.g. 0.5 ppm in the untreated state, the total hydrogen content after treatment in endogas increases to about 2 ppm, and in N_2/CH_3OH it reaches values that are higher by a factor of 4. About 50 to 75 % of the diffusable hydrogen is usually lost during tempering (Chapt. A.3, p. 58), although the growth of an oxide layer hinders effusion. The large amount of carbon in the carrier gas required for high loading densities in the furnace is provided by injecting liquid carburising agents into the furnace, such as isopropanol (C_3H_7OH), acetone (CH_3COCH_3) or ethyl acetate ($CH_3COOC_2H_5$), and each molecule provides three or four carbon atoms. A drawback of gas carburising is internal oxidation of the alloying elements, particularly along the austenite grain boundaries. The oxide deposits weaken the grain boundaries and bind e.g. Cr and Mn so that the hardenability is reduced and pearlite can grow. This depletion of the matrix induces secondary diffusion of the alloying elements, thus increasing their total content at the surface (Figure B.3.11 a). Surface oxidation, which is often 5 to 20 µm deep, damages the surface and adversely affects the endurance limit. This can be remedied by introducing residual compressive stresses by means of shot peening.

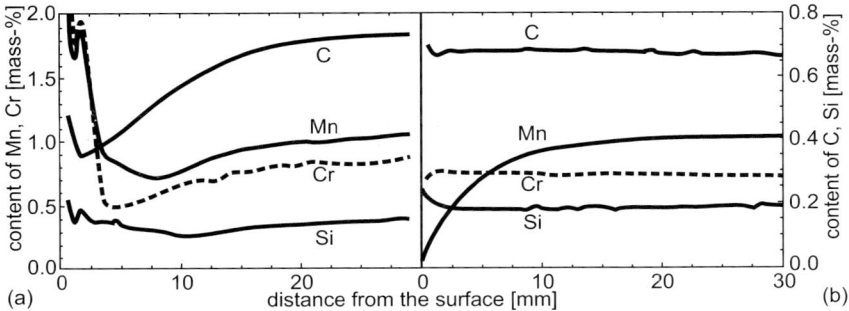

Fig. B.3.11 Near-surface chemical composition in the surface layer after case hardening from 930°C for CHD = 0.4 mm. (a) Gas carburising in nitrogen-methanol, (b) low-pressure carburising in 8 mbar propane (from S. Laue et al.).

(b) Vacuum carburising

Vacuum engineering, which was developed as a scale-free method of heat-treating workpieces, is also exploited in case hardening. One motive was to avoid surface oxidation, a further was to shorten the processing time. The use of pure hydrocarbons with a high carbon content, such as ethylene (C_2H_4), propane (C_3H_8) or acetylene (C_2H_2), avoids oxidation and furnishes a higher carbon potential. A pressure of e.g. 5 to 10 mbar is generally sufficient to lower C_P to a useful level and largely prevent sooting. This low-pressure treatment is carried out in a vacuum furnace, which is why it is known as vacuum carburising – which appears to be nonsensical from the true meaning of the words. In spite of the lower carbon mass-transfer coefficient compared to gas carburising, the higher C_P value during low-pressure carburising leads to a higher carbon-transfer rate $J = \beta \, (C_P - C_{S,t})$ into the surface of the workpiece. In this way, the gradually increasing carbon content at the surface $C_{S,t}$ reaches the required content at the surface C_S more quickly and is even increased beyond it. To avoid the growth of a thick and irreversible carbide layer, the furnace has to be evacuated in good time and left for a certain period to allow diffusion to take place. The pulsed process consists of carburising and diffusion periods. The momentary carbon profile in the surface layer is calculated from previously determined, characteristic carburisation values. Although there is no sensing element corresponding to the oxygen probe that can be used for control purposes under equilibrium conditions, experience shows that it is indeed possible to control a low carburisation depth of e.g. 0.1 mm in sheet components as well as a depth of several millimetres in heavy gear wheels. The vacuum carburising process not only avoids surface oxidation, but it also allows blind holes and narrow through-holes to be uniformly carburised, in which case, acetylene is more effective but more expensive than propane. On the other hand, the geometry-related hindrance of the supply of fresh gas during gas carburising tends to lower the uptake of carbon. Owing to its low gas pressure, the vacuum carburising process is suitable for plasma activation. This process is called plasma carburising. Further parameters that can be used to influence the carbon-transfer rate are the voltage, the amount of current and the pulse duration. This increases the dissociation of the gas and β. Vacuum carburising, with and without a plasma, promotes the evaporation of manganese from the surface with the corresponding depletion up to a depth of 20 µm (Figure B.3.11 b). It is possible to carburise at atmospheric pressure in a microwave furnace. Rapid heating is carried out e.g. in an argon plasma, to which acetylene is added once the carburisation temperature has been reached.

Vacuum furnaces are generally designed for high temperatures, so that high-temperature carburising above 1000°C can be used to shorten the processing time. To avoid exposing the furnace fixtures and workpieces to air oxidation during hardening, the material is subjected to high-pressure gas quenching in the furnace, which is equipped with a gas circulation and cooling system. In addition to quenching with nitrogen at a pressure of 6 to 20 bar, the cooling rate is increased by adding helium. Compared to a single-chamber vacuum furnace, the more modern multi-chamber furnaces offer a higher availability because gas quenching takes place in an exchangeable second chamber while the newly loaded chamber is being evacuated and coupled to the vacuum furnace. The furnace stays hot and the quenching chamber remains cold, and this increases the quenching rate during hardening. Liquid nitrogen is used for cooling to below room temperature, thus transforming retained austenite to martensite. The almost constant cooling rate during gas quenching reduces distortion of the workpieces compared to the very inhomogeneous cooling rate produced by quenching in a liquid with a maximum at about 600°C. This advantage of gas quenching is somewhat diminished by the fact that the $t_{8/5}$ time is often insufficient to achieve a sufficient core hardness in low-alloy case hardening steels with thicker workpiece cross-sections. The injection of liquid nitrogen between approx. 700 and 600°C can help to suppress the formation of pearlite.

(c) Carbonitriding

The addition of up to 10 % ammonia to the furnace atmosphere during gas carburising increases the amount of nitrogen diffusing into the surface as the temperature decreases. This lowers the case thickness. A process temperature of 780 to 870°C should result in 0.2 to 0.4 % N and $<1\%$ (C+N) in the surface layer. NH_3 in the carrier gas has a higher dissociation rate than N_2 on the surface of the metal. At elevated temperatures, it is possible that NH_3 prematurely decomposes on the furnace fixtures thus decreasing the nitriding effect and increasing the effusion of nitrogen that has already dissolved. Enrichment of the surface layer with C+N is known as carbonitriding. Nitrogen, in particular, increases the hardenability of the surface layer of unalloyed steels. A two-stage process is often used to increase the thickness of the surface layer: the material is first gas carburised at about 950°C and subsequently carbonitrided at e.g. 850°C. After an appropriate holding time, the workpiece is then hardened in oil.

A two-stage process is also used for vacuum carburising. The first step is carburising with a low hydrocarbon partial pressure and the second step is nitriding with added NH_3 at a lower temperature. A NH_3 pressure of almost 1 bar is required to achieve the specified N content of between 0.2 and 0.4 % as a result of a balanced inward and outward diffusion.

(d) Solution nitriding

As described above under sections (a) - (c), case hardening can be accomplished by carburising or carburising + nitriding. Nitriding on its own is recommended for stainless steels such as X20Cr13 because the maximum solubility of carbon in austenite is lowered by the presence of chromium to such an extent that carburising would not produce a sufficient increase in hardness (Figure B.6.13 b, p. 326). On the other hand, carbon (in steel) and nitrogen (from the gas) extends the phase field of austenite (Figure B.6.13 c) with satisfactory solubility of C+N. Because chromium hinders the diffusion of interstitial elements, a temperature between 1050 and 1150°C is used to nitride the surface within a reasonable time. In this temperature range, there is sufficient dissociation of N_2 gas on the surface of the steel to enable a nitriding equilibrium to develop between the gas and the surface of the steel. At a given temperature and steel composition, the nitrogen content on the surface of the steel depends solely on the N_2 pressure (Figure B.3.12). The SolNit case hardening process is carried out under controlled conditions in a vacuum furnace at a pressure of 0.2 to 2 bar N_2 followed by high-pressure gas quenching.

B.3 Materials for surface layer treatments

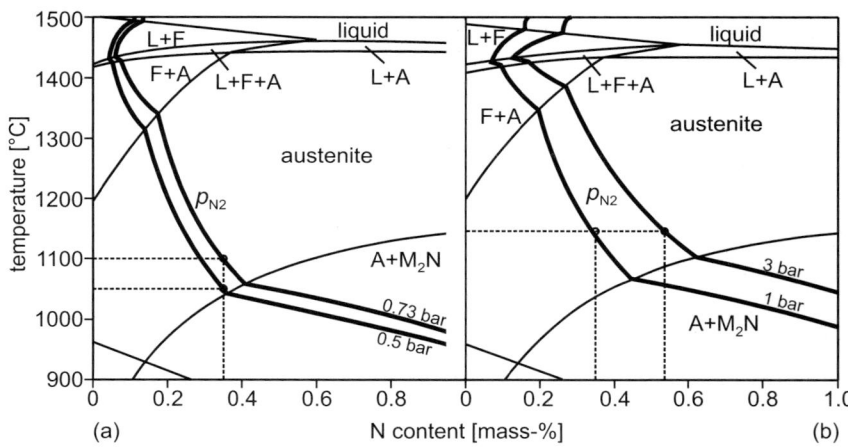

Fig. B.3.12 Case hardening of stainless steels with nitrogen: The Fe-13%Cr-0.2%C-N phase diagram describes steel X20Cr13, which has a surface nitrogen content according to the equilibrium pressure p_{N_2}. (a) Solution nitriding between 1050 and 1100°C to a surface nitrogen content of 0.35 %. Together with 0.2 % C, this produces a hardness of 58 to 60 HRC after direct hardening. (b) Solution nitriding at 1150°C, with optional overnitriding at $p_{N_2} = 3$ bar and homogenisation at 1 bar, intermediate annealing at 780°C and reheat hardening according to (a) (from F. Schmalt).

Table B.3.2 Steels for case hardening according to EN 10084. Mean values for J calculated from the hardenability scatter band, $(HL_{max} + HH_{min})/2$.

Steel	blank-hardened [1]				surface [2]
	J9		J20		hardness
	HRC	MPa	HRC	MPa	HRC
16MnCr5	31.5	1008	23	820	60 to
20MnCr5	36.5	1155	30	970	64 HRC
20MoCr4	31	995	22	800	or
17NiCrMo6-4	37.5	1185	31	995	700 to
14NiCrMo13-4	41	1300	37	1170	800 HV10

[1] According to Figure A.3.5, J9 and J20 correspond to a position close to the surface or the core, respectively, of a round bar, ⌀ 60 mm, hardened in oil. Tensile strength in MPa from the hardness in HRC according to EN ISO 18265

[2] After case hardening and tempering at 150 to 200°C

B.3.3.2 Materials and the surface layer

(a) As-delivered condition

Case hardening steels regulated by EN 10084 include unalloyed grades, such as C15E, and alloyed grades, such as 16MnCr5 or 17NiCrMo6-4 (Tab. B.3.2). Steels with 0.02 to 0.04 % S have an improved machinability (C15R, 16MnCrS5, 17NiCrMoS6-4). The as-delivered condition is expressed by an appended letter: U = untreated, S = treated to improve shearability, A = soft-annealed, TH = treated to hardness range, FP = treated to achieve a ferrite-pearlite microstructure and hardness range. The surface finish is indicated by further letters: HW = as hot-worked, PI = HW + pickled, BC = HW + blast-cleaned. Thus 16MnCr5 + A + BC indicates that the steel is to be delivered soft-annealed and blast-cleaned. The required hardenability is expressed as a predefined scatter band from the Jominy test according to EN ISO 642. The hardness ranges e.g. for 16MnCr5 + H from 37 to 21 HRC at a distance of 11 mm, which is expressed by J 11/37-21. In addition, narrower scatter bands in the upper or lower range can also be agreed upon e.g. 16MnCr5 + HH with J 11/37-26 or 16MnCr5 + HL with J 11/32-21.

The susceptibility to grain coarsening is tested by measuring the size of the former austenite grains after case hardening. As the surface layer generally remains more fine-grained than the core, a lengthy blank hardness test with subsequent determination of the quenched grain size is usually sufficient. The austenite grain boundaries are emphasised by a special etching process using picric acid with an added wetting agent. In contrast, the McQuaid-Ehn test is carried out after overcarburising at 925°C to precipitate cementite and thus visualise the austenite grain boundaries. However, this impairs their mobility, so that this method generally shows less grain growth. The materials are usually evaluated against a series of rating images whose code letter G shows the number m of grains per mm^2. A value of m = 16 is chosen for G = 1 so that it complies with the US standard ASTME 112. The mean grain diameter for G = 1 is 250 µm, which is relatively coarse. Fine-grain case hardening steels have G values of 5 and above. This corresponds to a mean grain diameter of < 62.5 µm. Special cases must be agreed upon with respect to acceptance for mixed grains with a basic size of e.g. G = 6 and a specific proportional area of e.g. G ≈ 2. According to the relationship $m = 8 \cdot 2^G$, G can also have negative values for extremely coarse grains. Details on grain size determination are given in EN ISO 643.

(b) Alloying and heat treatment

To limit grain growth during the several hours of heating required for high-temperature carburising, case hardening steels contain approximately 0.02 to 0.04 % Al and about 0.01 % N. At a soaking temperature of ≈ 1250°C, both elements are dissolved and only start to precipitate as AlN towards the end of hot-working and during subsequent annealing. If the precipitates have the right size (e.g. 20 to 40 nm) and a homogeneous distribution, they can effectively restrain the austenite grain boundaries. As the carburisation temperature increases, the AlN precipitates grow and gradually lose their effectiveness, which occurs at about 1000°C. Chromium and molybdenum, added to increase the hardenability, promote a fine-grained microstructure (example 20MoCr4). If the selected hot-forming and annealing parameters are not optimal, coarse grains may even appear below 950°C. The higher temperature used for ferrite/pearlite annealing (FP) usually has a less favourable effect on the AlN dispersion compared to spheroidising (AC). High-temperature carburising trials have been carried out between 1000 and 1100°C using more temperature-resistant MX precipitates, obtained by adding at least 0.045 % Nb and 0.01 % Ti, supplemented with more than 0.017 % N and 0.03 % Al. An increase in the carburisation temperature from 950 to 1050°C would more than halve the carburisation time, which is particularly attractive for heavy machine parts with case depths of up to 10 mm.

The above deliberations on ways of obtaining a fine-grained microstructure refer to hardening from the carburising temperature. This is known as direct hardening (Figure B.3.13). Another method is to allow coarse grains to grow during carburising and subsequently refine them. This is achieved by cooling from the carburisation temperature to e.g. a holding temperature below A_1 in order to induce a ferrite/pearlite transformation. Hardening occurs after reheating to a temperature below the carburisation temperature. This two-step

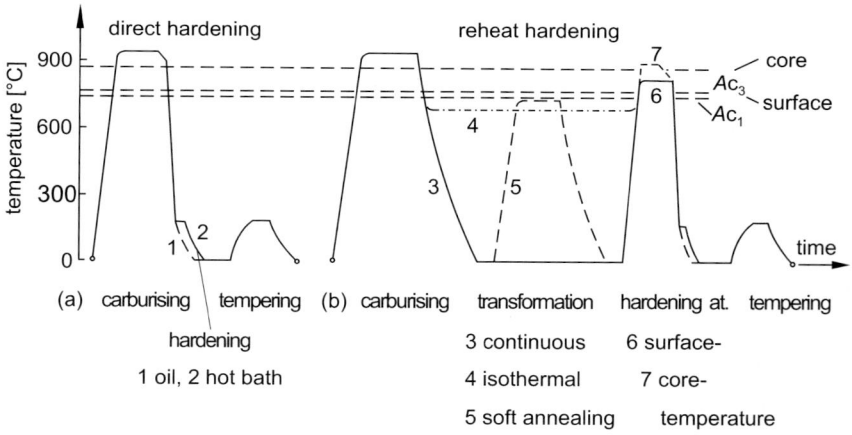

Fig. B.3.13 Case hardening sequence: (a) Direct hardening from the carburising temperature (or after cooling to the hardening temperature) in oil or a hot bath. (b) Reheat hardening with refining of the grain size by cooling below Ac_1 and subsequent hardening from the surface or core temperature.

treatment is known as reheat hardening. The diffusion time can be shortened by high-temperature carburising in a low-pressure process at e.g. 1100°C so that - in spite of the reheat hardening treatment to refine the grains - it is still faster than gas carburising at 930°C.

If the wear resistance is more important, the hardenability of unalloyed steels can be increased by the uptake of nitrogen to produce a carbonitrided, martensitic surface layer above a ferritic-pearlitic core. Higher mechanical loads require a greater core strength that can be attained by means of a martensitic-bainitic transformation. However, this requires a sufficient degree of hardness penetration, which can be achieved by alloying. The alloying content must increase with the thickness of the workpiece: C16E → 16MnCr5 → 17NiCrMo6-4. This particularly applies if the cooling rate decreases from gas carburising with oil quenching compared to low-pressure carburising with gas quenching. The standard specifies CrNiMo grades for gas quenching that have adequate hardenability; however, their alloying costs are comparatively high. Certain steels have been developed as a compromise, e.g. 23MnCrMo5. During through-hardening, the core hardness increases with the carbon content in the steel: C10E → 17Cr3 → 28Cr4. The carbon content in the surface layer usually lies between 0.6 and 0.8 %; however, in more highly alloyed steels it is kept closer to the lower value to limit the amount of retained austenite. In steels containing more than 2.5 % Ni, the surface layer of bending specimens exhibits dimple fracture rather than cleavage fracture. This ductilising effect of nickel is attributed to strengthening of non-directional interatomic bonds by its free electrons (Chapt. A.4 p. 85). Adding 0.001 to 0.005 % B improves the toughness if there is sufficient dissolved nitrogen present to precipitate it as boron nitride, e.g. 16MnCrB5. The addition of titanium to improve stability of a fine-grain microstructure could prevent precipitation of BN; however, free boron would then be available to increase the hardenability (Chapt. B.2, p. 192).

(c) Properties

In contrast to surface-hardening, the entire volume of the workpiece is affected by heat treatment during case hardening, and it is cooled more rapidly from a higher temperature compared to nitriding. Both adversely affect the *dimensional stability*. During direct hardening, the workpiece is often subjected to mild cooling from the carburisation temperature to the hardening temperature to reduce the temperature span of the rapid cooling regime and thus reduce distortion (Figure B.3.13). This becomes increasingly important as high-temperature carburisation progresses. For reheat hardening, the workpiece can be completely cooled and remachined after carburising to locally remove the case and avoid hardening where it is not wanted. This can be combined with hardening from a temperature below the carburisation temperature to reduce distortion. High-pressure gas quenching usually causes less distortion than quenching in oil. This reduces the high costs of finish-machining hard surfaces, which compensates the higher alloying costs that are required to maintain the core hardness. In nickel-alloyed case hardening steels with high hardenability, slow intermediate cooling can lead to cracks in the carburised surface layer, which undergoes a pearlitic transformation and develops tensile stresses due to the subsequent bainite/martensite transformation in the core. This can be avoided by holding for several hours in the pearlite stage during cooling so that the core also transforms. This holding time can be shortened by alloying with $\approx 0.1\,\%\,\mathrm{V}$.

In addition to improving the wear resistance, case hardening also aims to increase the *fatigue strength* of components such as toothed wheels or piston pins. Surface damage due to surface oxidation after gas carburising, carbide growth at the edges or loss of manganese after low-pressure carburisation, usually has a negative effect in unmachined areas, e.g. at the tooth root. This can be compensated by shot-peening to introduce residual compressive stresses. An undamaged hard surface layer without a soft skin has a greater tendency to form cleavage cracks at the surface. Coarse grains in a mixed-grain microstructure shift the formation of cleavage cracks into the transition zone to the core, thus reducing the fatigue strength. Introduction of residual compressive stresses in the case hardened surface layer leads to a considerable increase in the fatigue strength under bending and torsional loading conditions. The crack-initiating tensile half-cycles of the applied load are noticeably reduced by compressive prestressing until a lower load stress level is reached below the surface layer. As shown in Figure B.3.14, carburisation delays the transformation during direct hardening and also lowers the M_s temperature.

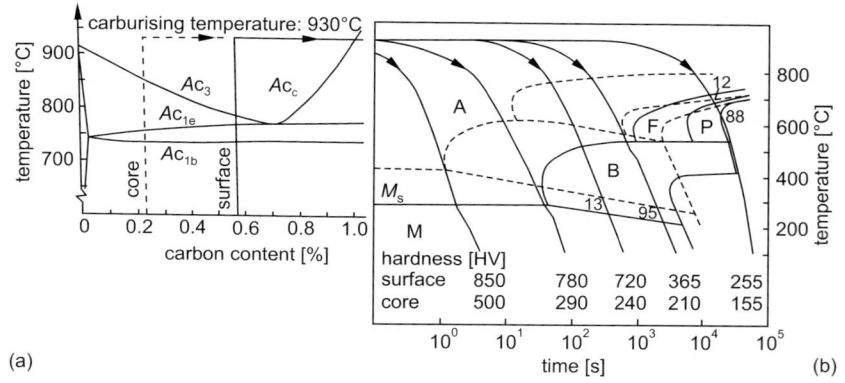

Fig. B.3.14 **Progression of the transformation in case hardening steel 20MoCr4:**(a) near-equilibrium temperature profile during carburising according to the phase diagram for steel with 0.56 % Cr and 0.44 % Mo, (b) TTT diagram for a core carbon content of 0.22 % (dashed line) and surface carbon content of 0.56 % (solid line), evaluated according to A. Rose and H. Hougardy.

Let us consider the second fastest cooling curve as an example: the transformation starts below the surface layer after a good 10 s and advances towards core. The increase in volume plastically strains the surface layer, which is still austenitic. After a delay, the surface layer transforms from the inside towards the surface until after $\approx 80\,\text{s}$ the surface is bainite-free martensitic with a hardness of 780 HV. The associated volume increase introduces the desired residual compressive stresses in the surface layer. These residual compressive stresses do not reach their maximum value if the cooling rate is too harsh, the core carbon content is too high or the core cross-section is too small, and this has a negative effect on the fatigue strength.

If austenite is retained in the surface layer, the volume increase is lower and so are the residual compressive stresses; this results in a lower increase in the fatigue strength. Deformation-induced transformation of retained austenite by shot-peening can considerably increase the sub-surface residual compressive stresses and improve the fatigue strength. A certain amount of transformation of the retained austenite can also be expected on cyclic loading, which increases with the level of stress increasing from the endurance limit to the low-cycle fatigue range. We must differentiate between crack initiation and propagation when considering the influence of retained austenite on the type of failure. Its ductility appears to prevent the formation of brittle cleavage cracks in the superhard martensite. If the proportion of retained austenite in the surface layer increases to more than 50 %, its lower yield point favours earlier crack initiation and a lower endurance strength. The more ductile retained austenite represents an obstacle with respect to the growth of microcracks. In addition, it is possible that mechanically induced transformation of retained austenite into martensite takes place in the plastic zone in front of the crack tip. The associated volume increase introduces local compressive stresses that lower the amplitude of tensile stresses due to externally applied loads, thus inhibiting crack propagation. This phase transformation remains in the flanks of the fatigue crack; however it is difficult to measure, owing to the shallow crack depth before instant fracture. The transformable volume increases with the proportion of retained austenite to a maximum value and then decreases again owing to the increasing stability of austenite. Alloying increases this stability. Furthermore, nickel increases the ductility of martensite, thus hindering the formation of cleavage cracks and their propagation as well as the mechanically induced phase transformation. Investigations of case hardened steels and tool steels indicate that 25 - 35 % retained austenite affords the greatest increase in the fatigue strength.

As discussed in Chapt. A.4, p. 106, the *wear resistance* increases with the material hardness for many types of wear. A high case hardness usually has a favourable effect on shape retention of workpieces subject to wear. The highest carbon content and thus the greatest amount of retained austenite can be expected close to the surface. This retained austenite can be transformed into particularly hard martensite by loading during wear. Thus surface layers with up to 30 volume-% of retained austenite often have a greater resistance to grooving wear. In special cases, overcarburisation with a high C content can produce a carbide-rich surface layer with a high resistance to adhesion and abrasion (Chapt. B.4, p. 257).

B.3.3.3 Applications

The industrial importance of case hardening can be seen from its market share of one third of all heat treatments used for hardening. An important application field of case hardening steels is in gear units. Toothed wheels in motor vehicles are generally case hardened. Typical steels are 16MnCr5 and 20MoCr4. According to DIN 3990, case hardened gearwheels have the highest endurance limit in the roots and flanks of the teeth, followed by nitrided and surface-hardened components. Of course, this statement depends on the ratio between the surface layer and the thickness of the component. Bending and torsion stresses decrease towards the core all the more slowly the larger the diameter. In contrast, friction-free Hertzian contact stresses increase under the surface. Their maximum shifts to increasing depths as the radius (i.e. the size of the component) increases. Variations in the load-induced stresses, residual stresses and hardness in the surface layer interact. Carburising and surface layer hardening allow greater hardening depths with lower residual stresses compared to nitriding. If the hard surface layer acts primarily as wear protection, the hardened depth should correspond to at least the permissible amount of wear. This may reach values of up to 1 mm in case hardened piercers for stone pressing moulds or gate valves in piston pumps for concrete, so that a nitrided surface layer would be too thin and a surface-hardened layer not hard enough. In addition, a case hardened layer is less susceptible to spalling under impact and shock loading conditions.

The high power density reached in case hardened automotive gear units is finding increasing use in large gears in order to save space and weight. A welded construction has been used successfully for e.g. large spur gears in ships. For example, a ring-rolled rim of 17NiCrMo6-4 with a diameter of more than 2 m is welded to a hub of unalloyed structural steel, e.g. S355, by circular blanks and cross-ties. After stress-relief annealing and premachining, all surfaces, except for the toothed rim, are covered with a layer of copper-donating paste to stop them absorbing carbon and to minimise distortion during case hardening. If this distortion has an out-of-plane and run-out deviation of only a few tenths of a millimetre, the final shape can be obtained cost-effectively by grinding. A CHD = 2 mm is sufficient for grinding. However, because the distortion increases with the wheel diameter, the limitation of this process is the cost of hard machining.

Piston pins are another practical application of case hardened steels. The key properties in this case are fatigue strength and wear resistance. Owing to its low annealed strength, steel 17Cr3 is suitable for cold extrusion of thin pins; a boron-alloyed grade can be used for thicker walls. Steel 17CrNi6-6 is used for thicker pins that have to be machined, e.g. in stationary diesel engines. At service temperatures of up to 180°C, the piston pins are exposed to long-term tempering during operation, which may have a negative effect on the microstructural and dimensional stabilities. Nitrided pins of 31CrMoV9 do not exhibit this phenomenon. Shift actuators in car gearboxes can also be made of sintered steels, such as Sint-D10 (Chapt. B.1, p. 142). These parts are carbonitrided.

The increase in toughness due to $>2.5\,\%$ Ni is currently restricted to heavy workpieces for reasons of cost; nickel also improves the depth of hardening. Such components include e.g. large self-aligning roller bearings of 18NiCrMo14-6 (regulated by EN ISO 683-17 for roller bearing steels) that have diameters of more than 2 m. X19NiCrMo4 is suitable for thick plastic moulding tools with a high closing edge stability and wear resistance to fillers. Steel 14NiCrMo13-4 is used for heavy eccentrics or shift gates.

Wire and bars used for manufacturing pins, bolts and short axles are often alloyed with sulphur if they are to be shaped by machining. They are given an FP treatment. Low-sulphur grades with an AC treatment are used for cold-worked parts. This also applies to stamped parts of unalloyed strip with a thin carbonitrided case. Precision-cast wear parts for knitting and packaging machines are case hardened to increase their service life. Injection nozzles for diesel engines should have a hard axial borehole to avoid being worn by the metering needle moving within them. Initially, the full length of the bore was often not reached during gas carburising, which meant that they had to be made of the through-hardenable grade 100Cr6. The injection pressure was then increased from 800 to 1600 bar, and a further increase to well over 2000 bar is in sight, which will improve atomisation and thus combustion. As the internal pressure increases, the tensile stresses around the feed channels and the dome also increase. These areas thus require residual compressive stresses, which cannot be introduced in a through-hardenable steel. In this case, low-pressure carburising with acetylene offers a way of producing a case hardened layer in narrow boreholes.

If the injection nozzles are susceptible to corrosion by the diesel fuel, the use of a chromium-rich steel may be appropriate. In this case, residual compressive stresses can be introduced by case hardening with nitrogen. N_2 continuously dissociates during nitriding, and any N atoms not taken up by the surface recombine to N_2. The gas thus stays fresh and available as long as the equilibrium pressure is maintained. This allows nitriding of narrow boreholes and slender gaps. Nitridable grades include the ferritic-martensitic steel X12Cr13 or the martensitic steel X20Cr13 (Figure B.3.12), depending on the required core strength. Solution nitriding increases the wear resistance of stainless steels, including austenitic and duplex steels, whose surface layer is not martensitic but high-strength austenitic instead. This also usually increases the resistance to pitting corrosion, which is exploited in numerous applications in the textile, food and chemical industries. Stainless steel moulds for polymer processing are made of e.g. X46Cr13, which is vacuum-hardened to 52 HRC. On the other hand, X12Cr13 facilitates machining, lowers the susceptibility to hardening cracks and has a surface hardness of $\geq 58\,\text{HRC}$ after solution nitriding in a vacuum furnace.

Hammer that has been used to break up minerals
(SCHMOLZ + BICKENBACH Guss GmbH & Co. KG, Krefeld, Germany)

B.4 Tools for processing minerals

B.4.1 Loading and material concepts

Tools have a very long tradition throughout the history of mankind. As early as 400 000 B.C. Homo erectus, a predecessor of Homo sapiens, was using wooden hunting tools that were produced with stone tools such as hand axes, scrapers and knives. In Iron Age Europe, about 800 B.C., tools were being made of iron, which allowed mankind to take an enormous leap forward (Chapt. C.2, p. 392). Tools and the knowledge of the appropriate materials to make them are still key factors driving technical advances.

The following discussions deal with some of the fundamental relationships with respect to wear loading of tools, and the material engineering countermeasures. These principles apply equally to tools for processing minerals as well as those for processing metals, which are discussed in the next chapter.

Iron is predestined for making tools. Fe-base tools are mainly used for processing ores, minerals and rocks as well as for shaping metals, polymers and ceramics. During service, they may be subjected to very complex loading conditions as a result of high multi-axial and cyclic stresses, as well as friction, wear, corrosion and the effects of temperature. In low-temperature applications, they undergo embrittlement below the transition temperature if the hardness is high, whereas in high-temperature applications they are susceptible to high-temperature corrosion and creep as well as thermal fatigue. The surface of the tool is frequently subjected to particularly high loads because it has repeated contact with the material being shaped or it must withstand the loads exerted by mineral particles, sometimes in conjunction with a high pressure and large relative movements. This requires a systematic consideration of a particular tool application as well as an analysis of the active wear mechanisms (Chapt. A.4, p. 106) so that long-lasting materials can be selected.

An increase in the hardness, e.g. by martensitic hardening, is the most important material engineering countermeasure against wear mechanisms frequently encountered with tools, namely abrasion, adhesion and contact fatigue. Furthermore, hard phases in the microstructure have proven successful against abrasion and adhesion. Thus, small and well-dispersed carbides with a fraction of 20 volume-% increase the resistance of a tool edge used for stamping and cutting, whereas gravel chutes are optimally protected by coarse carbides or borides with a fraction of around 50 volume-%. Inversely, coarse carbides would quickly lead to edge breakouts whereas very fine carbides can be easily gouged out by coarse gravel. Thus the selection of the appropriate material requires a case-by-case consideration of the necessary amounts, shape and distribution as well as the type of hard phases and their incorporation into a hard, tough and highly supportive metal matrix, which all act together to determine the wear resistance.

B.4.1.1 Hard phases

The most important properties of a hard phase (HP) with respect to grooving loads exerted, e.g. by a mineral grain, are its hardness (Figure A.4.16, p. 103) and fracture toughness. The HP have to be harder than the abrasive material ($H_{\text{HP}} > H_{\text{AB}}$) and at least as large as the width of the groove, if possible, to act as a obstacles to the scratching abrasive. If the hard phases are too small, they will be chipped out (Figure B.4.1). This lowers their resistance to grooving, unless they are present in large quantities, i.e. close together, thus hampering penetration of the grooving abrasive grain into the softer matrix. Net-like microstructures, as formed in cast alloys with a hypoeutectic composition, allow heavy grooving of the unprotected primary metal cells (Figure B.4.1 c). In addition to a high hardness, the hard phases should also have a high fracture toughness. If $K_{\text{Ic,HP}} > K_{\text{Ic,AB}}$ the mineral grain will tend to fracture rather than the hard phase. Figure B.4.2 a shows the optimum mode of action of hard phases using images of a worn surface and exemplified by monolithic fused tungsten carbide (FTC) that has a microhardness of $H \approx 2200\,\text{HV}0.05$ and a fracture toughness of $K_{\text{Ic}} \approx 6\,\text{MPa}\sqrt{\text{m}}$ and thus fulfils both criteria. Indentation testing devices provide information on further properties, such as Young's modulus, hardness and scratch resistance of the hard phases, even at elevated temperatures. The ratio between the tangential force measured in a single-scratch test under a constant normal force and the respective cross-section of the groove gives the specific scratch energy e_S as a measure of the resistance of a phase to grooving abrasion. Figure B.4.3 shows that the scratch energy of the hard phases is significantly higher than the scratch energy of various Fe matrices at all test temperatures. The hard phases thus also contribute to the wear resistance if they can be grooved by the abrasive. If they are harder than the abrasive, they lower the scratchable matrix path length as their volume fraction increases and thus lower abrasive wear by micro-cutting. However, if the fraction of hard phases is too high, wear increases again due to breakouts (microcracking).

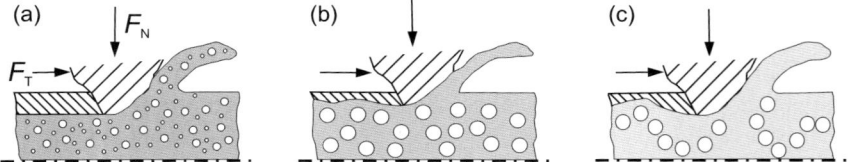

Fig. B.4.1 Abrasion: Schematic representation of wear due to grooving by an abrasive grain, (a) HP too small, (b) HP with an effective size, (c) HP unfavourable (net-like) distribution, F_N = normal force, F_T = tangential force

B.4.1 Loading and material concepts

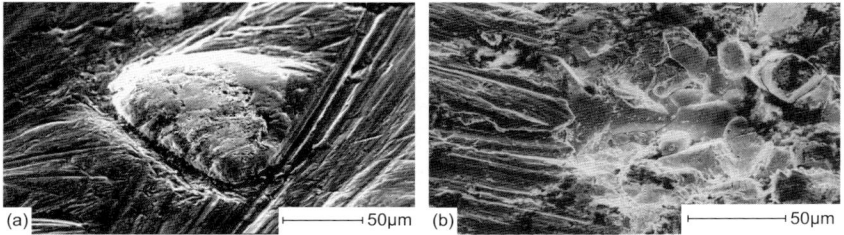

Fig. B.4.2 Mode of action of hard phases during abrasion: (a) Tough monolithic FTC ($H_{HP} \approx 2200\,\text{HV0.05}$) is not grooved by flint ($H_{AB} \approx 1200\,\text{HV0.05}$), (b) brittle, agglomerated WC is not grooved either, it breaks off instead.

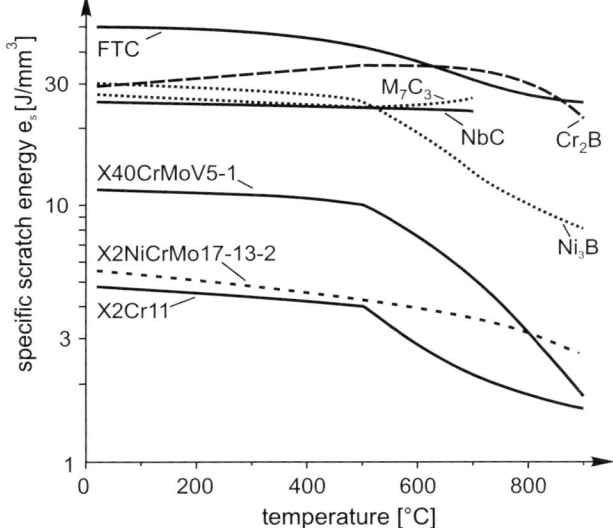

Fig. B.4.3 Specific scratch energy (according to J. Kleff): Temperature-dependence of the specific scratch energy $e_S = F_T/A_v$ (F_T = tangential force, A_v = groove cross-section) in the single-scratch test for selected hard phases and metal matrices (CBN indenter, contact angle $\alpha = 90°$, flank angle $2\theta = 115°$, $F_N = 0.3\,\text{N}$, scratch velocity $v = 2\,\mu\text{m/s}$)

B.4.1.2 Metal matrix

The metal matrix of ferrous alloys containing hard phases is often softer than the abrasive and is thus grooved by it (Figure A.4.16, p. 103). The groove cross-section decreases as the matrix hardness increases so that martensitically hardened matrices help to improve the wear resistance. Furthermore, the matrix has the task of supporting the hard phases and incorporating them in the microstructure. Therefore, the matrix should have not only a high hardness but also a high yield point and a certain degree of ductility. If the yield point is exceeded at room temperature due to mechanical loading, the additional strengthening induced by plastic deformation has a favourable effect on the wear resistance. An indication of this is the higher hardness of a worn tool surface compared to the core. The transformation of retained austenite in hardened metal matrices is particularly effective. It undergoes deformation-induced transformation to martensite inside a thin surface zone. The wear process thus self-generates a hard surface layer on top of a tougher zone that continuously renews itself as wear progresses. In ferrous alloys containing hard phases, this microstructural transformation is associated with a further advantage compared to pure strengthening by dislocations: residual compressive stresses are generated in the surface layer by the transformation of retained austenite to martensite on account of the associated increase in volume (Chapt. A.2, p. 42). These stresses provide greater support to the hard phases and also oppose the propagation of cracks within the metal matrix. A number of tool applications in the processing of metals and minerals take advantage of this effect (p. 257 and Chapt. B.5, p. 291).

As the service temperature increases, the hardness of the metal matrix decreases and recovery as well as recrystallisation processes take place. Up to $\approx 500°C$, the scratch energy of a martensitic metal matrix (X40CrMoV5-1; 1060°C oil 620°C) is about twice as high as that of a ferritic (X2Cr11) or an austenitic (X2CrNiMo17-13-2) matrix (Figure B.4.3). However, at higher temperatures, the scratch energy of a bcc matrix decreases to a greater degree than the fcc matrix, which exhibits a higher scratch resistance above 800°C.

In addition to abrasive action on the surface, tools must also withstand complex multi-axial stress conditions on the inside. This means that the mechanical properties of the tool materials are important, and if the material contains hard phases, they have to be determined as integral properties of the hard-phase/metal-matrix composite. In order to retain their shape for as long as possible, tools usually have to have a high hardness, which is increased by hardening the metal matrix as well as by the volume fraction of hard phases. The bending strength is regarded as a measure of the fracture resistance of tools, and is essentially influenced by the size and distribution of the hard phases (Figure B.4.4). Coarse hard phases lower the bending strength because they break or lose cohesion, even under low loads. In contrast, a dispersion of coarse hard phases leads to a higher fracture toughness than a dispersion of fine phases with the same volume fraction because, in the area of stress concentration in front of the crack tip, fewer particles break or detach and crack deflection increases. Although the transition from a dispersion to a network structure lowers the bending strength, the fracture toughness may increase due to crack deflection (Figure A.2.3, p. 22).

Tools for processing minerals are used to extract, treat and process minerals or mineral-containing materials e. g. in mining and mechanical processing applications, where they are parts of assemblies and machines used for crushing, grinding, classifying, transporting and storing mineral goods. This group of materials includes the previously discussed QT steels, whose developmental strategies are extended by increasing the carbon content with the aim of increasing the amount of hard phases. This has produced materials with a good combined resistance to wear and fracture. The properties can be varied within a broad range from metallic tough to ceramic hard by adjusting the quantity, type, size, shape and distribution of the hard phases. Materials used for toolmaking include rolled steel bars and forgings with < 20 volume-% carbides and castings with < 40 volume-% carbides, as well as products with thick coatings (hard-facing, composite casting and hot-isostatic pressing) containing hard-phase fractions of up to 70 volume-% that are applied to a cost-effective ductile substrate. Further materials include steels for surface hardening, nitriding and case-hardening, and, more rarely, boriding (Chapt. B.3, p. 217, 224 and 234).

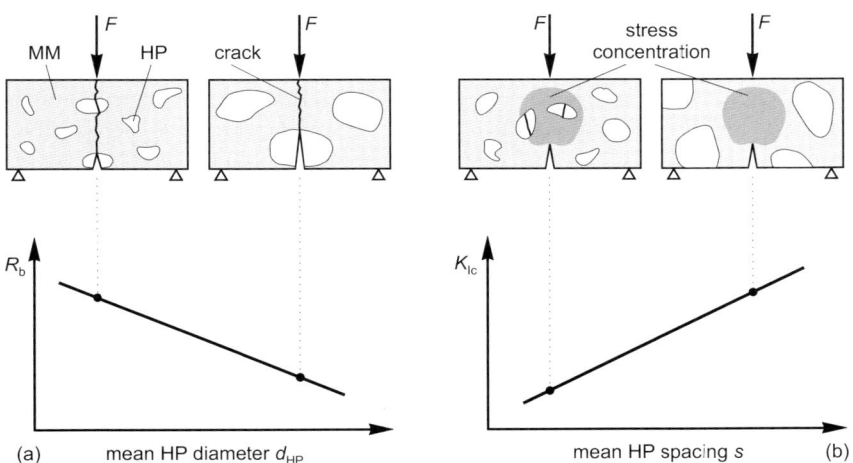

Fig. B.4.4 Fracture strength and fracture toughness: Schematic representation of (a) the bending fracture strength R_b and (b) the fracture toughness K_{Ic} as a function of the size and spacing (constant hard phase content)

B.4.2 Tools made of hot-formed steel

In this group, we find representatives from previously mentioned groups of steels: normalised low-C structural steels, pearlitic steels (e.g. 90Mn4), spring-tempered QT steels (e.g. 50Mn7), surface-hardened steels (e.g. 16MnCr5) and ledeburitic chromium steels (e.g. X210Cr12).

Medium and thick plates of wear-resistant special structural steels are suitable for components to store and transport mineral goods (silos, slides, chutes, screens, loading troughs and shovels). QT steels with a $< 0.4\,\%$ C and containing up to $1.5\,\%$ Cr, $0.5\,\%$ Mo, $1.5\,\%$ Ni and $0.005\,\%$ B have proven suitable in applications and show good welding, forming and machining properties. They are subjected to water quenching and tempering to produce differing levels of hardness between 300 and 600 HB. After laser or plasma cutting, they can be welded using all common procedures and mild preheating. Because the HAZ may have softened close to thermal cuts, rehardening after welding may be necessary. These steels are also suitable as welded-on tips of bucket teeth of mobile hydraulic excavators. These tips can be alternatively made of 50Mn7, which has a higher hardness at the tip compared to the inserted end.

The drill head on a full-face tunnel boring machine (TBM) is equipped with a number of disk cutters. The high pressing force (approx. 16000 KN for a shield diameter of 10 m) is transferred to the specially designed disk cutters that break and remove the rock as they pass over the surface by exceeding the withstandable local contact pressure, thus fracturing it. Owing to this specific load, hard but nevertheless tough materials with a high strength are required for the rings of the cutting disks. In the standard design, they are made of martensitic, low-carbide steel X50CrMoV5-1 and are tempered to a hardness of approx. 600 HV (tensile strength $R_\mathrm{m} \approx 2000$ MPa). Despite their high alloy content, rings with an external diameter of > 400 mm and piece weights of up to 25 kg can be die-forged. Alternatively, cast rings of X155CrVMo12-1 are heat-treated so that their hardness decreases from the cutting edge to the core.

Moulding frames used to produce building blocks and refractory bricks as well as cement slabs and grinding bodies are made of nitrided, case-hardened or borided steels, depending on their susceptibility to distortion. During boriding at 800 to 1050°C, boron diffuses out of a powder or a paste into the surface of the steel to precipitate an iron boride of type Fe_2B with a microhardness of ≈ 1800 HV0.05. Whereas the layer thickness in unalloyed and low-alloy steels can be as much as 200 µm, it may be only ≈ 20 µm in highly alloyed steels because the diffusion of boron is limited. In this case, it is possible that less boron diffuses inwards than is available at the surface so that the monoboride FeB may grow on the surface. Because these two borides have very different coefficients of thermal expansion, they cause considerable thermal stresses as the material cools and may even result in spalling of the outer FeB layer. An effective way of countering this is to allow post-diffusion without external boron so that the boron in the surface layer can diffuse towards the core and

FeB transforms into Fe$_2$B. During plasma-assisted boriding, the steel surface is first covered with a thin Ni layer (PVD) and then exposed to a boron-containing gas (trimethylborate, B(OCH$_3$)$_3$). Boron initially forms a M$_2$B layer with the nickel, underneath which nickel diffuses inwards and iron outwards. This means that the Ni$_2$B at the boundary to the steel is increasingly converted into the desirable and harder Fe$_2$B. The external Ni/Ni$_2$B layer prevents diffusion of oxygen to the iron and thus its oxidation. The boride layer is kept free of pores and the boriding time of 16 hours at 1000°C for a layer thickness of 160 µm is relatively short.

Plates (20 mm thick) for processing lime-sand bricks are produced cost-effectively from a case-hardenable steel with special case-hardening treatment. The starting point is a steel alloyed with a little boron as well as additional chromium (16MnCrB5) to increase the hardenability. This steel is case-hardened so that the surface layer is deliberately supercarburised. Injection of methanol/acetone into the furnace produces a carbon content of up to 3 % C in the surface layer. Carburising for up to ten hours at the usual temperature (940°C - 980°C) produces a surface layer with fine and densely precipitated Fe$_3$C carbides to a depth of about 0.8 mm (Figure B.4.5 a). After reheat hardening, they are embedded in a martensitic matrix with about 30 volume-% retained austenite. A conventionally case-hardened microstructure with a decreasing C content and a hardness penetration of CHD > 2 mm is formed below this zone. The as-quenched hardness of 66 HRC is reduced to 62 - 64 HRC by tempering at 180°C. Any retained austenite that has not yet transformed has a positive effect not only on the straightening step – which is still necessary in spite of hardening by press die quenching – but also during service due to its stress-induced transformation to martensite. Such plates are also used in vertical mould-making machines in which moulding sand is compressed to make a casting mould with narrow tolerances without using a flask. Ledeburitic chromium steels such as X210Cr12 are also used for press-forming. After slight overheating during hardening, they also contain elevated amounts of retained austenite that increase the resistance to grooving abrasion (Figure B.5.4, p. 278).

Depending on the operating conditions, briquetting and compacting rollers for potash, fertilisers, gypsum, dusts and fine-grain ores have tyres made of case-hardened high-strength QT steel (56NiCrMoV7) or, in corrosive environments, they are made of stainless martensitic chromium steels (X39CrMo17). The latter steels are sometimes given additional surface-hardening up to a depth of 30 mm on computer-controlled flame-hardening units, followed by tempering to a surface hardness of 52 - 54 HRC. Briquetting of blast-furnace dusts and fine-grain ores requires a more wear-resistant surface with hard phases. Forged rings of ledeburitic cold-work tool steel X153CrMoV12, surface-hardened to 60 HRC are suitable for this.

Grinding rods for rod mills are often made of steel bars of naturally hard 100Cr6. The hammers in hammer mills are made of e.g. flat steel with a limited hardness penetration. They can be made of steel 105Cr4 with dimensions

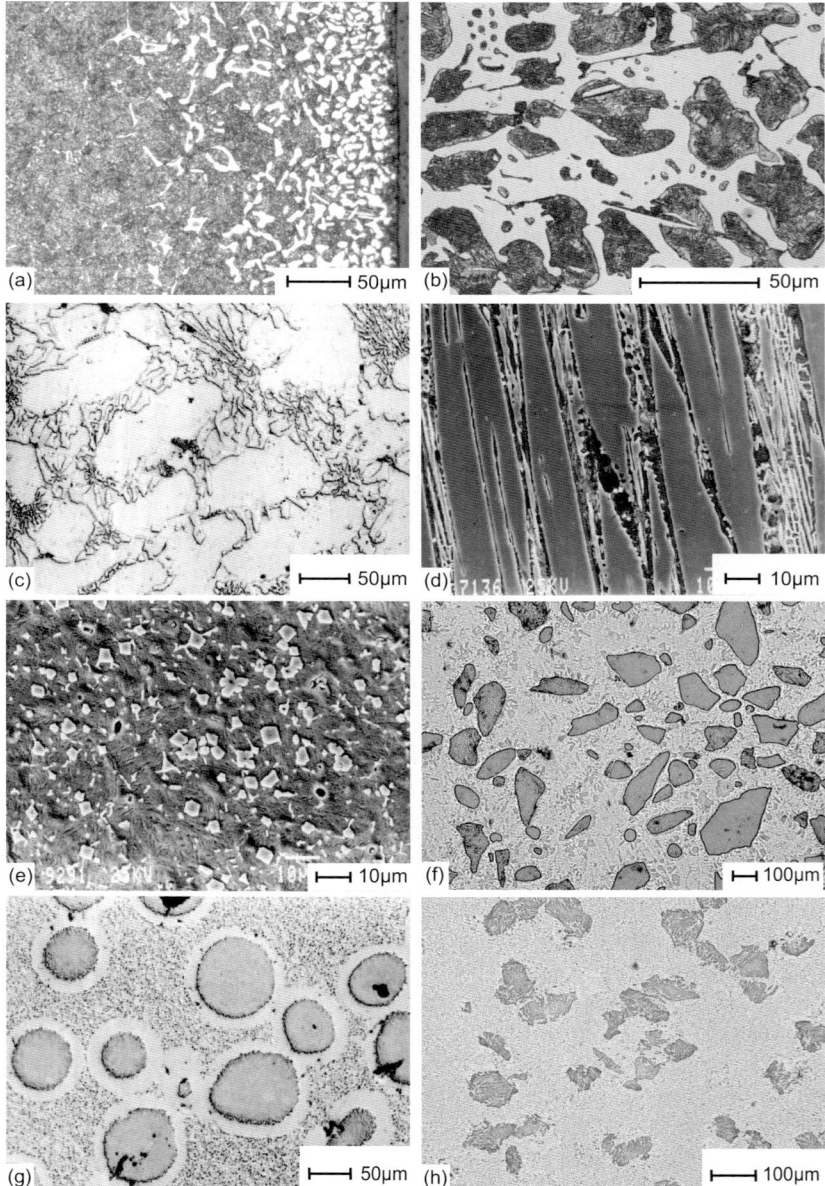

Fig. B.4.5 Microstructures of wear-resistant ferrous materials:
(a) 16MnCr5 supercarburised surface layer: hard phase M_3C; (b) pearlitic white cast iron: hard phase M_3C; (c) white chromium cast iron GX-300CrMo15-3; hard phase M_7C_3; (d) welded 500Cr27; hard phase M_7C_3; (e) welded 120NbCrNiMoV6-5-2-1: hard phase MC; (f) welded 50NiCrMo7 with WC/W_2C and $(W,Fe)_6C$; (g) HIP-MMC of X230CrVMo13-4 and WC/W_2C; (h) sintered MMC of X250CrNiMoV12-2-2 with WC/W_2C

of e.g. 350 x 120 x 40 mm and are hardened in water to produce a hard shell with a depth of approx. 10 mm. Initially, the entire hammer is immersed and shortly afterwards slowly pulled out of the water starting with the end that is to be mounted on the rotor. Owing to the residual heat, the mounting end experiences a higher tempering temperature than the hammering end. This single heat-treatment step produces a hard working end (64 HRC) and a tough mounting end. Less hard hammers can be made of 50Mn7 and partially hardened at the hammering end. Hadfield manganese steel X120Mn12 is also used, particularly if the hammer has tenons. This group of steels has an austenitic microstructure after solution-annealing and quenching. The dislocations with stacking faults produced by impacts and collisions along with tiny amounts of finely dispersed martensite can lead to cold-work hardening of this surface layer to 600 HV. This produces a hard, self-regenerating case on a tough core.

B.4.3 Cast tools

Some of the previously discussed wear-resistant steels are also used for cast tools. These include high-strength hammers in hammer mills (e.g. G55NiCrMoV6), grinding balls (e.g. G100Cr6) as well as small crushing and compacting tools made of ledeburitic cold-work tool steels (e.g. GX210Cr12) and cast Hadfield manganese steel (e.g. GX140MnCr17-2). It is used in jaw and cone crushers for coarse crushing of ores and natural rock as well as in the production of gravel and grit. Because Hadfield manganese steel is only weakly magnetisable, any material that wears off cannot be removed from the milled material by magnetic separators. Certain applications in the production of porcelain, glass or grinding media thus use a low-alloy QT casting steel with approx. 0.7 % carbon and strengths of 1000 to 1800 MPa. White cast irons are the workhorses within the group of wear-resistant casting alloys (Chapt. A.2, p.33). They are members of the Fe-C or Fe-Cr-C systems and contain up to 30 % Cr and 4 % C, and generally solidify near-eutectically. A silicon content $< 2\,\%$ in combination with manganese and chromium ensures that white solidification also occurs in large cross-sections. Primary solidification of the austenitic metal cell is followed by precipitation of the eutectic mixture consisting of a eutectic hard phase (M_3C or M_7C_3) and a metal matrix. Depending on the levels of alloying elements and the cooling rate, the metal matrix is pearlitic, martensitic/bainitic, austenitic or a mixture of these. White cast irons are classified according to their microstructure into 'pearlitic white cast iron', 'martensitic nickel white cast iron' and martensitic 'chromium white cast iron'

B.4.3.1 Pearlitic white cast iron

White cast iron is an unalloyed or low-alloy cast iron whose microstructure consists of an M_3C eutectic in a pearlitic matrix (Figure B.4.5 b).

White cast iron is now designated GJN-350 HV according to EN 12513 (Chapt. C.1, p. 388). The actual macrohardness may lie well above this value if the carbon content is at the upper limit (3.6 % C) and the volume fraction of M_3C from the melt is high (35 %, Tab. A.1.3, p. 13). An elevated manganese content, alloying with chromium and a high cooling rate produce a finely striped pearlite with a solid-solution-hardened ferrite. This microstructure provides good support for the eutectic cementite. Casting in metal moulds instead of sand produces a much finer solidification microstructure and also a macrohardness of up to 500 HV.

Pearlitic white cast iron can be produced as full-mould or chill castings. By casting against a metal plate inserted into the sand mould, the Fe-C melt with a specially tailored silicon content undergoes white solidification due to the rapid removal of heat (Chapt. A.2, p. 37). The white surface zone gradually changes towards the core to a grey/white mixed zone and then to grey solidification, which bestows good machinability on the remainder of the casting.

Whereas pearlitic, fully white cast iron is used e.g. for segmented chain rollers of bucket ladder excavators and bucket elevators as well as agricultural machinery, chill castings are used as a cost-effective material in grinding applications, e.g. as mill linings or grinding tools. Hammers for coal mills in power stations and cement works are made with a carbidic head and a soft, grey-solidified mounting end.

B.4.3.2 Martensitic nickel white cast iron

The wear resistance of pearlitic white cast iron can be significantly improved by a martensitic metal matrix. Martensitic nickel white cast irons (ASTM A532 Class I) thus contain not only iron, carbon and chromium, but also significant amounts of nickel in order to extend the austenite field, delay pearlite transformation and lower the transition temperature for cleavage fracture. Because nickel does not participate in forming carbides, it can exert its full effect on the matrix. To achieve a metal matrix consisting of martensite and retained austenite directly after cooling in the mould requires 3.5 to 5 % Ni, whereby the amount of Ni increases with the cross-section. The resulting tendency to form undesirable graphite is countered by adding 1.5 to 2.5 % Cr. The contents of Si and Mn (each up to 0.8 %) are lower than in pearlitic white cast iron on account of graphitisation on the one hand and stabilisation of retained austenite on the other. The M_3C carbides of the ledeburitic-martensitic cast iron are increased in hardness (up to 1100 HV0.05 compared to 850 HV0.05 for Fe_3C) by dissolution of Cr and their amount increases with the C content, which lies between 2.5 and 3.5 %.

Tempering at 275°C causes the retained austenite to decompose to lower bainite and the hardness of the as-cast condition of e.g. 550 HV30 increases by approx. 100 HV30. Compared to unalloyed white cast iron, this results in an improvement of the supportive effect of the matrix for the carbides and an increase in the wear resistance. Casting in a metal mould rather than in sand produces a finer microstructure and a further hardness increase of $\approx 50\,\text{HV}30$

As early as 1920, two Ni- and Cr-alloyed cast irons were developed that were marketed under the trade names Ni-Hard 1 and 2 (Tab. B.4.1). Ni-Hard 1 was a GX330NiCr4-2 according to the former DIN designation, and is now standardised in EN 12513 as GJN-HV 550, which specifies the minimum hardness; Ni-Hard 2 (GX260NiCr4-2) is now GJN-HV 520.

Ni-alloyed martensitic cast irons cover a broad range of applications. They are used as linings for ball and hammer mills and as crusher jaws and rollers as well as wear plates in mixing and conveying equipment. Because they are less wear-resistant than chromium cast iron, their applications are limited to handling moderately hard materials such as coal, lime and phosphate.

B.4.3.3 Martensitic chromium white cast iron

By increasing the Cr content, the harder M_7C_3 carbides (1300 - 1500 HV0.05) can be precipitated instead of the somewhat softer M_3C (Figure A.4.16, p. 103). At $\approx 5.5\,\%$ Ni and $\approx 2\,\%$ Si, only $8.5\,\%$ Cr is sufficient to achieve this and exclude the formation of graphite. Lowering of the Ac_1 temperature and suppression of pearlite formation by adding nickel allow low-stress air-hardening from a low hardening temperature ($\approx 750°C$) that can produce a macrohardness of $> 700\,\text{HV}30$. A Ni-alloyed cast iron with $9\,\%$ chromium and $5\,\%$ nickel is marketed as Ni-Hard 4 (ASTM A532 Class I) and belongs to the family of nickel-chromium cast irons. The European standard specifies that the minimum hardness and the approximate chemical composition must be given in the designation for this high-alloy cast iron. Alloy GX300CrNiSi9-5-2 is thus unfortunately only incompletely renamed to GJN-HV 600(XCr11) (Tab. B.4.1).

Table B.4.1 Cast tooling materials for processing minerals: Microstructure, recommended values for the as-installed hardness, type of treatment

Alloy	Name according to EN 12513	Hard phase type	Hard phase proportion [volume-%]	Metal matrix	Installed condition	Service hardness [HRC] from - to	Example application
G55NiCrMoV6		–	–	M	Q + T	55 - 56	impact hammer
GX210Cr12		M_7C_3	17	M	Q + T	55 - 65	compacting tool
GX140MnCr17-2		–	–	A	AT	< 20	crusher tools
GX300MnCr	GJN-HV350	M_3C	28	P	U	35 - 50	mill lining
GX260NiCr4-2	GJN-HV520	M_3C	32	M	(Q) + T	48 - 56	=
GX330NiCr4-2	GJN-HV550	M_3C	42	M	(Q) + T	52 - 62	=
GX300CrNiSi9-5-2	GJN-HV600	M_7C_3	32	M+(RA)	(Q) + T	59 - 63	grinding tool
GX300CrMo15-3	GJN-HV600(XCr14)	M_7C_3	33	M+(RA)	Q + T	60 - 65	=
GX260CrMoNi20-2-1	GJN-HV600(XCr18)	M_7C_3	28	M+(RA)	Q + T	62 - 65	=
GX260CrMo27-2	GJN-HV600(XCr23)	M_7C_3	35	M+(RA)	Q + T	62 - 65	blow bars
GX130NbCrMoW-6-4-2-2	no standard	MC	13	M	Q + T	58 - 62	briquetting tools

U = untreated, AT = solution-annealed, Q = quenched, T = tempered, M = martensite, P = pearlite
A = austenite, RA = retained austenite

Chromium decreases the size of the austenite field and thus increases the hardening temperature. Because the majority of the chromium is forming M_7C_3 carbides, the hardenability must be raised for thicker cross-sections by adding 1 to 3 % of other alloying elements such as Cu, Mn, Mo, Ni. Of these, Mn and Mo partially dissolve in the carbide, whereas Cu and Ni remain fully effective in the metal matrix. Although Cu is cheaper than Ni, it is not wanted in scrap metal. With respect to the desirable increase in the hardenability and the undesirable stabilisation of retained austenite, Mo is more favourable than Mn, but it leads to higher alloying costs. In unalloyed white cast iron, the eutectic point lies at $\approx 4.2\,\%$ C; however, it drops to $<3\,\%$ if the chromium content is increased to the usual upper limit of 28 %. Figure B.4.5 c shows the microstructure of cast alloy GX300CrMo15-3 (GJN-HV 600 (XCr14)).

An increase in the C content above 3.3 % leads to the precipitation of coarse columnar primary M_7C_3 carbides that offer high resistance to coarse grooving; however, they also have an adverse effect on the toughness. An increasing Cr/C ratio increases the Cr content in the carbides and in the matrix so that the carbide hardness and the hardenability both increase. This means that the molybdenum content can be reduced from 3 to 1 %. Despite the high Cr content, high-chromium cast irons are not usually resistant to wet corrosion because too little Cr remains dissolved in the metal matrix on account of the large fraction of carbides (Chapt. B.6, p. 344).

After cooling in the mould, the matrix exhibits a mixed microstructure whose proportion of retained austenite increases with the alloying level. High-chromium white cast irons (ASTM A532 Class II and III) are thus given their service hardness by hardening and tempering. The hardening temperature increases with the Cr content from 950 to 1100°C. Depending on the alloying content, it dissolves the carbides or destabilises the retained austenite by precipitation of carbides. A hardness of between 750 and 900 HV30 is obtained after tempering at about 200°C. Tempering in the secondary hardening range is associated not only with a lower hardness (650 to 750 HV30), but also with the relief of residual stresses and is advantageous under wear conditions at elevated temperature.

Special cast irons have been developed for applications in which the hardness of the abrasive material exceeds the hardness of M_7C_3. They contain monocarbides such as VC (2900 HV0.05) or NbC (2400 HV0.05) as the main hard phase, and are produced by primary precipitation from the melt. Deoxidation with aluminium and late addition of a few tenths of a percent of titanium initially produce fine Al_2O_3 precipitates in the melt that act as nuclei for the precipitation of titanium carbonitrides. They in turn act as nuclei for the growth of MC carbides. This not only increases the hardness of the carbides, but also leads to their more homogeneous distribution, thus resulting in a greater wear resistance. Tab. B.4.1 includes one of these alloying variants with Nb-based monocarbides (GX130NbCrMoW6-4-2-2).

Chromium white cast irons are used in the same types of applications as martensitic nickel white cast iron; however, they generally exhibit a comparatively high wear resistance, particularly against harder, quartz-containing rock. An example of the fracture strength of such castings are the blow bars of impact mills used to crush basalt and diabase. They are attached to a rotor turning at e.g. 600 rpm and hit lumps exiting from a coarse crusher with a mesh size of e.g. 300 mm. The lumps are flung against an impact plate and they break into gravel. After this process, the material still contains a high proportion of 'cubic' material, as is required for railway tracks, in particular. The blow bars with a piece weight of more than 100 kg are made of near-eutectic cast iron, e.g. GX260CrMo27-2 by casting in a sand mould. They are then hardened in moving air and tempered at $\approx 180°C$ to achieve a hardness of 800 to 900 HV30. In this condition, their fracture strength is greater than that of the rock. However, prerequisites for safe operation is a planar seating on the rotor, the avoidance of notched contours and a feed material without any coarse pieces of metal. Chromium white cast irons are also used in many grinding applications, e.g. as grinding balls and linings in ball mills, as grinding tables and rollers in vertical mills and as roller segments in high-pressure roller mills that are used to grind cement and ore.

B.4.4 Coated tools

The thin CVD or PVD films or hard chromium coatings commonly used for metal working tools are rarely used for applications in which abrasive minerals are involved because greater wear depths are generally allowed and the point loads are often higher. Millimetre-thick coatings of materials with coarse hard phases are used instead and their hard-phase fraction can significantly exceed the technical casting limit of approx. 40 volume-%. These layers are applied by hard-facing, composite casting or by powder metallurgy, usually in combination with hot-isostatic pressing (HIP).

B.4.4.1 Hard-facing

Various hard-facing procedures can be used to apply 2 - 15 mm-thick wear-protection layers by multi-layer welding on a cost-effective substrate. The range of coating materials extends from simple martensitic steels with 0.3 - 0.5% C to hypereutectic alloys containing superhard phases and up to 6 % C, 2 % B, 40 % Cr and further metallic elements (Nb, V, W, Mo, Co). Examples of various types of materials are listed in Tab. B.4.2. Macrohardnesses of 70 HRC and more are obtained by precipitation of M_7C_3 carbides solidified from the melt with proportions of up to 50 volume-% in the third welded layer, which is hardly affected by dilution. Figure B.4.5 d shows the columnar M_7C_3 carbides growing in the direction of the temperature gradient in one of these materials. The wear resistance of such an alloy may even exceed that of white cast iron (Figure B.4.6).

Table B.4.2 Wear-resistant alloys for hard-facing

Composition	Hard phase type	proportion [volume-%]	Metal matrix 2)	Hardness [HRC]	Example applications
X20CrMo5	–	–	M	42 - 47	rollers
X50CrMoV5-1	–	–	M	54 - 57	shredder hammers
X120NbCrNiMoV6-5-2	MC	13	M+RA	55 - 59	comminution rollers
X180CrTiMo7-5-2	MC	19	M	56 - 58	mill hammers
X500Cr27	M_7C_3	55	M	59 - 62	chutes, screens
X540CrNb22-7 + (B)	M_7C_3 MC	50 15	M+RA	63 - 65	excavator bucket teeth
X440CrMoNbWV24-7-6-2	M_7C_3	42	M+RA	62 - 65	grate bars hot screens
X380VCrWMo17-13	MC	30	M	61 - 64	extruder screws
50NiCrMo7 + WSC	FTC[1]	60 60	M	63 - 66	overburden removal equipment

[1] FTC = fused tungsten carbide WC/W_2C
[2] M = martensite, RA = retained austenite

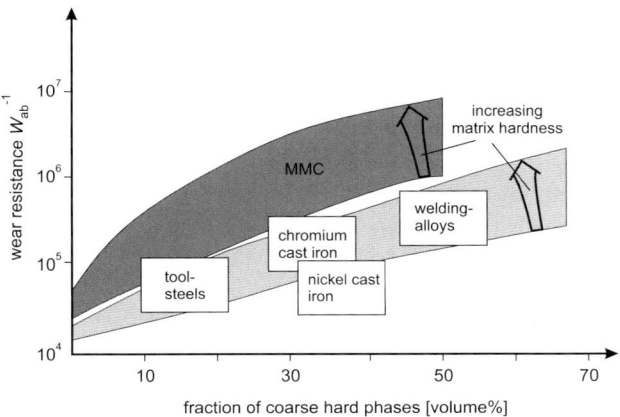

Fig. B.4.6 **Wear resistance and proportion of hard phases**: In the pin-on-disk test using abrasive paper with flint (80 mesh), which is softer than the hard phases in the sample, the wear resistance W_{ab}^{-1} increases with the proportion of hard phases. Tool steels, martensitic chromium white cast iron and welding alloys generally contain M_7C_3 carbide and lie within the same scatter band. Despite their high proportions of hard phases, martensitic nickel white cast irons lie underneath the band because they contain softer M_3C. MMC with 30-50 volume-% of the WC/W_2C and CrB_2 hard phases (Figure A.4.16, p. 103) have outstanding wear resistance.

Owing to the high proportion of hard phases, the layers are always pre-cracked and are thus not suitable for high mechanical loads, particularly impacting types. However, they can be used for purely abrasive loading conditions (e.g. as coated sheets for slides, chutes, screens, bins). Coated wear plates are commercially available that are produced by the plasma-transferred arc (PTA) process, which produces a low degree of dilution.

Worm screws are often protected by weld deposits. Whereas the shaft and the large blade surface of conveyor worms have multiple coatings, hard-facing of pressure screws can be used to shape the helix. Extruder and injection moulding screws in plastics processing equipment subjected to a particularly high degree of wear by the abrasive fillers (e.g. quartz powder, glass fibres). Their blades are thus built up of 3 PTA layers of high-V alloys with inductive preheating (Tab. B.4.2). In screw presses used to extract plant oils and to process animal carcasses, the screw blades, with sizes of up to 50 mm, are built up by more than 10 welded layers of slightly hypereutectic Fe-Cr-C alloys.

Whereas low-alloy steel substrates are welded after preheating to guarantee good adhesion of the layer, a different route is used to regenerate wear parts of white cast iron. The worn grinding plates and stones of vertical mills, which are not really suitable for welding, are automatically regenerated by multiple hard-facing layers using a flux-cored wire of e.g. X500Cr27 (Figure B.4.5 d). The welding parameters are selected so that the welding site remains relatively cold. This induces early formation of regularly spaced vertical cracks to the surface in this layer (segmentation cracks) and usually ensures good bonding to the substrate between the cracks and prevents spalling of the welded layer, even under high grinding pressures, and results in a long service life. Grinding rollers of high-pressure roller mills used to grind cement clinker are subjected to such high mechanical loads that the three layers of hard-facing must remain crack-free. Although carbide-free alloys that undergo a martensitic transform during cooling (e.g. X50CrMoV5-1), are used for hard-facing owing to the danger of spalling, their service lives are short, e.g. 3 months. Alloys containing < 20 volume-% monocarbides of the elements niobium, vanadium and titanium in a martensitic metal matrix have proven successful in such cases (Figure B.4.5 e). The MC carbides have the advantage of a high hardness (2000 - 3000 HV0.05) and solidify from the melt before the metal dendrites so that they are dispersed within the microstructure. A nickel content of about 2 % in 120NbCrNiMoV6-5-2-1 and natural air cooling produces retained austenite whose plastic deformability relieves thermal stresses and thus decreases the susceptibility to cracking. Tempering of Mo- and V-alloyed variants at about 500°C increases the hardness as a result of the transformation of retained austenite and precipitation of secondary carbides with simultaneous stress-relief of the entire workpiece.

Excavator buckets and overburden removal equipment in open-cast mining of lignite coal and oil shale are protected by layers of hard-facing to whose molten pool special hard particles are added. Tungsten carbides (WC or WC/W_2C) are generally used for Ni and Fe matrices. They are added to the plasma beam or to the molten pool during PTA welding in proportions of up to 60 volume-%. In Fe matrices, it is possible that some of the carbides melt or start to melt and are then reprecipitated as a brittle mixed-metal Fe-W carbide $(W, Fe)_6C$ (Figure B.4.5 f).

B.4.4.2 Powder metallurgical coatings

The production of thick layers using powder metallurgy offers the greatest flexibility in the design of wear-resistant layers and they have proven exceptionally successful in service. In the so-called HIP cladding process, which is based on hot-isostatic pressing, a powder mixture of a hard phase and a steel are compacted at e.g. $T = 1150°C$ and $p = 1000$ bar to the theoretical density and simultaneously applied to a steel substrate. Metal-matrix composites (MMCs) produced by powder metallurgy offer some decisive advantages compared to parts produced from melts. Whereas the microstructure of hard alloys from melts depends on the solidification sequence corresponding to their chemical composition, the type of hard phase and their size are freely selectable using PM. Furthermore, thick and crack-free PM coatings can be obtained with higher proportions of hard phases, and the microstructural homogeneity of these coatings is better than that obtained by hard-facing or thermal spraying, which is rarely used for ferrous materials.

The microstructure of the MMC is governed by the processes occurring during hot-compaction, namely interdiffusion of elements, as well as by the grain-size ratio of the powered components (Figure B.4.5 g). As a consequence of approaching a thermodynamic equilibrium, a diffusive shell develops in the steel matrix around the original hard phases during hot compaction, e.g. for fused tungsten carbide (WC/W_2C). This shell consists of W_2C as well as M_6C towards the outer regions and has a positive effect on embedding the hard phase in the steel matrix because it lessens sudden changes in the properties (E, HV, α) at the hard-phase/matrix interface. The distribution of hard phases can be influenced via the grain-size ratio and the volume fractions of hard phase and matrix powders. Hard-phase fractions of < 30 volume-% can form brittle networks if the hard-phase powder is significantly finer than the matrix powder. In contrast, a respective hard-phase fraction with a narrow range of grain sizes similar to or larger than that of the matrix powder produces a tough material with a dispersed hard phase.

The MMC layers, which can have almost any thickness, exhibit excellent wear resistance. For example, in the pin-on-disk test against flint, a MMC from powdered X230CrVMo13-4 and a hard-phase fraction of 30 volume-% WC/W_2C (150 μm) exhibited a wear resistance that was greater by a factor of 40 than that of the pure steel matrix and greater by a factor of 6 than martensitic chromium cast iron GX300CrMo15-3 hardened to 800 HV30 (Figure B.4.7). This is due not only to the high hardness of the tungsten carbides but also to their high fracture toughness. They have proven very resistant, even against Al_2O_3, and give the composite a significantly greater wear resistance compared to a hard-facing.

In addition to the possibility of producing thick, crack-free layers, a further advantage of MMC over a weld deposit is its greater toughness and fracture resistance because the matrix, unlike hypereutectic hard-facing alloys, is not embrittled by an unfavourable arrangement of eutectic hard phases.

In spite of their high production costs, PM coatings are capturing the market in mineral-processing equipment. The first applications of MMC as protective layers on roller surfaces in high-pressure roller mills used for grinding of kimberlite and kaolin exhibited significantly longer service lives than

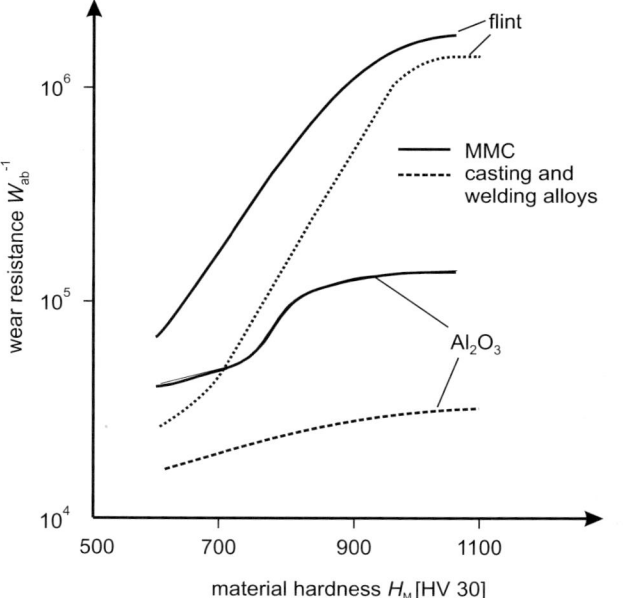

Fig. B.4.7 Wear resistance and hardness: As demonstrated by the wear test, the wear resistance W_{ab}^{-1} increases with the material hardness H_M. It increases with the hardness of the metal matrix (H_{MM}) and with the volume fraction of hard phases. If these phases are softer than the abrading material ($H_{HP} < H_{AB}$), the increase is small, as shown by the curve for casting and welding alloys abraded with Al_2O_3. Against flint, $H_{HP} > H_{AB}$, so that W_{ab}^{-1} lies in the upper shelf region (2 orders of magnitude higher). The latter also applies to MMC, whereby the lower shelf is only reached at lower hardnesses owing to the coarse hard phases ($> 100\,\mu m$). The special potential of MMCs is particularly noticeable against Al_2O_3: in this case, only 30 volume-% of WC/W_2C ($H_{HP} > H_{AB}$) in a martensitic matrix is sufficient to increase W_{ab}^{-1} compared to welding and casting alloys that contain mainly M_7C_3 ($H_{HP} < H_{AB}$).

cast materials or hard-facings. In the meantime, there have been a number of 10-fold increases in the service life of multi-component roller surfaces used to grind cement. In analogy to a dispersion of hard materials in a metal matrix at the microstructural level, hexagonal plates (approx. 35 mm wide, approx. 20 mm thick with 5 mm gaps) have been attached to the surface of the rollers. The material in the gaps wears away more quickly because it is softer than the hexagons. The material being milled is pressed into these depressions where it provides autogenous wear protection. At the same time, this means that part of the roller surface is rough so that the pull-in and consequently the throughput noticeably improve.

The forerunner in HIP cladding technology is the plastics processing industry. The working surfaces of screws and housings of extruders and injection moulding equipment have been protected by PM layers for a long time. Here too, various high-alloy tool steel powders are used, sometimes with added hard-phase powders, depending on the application.

In spite of the excellent service life of the MMC, HIP cladding is not always the first choice for reasons of cost: the high investment costs of the HIP equipment, the complicated welding technology required to produce pressure-tight weld seams on metal capsules as well as the costs for their subsequent removal by machining. An answer to this problem is the newly developed process to densely sinter Fe-based MMCs with the assistance of a liquid phase. The starting point are gas-atomised cold-work and high-speed steel powders whose low-melting eutectics produce a pore-free microstructure during vacuum or protective-gas sintering just above the solidus temperature. Sintering under nitrogen leads to the uptake of nitrogen, particularly by high-V-containing powders that bind it in VC carbides, which are frequently sub-stoichiometric. These release carbon to the matrix so that the solidus temperature and the associated optimum sintering temperature is lowered by $\approx 20°C$ compared to vacuum sintering.

Whereas thermodynamically stable hard phases, such as TiC, can be sintered into the material without problems, the liquid phase may cause violent diffusion reactions between WC/W_2C and the matrix powder. This can be countered by controlling the temperature so that the liquid phase only just starts to form and the sinter temperature is decreased well below T_{sol}. Figure B.4.5 h shows the microstructure of a MMC containing 10 volume-% WC/W_2C in a cold-work steel matrix of X250CrNiMoV12-2-2 after liquid-phase sintering. This method can also be used to coat a steel substrate in which the interface has a quality similar to that produced by HIP. The 2 % nickel in the above-mentioned steel matrix means that even thick-walled layer composites can be hardened in static air.

B.4.4.3 Composite casting

Thick wear-protection layers can also be produced by composite casting. For hammers used in hammer mills, a white cast iron, e.g. GX300CrMo15-3, is poured onto a cast steel substrate that has just solidified using a relatively complicated system of runners and a contact agent, which provides local lowering of the melting point. Rotationally symmetrical composite castings can be manufactured more simply by spin casting. A shell (with a white cast iron or a cold-work steel composition) is first cast into a horizontally or vertically rotating mould where it solidifies from the outside inwards. Just before the shell has solidified completely, a core material of spheroidal graphite cast iron with a high degree of saturation is cast (low casting temperature, see Eq. A.2.2, p. 34). The bonding between the two alloys, which are still in a semi-liquid state, produces a high interfacial strength that can withstand the loads applied to comminution and metal-working rollers.

The insides of cylinders for extruders and injection-moulding machines are coated by spin-casting with a wear-resistant layer containing additional hard phases. An alloyed steel powder with added hard phases is filled into a hollow, drilled cylinder of QT steel held horizontally. It is then rotated and induction-heated to $\approx 1200°C$. The use of a steel powder with more than 3 % C and 3.5 % B produces low-melting eutectics that reduce the solidus temperature of the alloy below 1200°C. The steel powder melts and is pressed by centrifugal force onto the insides of the cylinder where it solidifies when the inductor is switched off. The addition of spheroidal fused tungsten carbide (up to 50 volume-%) produces naturally hard coatings that are sufficiently wear-resistant, even without a subsequent heat treatment, owing to their high proportion of hard phases.

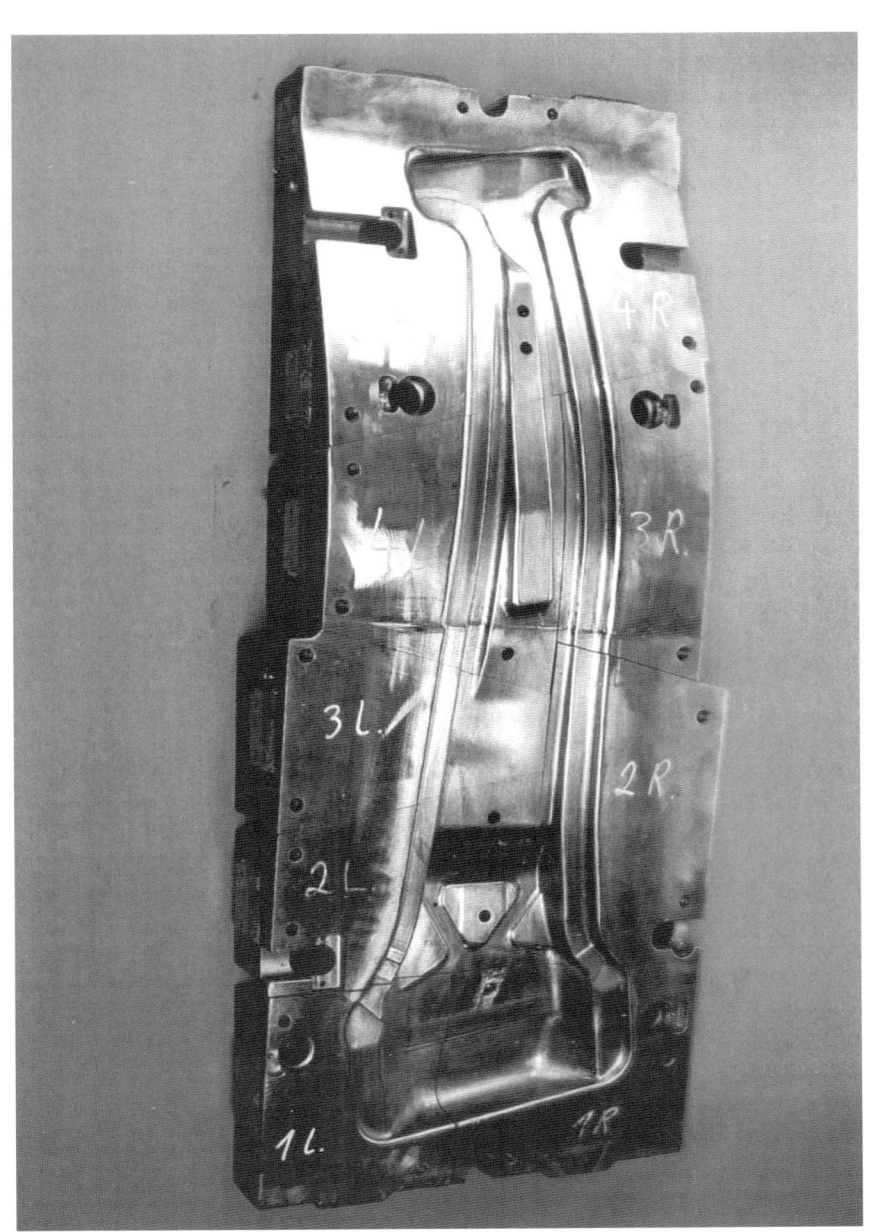

Segmented deep-drawing tool of ledeburitic cold-work tool steel
(Deutsche Edelstahlwerke GmbH, Witten, Germany)

B.5 Tools for processing materials

Tools for processing materials are the key elements in hot and cold forming processes i.e. from the melt (e.g. pressure die casting and powder metallurgy), by hot working (e.g. forging and extrusion) and separation (e.g shear cutting and chipping). The shape of the tools is preserved by using different hardnesses that depend on the type of application. In practice, this leads to the following reference values for the service hardness: pure polymer processing 30 - 35 HRC, metal processing 40 - 50 HRC (hot) and 55 - 65 HRC (cold). Greater hardnesses or hard layers may be required if the wear behaviour is of particular importance.

Tools for processing materials are generally made of tool steels, whose properties can be appropriately tailored by the production method, alloying and heat treatment. In a few cases (usually large tools), grey cast iron is used instead. In addition to conventional production by ingot or continuous casting, the cleanness and the degree of segregation can be improved by remelting processes, such as electroslag remelting (ESR, Figure A.2.16, p. 38) or vacuum remelting. Vertical or horizontal forging can be used to influence the orientation of core segregation in disk-shaped tools. Production by powder metallurgy or spray-forming refines the eutectic carbides and prevents macro-segregation. Low final temperatures during hot forming and accelerated cooling prevent the formation of secondary carbide networks. Soft-annealing leads to good machinability during tool manufacturing (Chap. A.3, p. 60). The service hardness

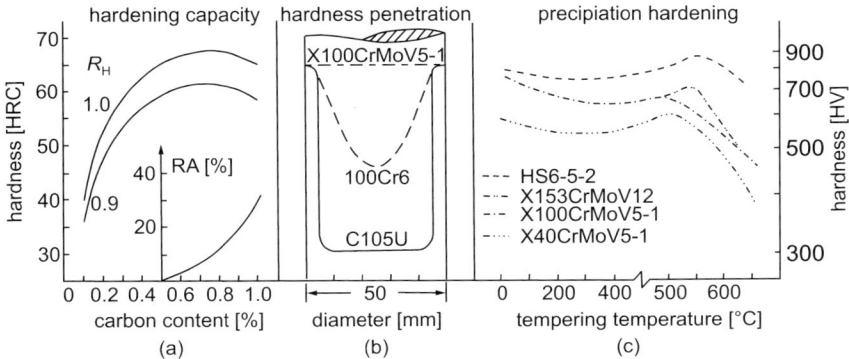

Fig. B.5.1 Hardening of tool steels: (a) Increase in hardness of an unalloyed steel due to dissolved carbon, and lowering of the hardness due to an increasing proportion of retained austenite (RA) above $\approx 0.7\,\%$ C or due to a slow cooling rate, for a degree of hardening $R_H < 1$, Eq. A.3.2, p. 67. (b) Hardness penetration due to dissolved alloying elements such as Cr, Mo and V. (c) Precipitation hardening (secondary hardening) by special carbides such as MC (V) and M_2C (Mo) during tempering (from H. Berns).

274   B.5 Tools for processing materials

is reached by hardening and tempering (Chap. A.2, p. 64 and 68): dissolved carbon increases the hardening capacity (Figure B.5.1), dissolved alloying elements increase the hardness penetration and formation of secondary-hardening alloy carbides during tempering results in precipitation hardening (Figures A.2.25, p. 49 and B.5.1 c).

Carbon contents above the approx. 0.7 % required for hardening remain precipitated as carbides. Their size decreases with the precipitation temperature: primary and eutectic carbides from the melt are 100 - 1 μm, secondary carbides from the solid solution are 1 - 0.1 μm, temper carbides are 0.1 - 0.005 μm. Coarse, hard carbides increase the wear resistance, whereas very fine carbides increase precipitation hardening and the creep resistance.

Most forgeable tool steels can be classified into three groups (EN ISO 4957): cold-work tool steels (CWS) with a high hardening capacity and optional carbides to improve wear protection, hot-work tool steels (HWS) with increased creep resistance through precipitation hardening and, as a combination of the aforementioned groups, high-speed tool steels (HSS) containing carbides in a creep-resistant, hardened matrix. The main application field of these three groups is the processing of metals and plastics.

B.5.1 Cold-work tools

Cold-work tools are generally used at room temperature, although temperatures of up to approx. 250°C are reached during polymer processing or due to frictional heating. The majority are made of cold-work tool steels, which are divided into three subgroups (Tab. B.5.1).

Table B.5.1 Cold-work tools: Group 1: 0.4 - 0.7 % C; Group 2: 0.8 - 1.3 % C; Group 3: > 1.5 % C

Group	Material	Carbide content Type	[%] [1)]	Annealed hardness [HB]	Service hardness from - to [HRC]	Fabrication	Tool application
1	C60U		<1	231	56 - 58	W	manual tools
	60WCrV8		<1	229	58 - 62	W	cutting, embossing
	X45CrNiMo4		<1	260	40 - 53	W, C	embossing, extrusion
	60CrMoV10-7		<1	250	50 - 60	C	working
	40CrMnNiMo8-6-4		<1	235	30 - 50	W	plastics moulding
	X54CrMoV17-1		<1	260	54 - 58	W	plastics moulding
2	C105U	Fe$_3$C	8[4)]	213	56 - 61	W	cutting, forming
	100Cr6	M$_3$C	8[4)]	223	58 - 62	W	impact extrusion
	X100CrMoV5-1	M$_7$C$_3$	5[4)]	223	56 - 62	W, C	cutting, forming
3	X153CrMoV12	M$_7$C$_3$	12	235	63 - 65	W, C	cutting, drawing
	X210Cr12	M$_7$C$_3$	17	255	55 - 65	W, C	cutting, forming
	X230CrVMo13-4	M$_7$C$_3$ MC	18 4	260	58 - 64	PM	embossing, cutting, impact extrusion, extrusion, plastics moulding
	X190CrVMoW20-4-1	M$_7$C$_3$ MC	21 2	280	58 - 62	PM	plastics moulding
	X245VCrMo10-5-2	MC	21	270	59 - 64	PM	embossing, forming, plastics moulding
	GJL-235HB	M$_3$C	2	235[2)]	52 - 60[3)]	C	forming, deep drawing
	GJS-265HB	M$_3$C	1	265[2)]	55 - 65[3)]	C	deep drawing, cutting

W = hot-worked, C = cast, PM = powder metallurgy
[1)] as hardened (volume-%) , [2)] as cast , [3)] in surface-hardened edges , [4)] quenched

Tough steels of group 1 with ≈ 0.5 % C do not reach their full martensite hardness. They are hardened at temperatures above Ac_3 (Figure A.2.9, p. 29) and are practically carbide-free. In contrast, the hard steels of group 2 with 1 % C contain small undissolved secondary carbides because their hardening temperature lies just above Ac_{1e} (Chap. B.2, p. 203). Wear-resistant chromium steels of group 3 contain coarser eutectic carbides with a higher hardness (Figure B.5.2). They also contain small undissolved secondary car-

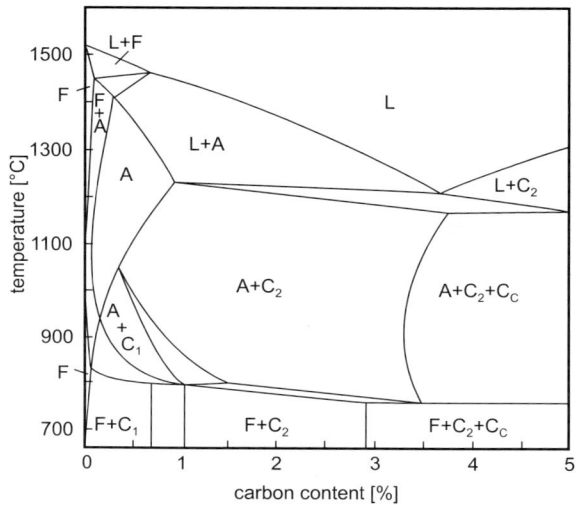

Fig. B.5.2 Fe - 13 % Cr - C phase diagram: In this isoplethal section, L = liquid, A = austenite, F = ferrite, $C_c = M_3C$, $C_1 = M_{23}C_6$, $C_2 = M_7C_3$ (from K. Bungardt et al., cf. Figure B.6.13 b, p. 326).

bides. Even higher carbide contents have an adverse effect on the hot workability of ingots and are thus limited to castings and PM steels. High-speed steels, which were originally developed for machining applications, are frequently used for cold-work tools as well (cf. Chap. B.4, p.288). They contain carbides of W, Mo and V (M_6C, M_2C, MC) that are embedded in a martensitic metal matrix (p. 299).

The cast irons listed in Tab. B.5.1 belong to group 3 on account of their C content; however, they have a different basic microstructure compared to the aforementioned steels. Their working surfaces are too soft because of the flake or spherulitic graphite precipitates in the pearlitic matrix so that they are often surface-hardened (nitriding, laser and induction hardening).

B.5.1.1 Properties

Segregations lead to microstructural banding in steel bars and are particularly noticeable in ledeburitic steels (Figure B.5.3 b). The size of the eutectic carbides increases with the solidification cross-section. These are chromium carbides of type M_7C_3, which are significantly harder than M_3C carbides (Figure A.4.16, p. 103) and more cost-effective than the even harder Mo, W, V and Nb carbides. Dissolved chromium also improves the hardenability (Figure B.5.1 b). This is why steels of group 3 contain approx. 12 % chromium. An increase in the hardening temperature increases the dissolution of secondary carbides in groups 2 and 3. This lowers M_f and the proportion of retained austenite increases (Figure B.5.4 a). If the steel contains sufficient amounts of e.g. Mo, and V, as in X100CrMoV5-1 and X153CrMoV12, significant

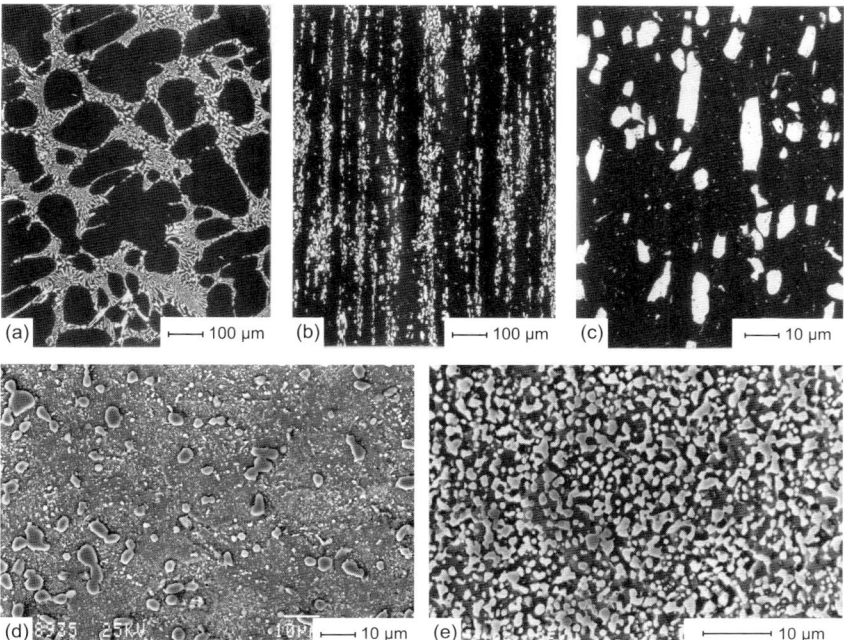

Fig. B.5.3 Carbides in ledeburitic tool steels: (a) to (c) X210CrW12, (d) X153CrMoV12, (e) X230CrVMo13-4. (a) Network of eutectic M_7C_3 carbides (light) around primary solidified dendrites in the ingot. (b) After hot-working, a banded structure of carbides is oriented along the longitudinal axis. (c) A high degree of plastic deformation produces a dispersion of oriented carbides. (d) After spray-forming and hot-working, the steel contains fine M_7C_3 carbides ($< 10\,\mu m$) and less banding owing to the lower degree of plastic deformation. (e) Dispersion of M_7C_3 carbides ($< 5\,\mu m$) after PM-HIP.

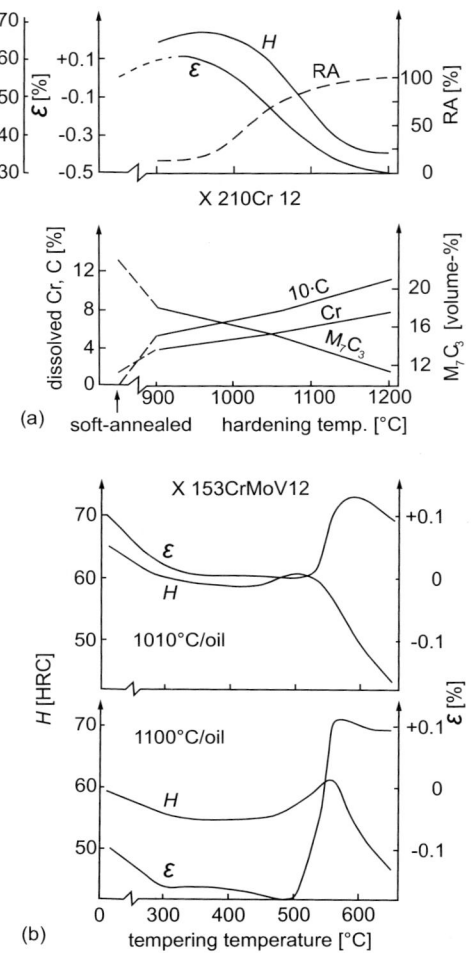

Fig. B.5.4 Hardening and tempering of ledeburitic chromium steels: (a) As the hardening temperature increases, the content of M_7C_3 carbides decreases and the content of dissolved carbon and chromium in the matrix increase. This initially increases the hardness H and the size change ε, which then drop again as the proportion of retained austenite RA increases (from H. Berns, J. Kettel). (b) After hardening from a low temperature, a tempering temperature of $\approx 300°C$ is sufficient to attain a hardness of $H \approx 60\,\mathrm{HRC}$ and a size change of $\varepsilon \approx 0$. Hardening from a higher temperature promotes secondary hardening and increases the proportion of retained austenite, which only decomposes at tempering temperatures above 500°C. A small size change at 60 HRC therefore requires tempering at about 540°C (from H. Berns).

secondary hardening occurs during tempering between 500 and 550°C. Service temperatures that lie well below the tempering temperature are tolerated for long periods without a loss of hardness (Eq. A.3.3, p. 69). Precipitation hardening thus increases the hot hardness during service. Furthermore, this effect is exploited during manufacture if already hardened tools are to be nitrided or PVD-coated at approx. 500°C without significantly affecting the core hardness. In addition, secondary tempering can be used to to lower the size change of the workpiece. The formation of retained austenite during hardening causes the tools to shrink (Figure A.3.7, p.73), whereas its transformation during secondary tempering causes them to expand. This leads to two methods of hardening with minimal size changes: a low hardening and tempering temperature (low hot hardness) or a high hardening and tempering temperature (high hot hardness, Figure B.5.4 b). The banding of the metal matrix and the difference in the coefficients of thermal expansion between the matrix and the elongated eutectic carbides gives rise to an anisotropy in the size change, which is larger in the direction of forming than in the transverse or perpendicular directions. This can be lessened by cross-rolling of slabs or largely avoided by using PM steels.

Under compressive loads, the fracture strength increases with the hardness. However, above a critical hardness, it drops as the multiaxiality increases (Figure A.4.14, p. 97). This drop is enhanced by crack-initiating defects: eutectic carbides and non-metallic inclusions in this case. Hard steels of group 2 and particularly carbide-rich steels of group 3 are thus suitable for tools subjected to predominantly compressive loads. If they are also subjected to notched tensile and bending stresses, the hardness must be lowered and the tougher steels of group 1 should be used instead, even if this means a lower wear resistance. Figure B.5.5 a, b shows a comparison of the mechanical properties in a static bending test. The strength and strain at fracture decrease from group 1 to group 3. The yield strength decreases as the proportion of softer retained austenite increases. The fracture strength and strain at fracture decrease as the carbide size in group 3 increases (Figure B.5.5 c). The drop in the yield strength is caused by the formation of retained austenite shells and an increase in residual microstresses around the coarser carbides. Figure B.5.6 shows that the CWS and HSS grades, which are suitable for metal-working tools, represent a good compromise between hardness and bending strength compared to other materials.

Under fatigue conditions, an increasing hardness has a favourable effect on the fatigue life as long as carbides and other defects remain small. In contrast, coarse carbides initiate premature cracks that propagate more quickly in a harder matrix, thus lowering the time to fracture (Figure B.5.7 a).

The stress-induced transformation of retained austenite is associated with a volume increase. This reduces the effective bending stress in the surface layer and crack initiation is delayed. Subsequently, the propagation of the fatigue crack is slowed down by the transformation of retained austenite in front of the crack tip (Figure B.5.7 b). The fracture toughness K_{Ic} decreases as the hardness, amount of carbides and banding increase. Even minor cracks, caused by fatigue, grinding or spark erosion, that have a depth of e.g. $a = 0.1\,\text{mm}$, can thus adversely affect the fracture strength $\sigma_c \approx K_{Ic}/\sqrt{\pi a_c}$ in a similar manner to the internal defects shown in Figure A.4.15 (p. 97). Under tensile or bending loads, an instant fracture may occur for a stress that lies well below that of a crack-free surface.

As described in Chap. A.4 (p. 106), a high hardness usually has a favourable effect on the wear resistance, whereas carbides can have differing effects,

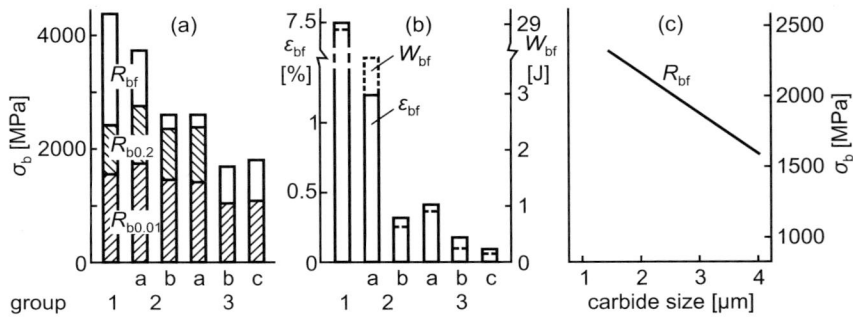

Fig. B.5.5 Mechanical properties of cold-work tool steels: Measured in a three-point bending test using longitudinal specimens with strain gauges on the tension side. (a) The bending strength R_{bf} and the 0.01 and 0.2 % bending proof strengths. (b) The plastic bending strain at fracture ε_{bf} and plastic work of bending at fracture W_{bf} for the steels and heat-treatment conditions given below. (c) Influence of the mean carbide size in ledeburitic chromium steels (group 3) with RA $\approx$ 10 %, CC $\approx$ 15 volume-% and hardness $\approx$ 770 HV (compressive strength $\approx$ 3000 MPa) (from H. Berns, W. Trojahn)

Group	Example		HT [°C]	TT [°C]	RA [%]	Hardness [HV]	CC [%]
1	56NiCrMoV-7		860	180	5.1	670	<1
2	100Cr6	a	820	180	4.0	748	8.8[1]
		b	880	180	12.0	803	6.4
3	X153CrVMo12-1	a	1020	180	4.0	698	12.5[2]
		b	1080	180	23.0	722	
		c	1080	540	4.0	748	

HT = hardening temperature, , TT = tempering temperature, RA = proportion of retained austenite, CC = carbide content
[1] mean carbide diameter 0.86 µm
[2] mean length and width of eutectic carbides 8.3 and 4.2 µm, respectively
(from H. Berns)

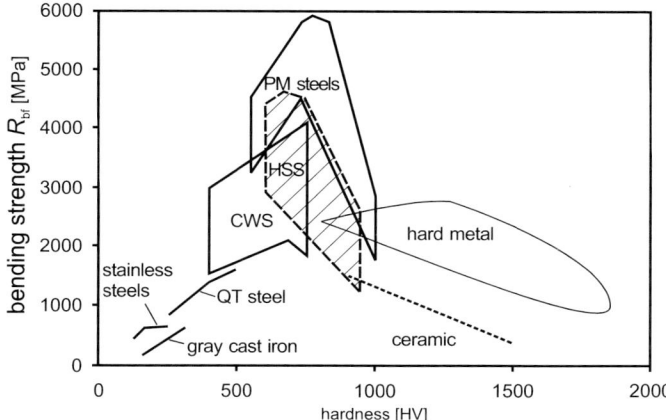

Fig. B.5.6 Comparison of tool materials: Bending strength of various tool materials plotted against their hardness. For a hardness between 400 and 1000 HV, CWS and HSS with hard phase contents of < 20 volume-% have the highest strengths. Hard metals and ceramics are harder, but not as strong. PM-CWS and PM-HSS are superior to conventionally produced steel variants (from L. Westin, H. Wisell)

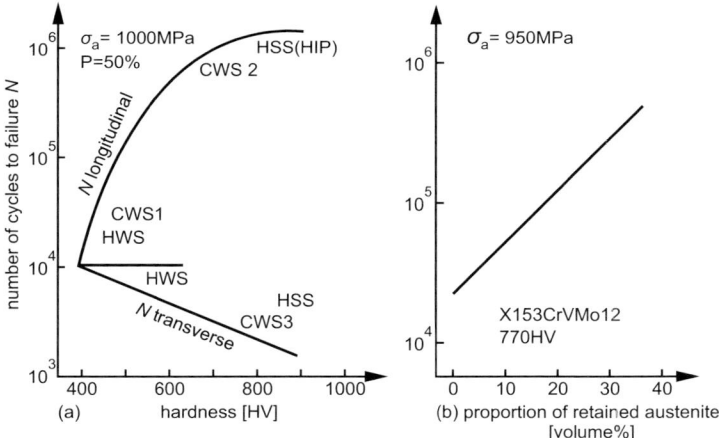

Fig. B.5.7 Fatigue strength of tool steels: (a) In a rotary bending test, the service life increases with the hardness if the carbides are small and round (upper limiting curve). It decreases if the effective carbide size increases, as in e.g. transverse specimens containing eutectic carbides (lower limiting curve for steels according to Tables B.5.1, B.5.4 and B.5.5) (from H. Berns). (b) The service life increases with the proportion of retained austenite (from H. Berns, W. Trojahn).

depending on the main wear mechanism, so that their amount and size must be tailored to the type of wear. Coarse, hard carbides in steels of group 3 have a favourable effect on the wear resistance for 3-body sliding abrasion and sliding wear, which occurs during e.g. powder compaction, bending and extrusion (Figure B.5.8). In contrast, the smaller carbides in steels of group 2 or PM steels of group 3 are more favourable in the case of contact fatigue, as experienced by e.g. rolling and roll-embossing tools.

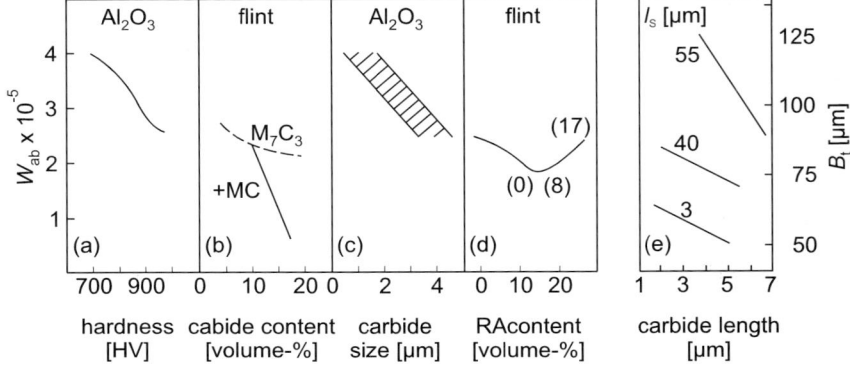

Fig. B.5.8 Microstructure and wear of ledeburitic chromium steels: containing 10 to 14 % Cr, 1.2 to 2.7 % C and some Mo, V, Nb and Ti. Wear loss W_{ab} against abrasive paper with 80 mesh; flank wear B_t in a slow cutting test on a structural steel shaft. (a) X210Cr12 hardened from 1000°C/oil, cooled to -196°C and tempered at 100 - 400°C. (b) W_{ab} decreases with the amount of primary and eutectic carbides, particularly the harder MC. (c) The wear protection provided by the carbides increases as the mean carbide size approaches the width of the groove. (d) Transformation of retained austenite slows the wear rate. RA will not completely transform if its stability is too high (values in brackets = RA contents in the wear surface). (e) Metal/metal sliding wear also decreases with the mean carbide size; however, it increases with the distance between the carbides or carbide bands because the scratch length l_S in the unprotected matrix increases (from H. Berns, W. Trojahn).

B.5.1.2 Coated tools

Owing to the fact that the service life of a tool is determined by the properties of its surface and/or near-surface zone, various methods are employed for surface finishing. In addition to surface layer treatments (Chap. B.3, p. 217), which have been used for decades, the deposition of thin coatings has become a very popular method in recent years to provide wear protection on tools used for processing materials. Thin ($< 15\,\mu$m) layers of hard materials have a high hardness, a high compressive strength, a low adhesion tendency and often low coefficients of friction. They can thus greatly increase the service life of the tool. They are deposited as mono- or multilayers from the gas phase either chemically (chemical vapour deposition, CVD) or physically (physical vapour deposition, PVD). VDI Guideline VDI 3198 provides information on the use of thin coatings on cold-forging tools.

(a) CVD coatings

The CVD process is based on the reaction of gaseous metal compounds (e.g. fluorides, chlorides and bromides) with reactive gases (e.g. CH_4, CO_2, N_2, H_2) in a closed reactor followed by deposition of the product as a thin hard coating on a steel surface. The reaction temperature in the popular high-temperature CVD process is about 1000°C. Tab. B.5.2 lists commonly used constituents for coating tools along with some important properties and application examples.

The activation energy for this reaction can be introduced by heating the substrate or the reactor wall as well as by plasma ignition, magnetic induction or laser beams. CVD coatings are uniformly thick and can even be applied to complex geometries without shadowing effects. Not only monolayers, but also multilayer coatings are frequently used. From the broad range of hard coatings, which includes oxides, carbides, nitrides and borides, approx. 6 - 9 µm-thick TiC layers on TiN have proven successful on steel tools (Figure B.5.9 a). TiN improves the adhesion to steel and is deposited as the first layer of multilayer coatings. An intermediate layer of TiCN with an increasing C content towards the surface produces gradients in the hardness and residual compressive stresses, which has a favourable effect on the adhesion. TiC is often used as the top layer owing to its high hardness. A number of alternating layers of TiN and TiC produces multilayer coatings with an overall thickness of up to 10 µm,. Their good adhesion and high toughness make them suitable for high local loads (e.g. embossing). Good adhesion is obtained by diffusion reactions between the coating and the substrate that may occur at high coating temperatures.

Table B.5.2 Thin films: Commonly used coatings for tools with special properties and application examples

Film material	Microhardness [HV0.01][1]	Coeff. friction[2]	T_{max}[3] [°C]	Coating structure	Area of app.[4]	Example applications
TiN	2300	0.4	600	monolayer	CT, CW, PP	metal working, injection moulding
TiC	3500	0.2 – 0.3	350	monolayer	CT	metal working, sliding surfaces
AlCrN	3200	0.35	1100	monolayer	CT, CW, HW	tool bits, forging, pressure die casting
TiCN	3000	0.4	400	gradient	CT, CW, PP	milling, metal working, stamping
CrN	1750	0.5	700	monolayer	CT, CW, PP, HW	die casting, injection moulding, metal working
Cr_3C_2	2200	0.35 – 0.4	700	monolayer	PP, HW	pressure die casting, injection moulding
Diamond	9000	0.15 – 0.20	600	monolayer	CT	cutting Al and Al-MMC
TiAlN	3300	0.25 – 0.35	900	nanostructured	CT, CW	milling tools (HSC, hard cutting)
TiCN + TiN	3000	0.4	400	multilayer, gradient	CT	milling, metal working, blanking
TiAlN	3300	0.4	900	monolayer	CT, CW, PP, HW	milling tools (HSC), pressure die casting
AlCr - based	3000	0.25	1100	multilayer	CT	drills (HSC, dry)
DLC (a-C:H)	2500	0.1 – 0.2	350	monolayer	CT, CW, PP	cutting plates for Al and Al-MMC
TiAlN + a-C:H:W	3000	0.15 – 0.20	800	multilayer, gradient	CT, CW	tap drills, dry cutting
TiNCrN	2100	0.5	700	gradient	CW (HW)	metal working (cold, semi-hot)

[1] mean hardness, [2] against steel (dry), [3] maximum service temperature

[4] CT = cutting   HSC = high speed cutting   CW = cold-working   HW = hot-working   PP = polymer processing

The typical slow rate of cooling from the coating temperature often means that the substrate has to be rehardened. With respect to dimensional tolerances, secondary hardening steels are particularly suitable for this because they can be tempered to specific dimensions (Figure B.5.4).

In plasma-assisted CVD processes (PACVD), a pulsed low-pressure glow discharge - with the substrate as the cathode - leads to a higher internal energy in the gas compared to the thermodynamic equilibrium. This allows the process temperature to be lowered to 250 - 600°C and the tools can be coated after secondary tempering. The process parameters must be carefully selected to match the substrate and the coating materials to achieve good adhesion, particularly at low processing temperatures. There are many ways of achieving this, e.g. by varying the composition and partial pressure of the gas as well as the glow discharge parameters. These control variables ultimately determine the stoichiometry and thus a variance of up to 20 % in the layer hardness. At deposition rates of $\approx 1\,\mu m/h$, the PACVD layers have a dense columnar structure with a smooth surface that provides the required low coefficient of friction without the need for final polishing.

PACVD can also be used to deposit superhard boride layers (thickness 1 - 3 µm): TiN is deposited first as an adhesive base coat, then the proportion of

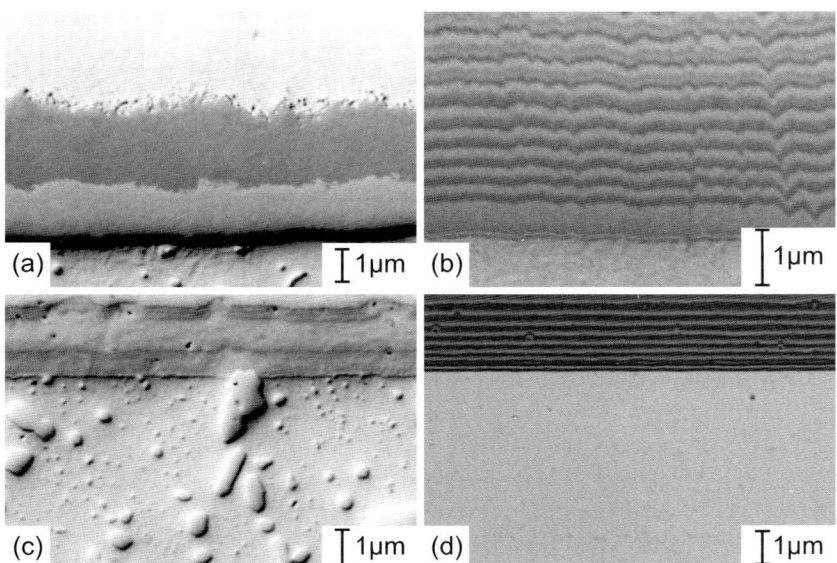

Fig. B.5.9 Thin films: ((layer sequence from the substrate to the surface) (a) CVD-TiN/TiCN/TiC sandwich, b) PACVD-TiN/TiB$_2$ multilayers in alternation (c) PVD-TiN-TiCN (d) PACVD-a-C:H multilayers, carbon layers, hard and soft in alternation (after C. Escher)

BCl_3 is slowly increased so that the composition gradually changes to TiBN and finally to a top coat of TiB_2, which is even harder. Multilayer coatings of TiB, TiN and TiBN are also used (Figure B.5.9 b). Boride coatings have a high thermal stability and their oxidation resistance is only exceeded by that of TiAlN coatings. A combination of plasma nitriding and subsequent PACVD is used to deposit load-bearing layers on substrates with < 60 HRC.

Crystalline diamond layers and diamond-like carbon (DLC) layers, which are included in VDI Guideline 2840 (draft), are now becoming increasingly popular, mainly because of their very low coefficients of friction when paired against various solid materials. They can be deposited by PACVD or PVD as highly cross-linked amorphous carbon layers (a-C films) at 150-250°C. The process-related inclusion of hydrogen (10-30 atom-%) to produce C:H layers allows the proportion of diamond-like bonds and thus the layer properties (e. g. hardness, adhesion, toughness, coefficients of friction) to be controlled via the processing parameters. Furthermore, DLC layers can be adapted to the respective application within wide limits by doping with non-metals (e. g. oxygen, nitrogen, boron, fluorine) or metals (e. g. a-C:H:Me where Me = W and Ti). They are frequently used as the top layer in mono- and multilayer coatings on account of their low coefficient of friction.

(b) PVD coatings

The physical deposition of hard coatings from the gas phase involves vaporisation of one or more solid sources to form a gas, which may react with a reactive gas (e. g. N_2, CH_4, C_2H_2, CO_2), and then condenses on the substrate as a carbide, nitride or oxide. Of the many different PVD processes known today, cathode sputtering as well as electron beam and arc evaporation have become established processes for producing hard coatings.

This process involves placing tools with a polished and degreased surface in a vacuum chamber that is evacuated and then filled with an inert gas, usually argon, to a pressure of 5 Pa. A negative voltage is applied to the substrate connected as the cathode. This voltage is gradually increased until a self-sustaining glow discharge develops around the substrate that cleans it by ion bombardment (sputter etching). Subsequently, with the glow discharge burning, the source material is vapourised, e.g. by an electron beam, and then condenses on the substrate, whereby the deposition rate must be greater than the sputter rate. During continuous bombardment, ions are able to penetrate the growing layer thus producing a higher lattice defect density, microstructural refinement, densification of the layer and good adhesion. An arc evaporator (arc PVD) produces a significantly higher ion density than an electron beam evaporator. In this case, the path of the arc is either left uncontrolled or it is controlled by an additional magnetic field so that it tracks over one or more arbitrarily located solid source targets that vaporise within the small focal spot.

Cathode sputtering involves the bombardment of Ar ions accelerated against the source material connected as the cathode by ignition of a glow discharge. This ejects particles that are guided by a magnetic field and deposited as a layer on an opposing substrate (magnetron sputtering). Ion bombardment of the substrate can also be increased in this case by glow discharge.

Although the adhesion and deposition rate are higher for arc evaporation than for cathode sputtering, they are still significantly lower compared to CVD coating. However, the structure and thickness of the sputtered layer are much more homogeneous compared to arc evaporation.

The aforementioned hard materials (Tab. B.5.2) are used to produce PVD coatings on tools as mono-, sandwich and multilayers (Figure B.5.9 c). Similar to PACVD (Figure B.5.9 d), PVD can be used to deposit amorphous diamond-like layers doped with metals. They are used as mono- and multilayer coatings; frequently as the top layer on the latter. Owing to their lower bonding to the substrate PVD coatings are generally thinner than CVD layers. The advantage of the low substrate temperature (250-500°C) and the resulting possibility of coating secondary hardening steels after tempering close to the secondary hardening maximum (as for PACVD) must be weighed against the drawback of poor coating results on surfaces out of the direct gas stream, e.g. in bores (shadow effect). Developmental work is aimed at lowering substrate temperatures. Owing to the high particle energies in an arc plasma, vacuum arc evaporation can be used to deposit dense, hard coatings, even at substrate temperatures below 200°C. The fact that bonding decreases with the coating temperature has led to the development of hybrid technologies such as a combination of plasma nitriding (Chap. B.3, p. 227) and arc PVD. Arc PVD equipment can also be used for plasma nitriding: the tools can be nitrided and then coated directly with TiN in a continuous process without flooding the recipient and without atmospheric contamination. The nitrided layer beneath the hard TiN layer increases. This means that low-alloy substrates can also be coated. The choice of suitable parameters allows the formation of a stable multiphase Fe_xN compound layer that lessens the sudden change in hardness between the diffusion layer (≈ 800 HV 0.01) and the coating (≈ 2300 HV 0.01). Furthermore, in-process ion etching after plasma nitriding significantly increases the adhesion.

(c) Applications of thin films

Tools with thin coatings often achieve considerably longer tool lives compared to uncoated ones. For example, a deep-drawing insert made of ledeburitic cold-work tool steel X153CrMoV12 was used for zinc-coated automobile sheets. In the uncoated state and a hardness of 61 HRC, it processed $\approx 15{,}000$ parts, whereas its counterpart with an additional CVD sandwich coating processed $\approx 600{,}000$ parts. The tool life of expansion mandrels of X165CrMoV12 (62 HRC) used for stainless austenitic steel pipes increased from ≈ 4000 parts in the uncoated state to $\approx 10{,}000$ parts with a PVD-CrN coating and to $\approx 180{,}000$ parts with a CVD-TiC coating. A PVD-CrN coating on punches of X153CrMoV12 (60 HRC) increased the tool life by a factor of 25 for processing austenitic sheets 0.4-mm thick. The same coating increased the tool life by a factor of 33 in a punching-drawing operation of an austenitic locking ring. It also allowed a 50 % higher stroke frequency and a reduction of the amount of lubricant per part to a sixth.

Amorphous DLC films have been very successful as mono- and multilayer coatings for deep-drawing tools and on thread rolls used to manufacture high-strength bolts under minimum lubrication conditions. They increased the service life by up to 200 %. They are particularly successful for forming austenitic steel plates and aluminium. Their high proportion of covalent bonding considerably reduces their tendency to adhere, which is responsible for cold welding, particularly with face-centred cubic materials. This means that aluminium beverage cans can be produced with a significantly improved cutting quality without burrs and a significantly longer service tool life. A modified CrN/a-C:H:W coating system, used as a non-stick coating on an automobile wheel mould, improved demoulding and cleaning.

B.5.1.3 Applications of cold-work tools

The force required to cut polymers, textiles, metals etc. increases with the strength and thickness of the workpiece, and the load on the cutting edges also increases. The rate of edge wear grows with the load and the number of workpieces (tool life). The example given in Figure B.5.10 for the shear cutting of unalloyed structural steel sheets shows that the tool life increases with the carbide content.

With increasing plate thickness the load on the tool edge is raised until the yield strength of the matrix is reached. A high content of coarse carbides would then lead to break-outs. After perforating, the workpiece springs back elastically and hinders retraction of the punch. The retraction stress increases with the ratio between the plate thickness and the hole diameter. The oscillating compressive load becomes a compression-tension load, which induces fatigue cracks and may tear the punch. This is why punches for perforating thick plate should not be made of steel with coarse carbides. Steel HS6-5-3 PM with a nitrided or PVD-TiN-coated surface has proven successful e. g. for high-speed cutting of electrical steel sheets as well as for thin ferritic and austenitic stainless steel plates. If the tool steel is to be nitrided or PVD-coated, it must be subjected to secondary hardening to impart sufficient tempering resistance.

Mechanical loading of the tool is lower for deep-drawing of sheet steel compared to punching. Tribological loading due to adhesion and abrasion is more important in this case. Low-carbide ferritic deep-drawing sheets and austenitic sheets tend to cold-weld to the tool. This tendency increases with the draw depth because it becomes more difficult to maintain an intact lubricating film. In addition to the problem of tool wear, the workpiece may be grooved by the punch. The high carbide contents in group CWS 3, coupled with a hardness of 61 to 63 HRC have a favourable effect. Further improvements are possible by nitriding or application of a CVD-TiC coating. The process temperature for ledeburitic chromium steels is close to that of the subsequent hardening step. Flat draw dies with medium tool lives are made of steels with a shallow hardness penetration, such as C105U, and are subjected to low-distortion hardening to 62-64 HRC using a water jet only in the bore hole in which a martensitic case of 2 to 3 mm depth is formed, backed up by pearlite.

Extrusion punches must have a high compressive yield strength to avoid bulges due to upsetting. They must also be wear-resistant. These requirements are fulfilled by ledeburitic steels with a low proportion of retained austenite, which can be achieved by secondary hardening of e.g. X153CrMoV12 (≈ 59 HRC) or X220CrVMo12-2 PM (≈ 61 HRC) or by the use of a high-speed steel (≈ 63 HRC). A further possibility is the use of chromium steels with a low degree of tempering (62 to 63 HRC) and removal of slight bulges by regrinding. After the retained austenite has undergone local strain-induced transformation during service, the punches can then usually withstand greater loads than after secondary hardening. Tangential tensile stresses develop in extrusion dies due to the internal pressure. These stresses may cause bursting of through-hardened steels. Water-jet cooling of the internal contour of a steel with a low hardness penetration, e.g. C105U (Figure B.5.1 b), produces a surface layer with residual compressive stresses, hardened to a depth of 2 to 3 mm. It is then tempered to 62 - 63 HRC. This prestressing relieves stresses

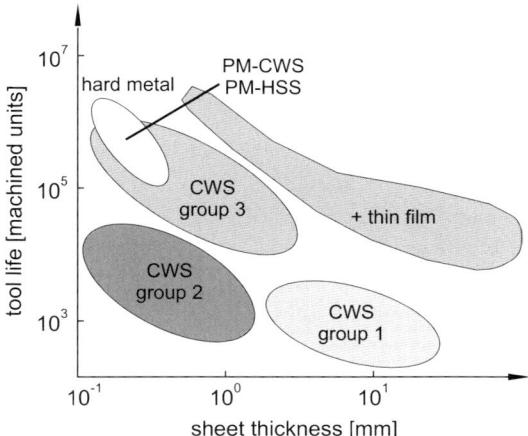

Fig. B.5.10 Selection of materials for shear cutting: Rough classification of tool materials used for cutting structural steel sheets. For thin sheet, the number of workpieces cut per tool (tool life) increases with the wear resistance due to the increasing carbide content. For thick sheet, the higher content of coarse carbides causes breakouts in the cutting edges. The application of thin films increases the tool life of the CWSs of group 2 much more than those of group 3 for thin sheets. PM steels close the gap between conventionally manufactured steels and hard metals.

arising during service, e. g. caused by pressing bolt heads and nuts. In contrast, a higher service life is achieved by the use of carbide-rich steels of group 3 or sintered hard metals that are prestressed via longitudinal or transverse press fits using one or two rings of high-strength QT steel. If a transverse press fit is used, the CWS insert must be secondary hardened otherwise its tempering temperature would be exceeded due to the shrink-ring temperature of $\approx 500°C$ and the hardness would drop. The tempering resistance is also of vital importance for the rings. Secondary-hardenable hot-work tool steels allow higher strengths than QT steels; however, they are more expensive. These problems do not arise with longitudinal i.e. tapered press fits. Rings of high-strength maraging steels (Chap. B.2, p. 202) permit oversizing by up to 0.7 % and thus high prestressing levels.

During embossing of flat materials, the workpiece material has to rise into the engraving. This usually requires a high pressure as well as through-hardened tools. Notch stresses develop in the root of the engraving that have to be plastically relieved by the tool steel to avoid premature crack initiation. Tough steels of group 1, e. g. X45NiCrMo4, are used for embossing tools e. g. in the cutlery and tool industries. If the engravings are to be carried out by hobbing, hot-work tool steel X32CrMoV3-3 can be used owing to its low annealed hardness. It can be converted into a cold-work tool steel by carburising during hardening.

Because the processing of large tools may account for more than half of the tool costs, near-net-shaped cast products are in demand. In the automobile industry, large cast tools are used for deep drawing, bending and cutting of bodywork and chassis parts from thin plate. Weldable materials are becoming increasing popular because contour changes are often necessary. The carbide-free grades G47CrMn6 and G59CrMoV18-5 are thus used as lost-foam castings for monoblock tools. A slightly lower C content and a higher Mo content makes these casting alloys secondary hardenable and suitable for PVD coatings. If the main requirement is not weldability, but wear resistance and CVD coatability, hard steels such as GX100CrMoV5-1 or GX153CrMoV12 with hardness of 58 to 62 HRC are used. Some parts consist of lost-foam cast segments that are assembled on a framework of steel or grey cast iron. The cast microstructure has the advantage of quasi-isotropic size changes thus allowing the segments to be fitted without gaps. The cutting edges can be flame- or laser-hardened in air . QT steel castings are used for flat deep-drawing tools and blank holders. Particularly loaded areas are surface-hardened.

To save material and manufacturing costs, grey cast iron is used as an alternative for large forming tools. Because highly loaded edges may be subjected to martensitic surface-hardening, only grades with a high pearlite content are suitable. Thus, lamellar graphite cast iron GJL-HB 235 alloyed with a little chromium and molybdenum and with ≈ 2 volume- % eutectic cementite can be easily machined and is used for deep-drawing tools, such as the die, punch and blank-holder. Also suitable for the same application is a lightly alloyed and more ductile spheroidal graphite cast iron GJS-HB 265, that has a tensile strength of $R_\mathrm{m} \approx 750\,\mathrm{MPa}$ and a strain at fracture of $A \approx 2\,\%$ and which is subjected to subsequent surface-hardening of the edges. It is used for deep-drawing tools for mass production. The ferritic grade GJS400-15 is used without edge hardening for thin-walled fatigue- or impact-loaded blank-holders or wipers. If there is a likelihood of cold-welding, the tools can also be plasma-nitrided (Chap. B.3, p. 241) or given a hard chromium plating.

Cold rolls for manufacturing thin sheets are made of e.g. 85CrMo7, which is forged, quenched, tempered and inductively surface-hardened. This results in a surface hardness of up to 64 HRC and a depth of hardening of up to $\mathrm{DS} \approx 50\,\mathrm{mm}$ in large-diameter rolls (Chap. B.3, p. 218). In contrast, bending and straightening rolls for pipe manufacture are made of group 3 steels. Rolls with a diameter of up to $\approx 500\,\mathrm{mm}$ are forged or made of steel bars. Those with a larger diameter are usually cast. To prevent heating of guide rollers in high-frequency pipe welding equipment, they are made of e.g. steel X210CrW12 that is hardened from $\approx 1200°\mathrm{C}$ to produce a non-magnetisable matrix of retained austenite. Although the resulting hardness lies below 40 HRC, the eutectic carbides and a thin friction-induced martensite layer provide good wear resistance.

Circular saw blades are made of high-speed steel, e.g. HS6-5-2, hardened and tempered to 63 - 65 HRC. Diameters of up to 3 m are produced from partially hardened toothed segments of high-speed steel riveted to a central disk made of e.g. 75Cr1, quenched and tempered to 40 - 45 HRC. Alternatively, hard metal inserts are brazed on. Stone saws have diamond-containing cutting elements. High-speed steels and hard metals are also necessary for coated wood as well as filled or reinforced polymer materials. Band saws are often made from strips of spring steel that are electron-beam-welded to a strip of high-speed steel whose width is slightly less than the tooth height. After the teeth have been ground, their roots lie in the tougher spring steel. Pure woodworking saws, e.g. gang saws, are usually made of low-alloy steels, such as 80CrV2, that are hardened and tempered to only 48 to 50 HRC so that the teeth can be punched and set. The tips can be inductively hardened.

The blade of stone chisels of C70U and metal chisels of 45CrMoV7 are given a higher hardness (stone: 58 to 60 HRC, metal: 54 to 56 HRC) than the hammered end (46 to 48 HRC) and the shaft. Wrenches are forged from 31CrV3 and then quenched and tempered to 35 to 40 HRC, whereas screwdrivers and impact wrenches are made of spring steel such as 50CrV4 with 45 to 55 HRC. Hammers are made of steels C45U or C60U, which do not undergo through-hardening in water if the thickness is appropriate. They are then tempered to 40 to 45 HRC. Files can be made of e.g. steel C125U with $\approx 65\,\mathrm{HRC}$.

B.5.2 Tools for processing plastics

The multitude of tasks involved in processing plastics means that tools require very different sets of properties, which can only be fulfilled by using steels from the various groups. Together they are known as plastic moulding steels, but they are not specified as a separate group in EN ISO 4957. They include through-hardenable and surface-hardenable QT steels, hot- and cold-work tool steels as well as stainless martensitic chromium steels.

Injection moulds for processing plastics are made of QT steels such as 40CrMnMo7. Sulphur is added to improve the machinability, and nickel is added for large moulds to improve hardness penetration (e.g. 40CrMnNiMo8-6-4). Most polymer parts have decorative and functional surfaces with extremely high demands on the quality of their surface finish. The production of large forged parts with low segregation and a good cleanness is thus a prerequisite for defect-free polishability and etchability during surface texturing. A strength of ≈ 1000 Pa is required to withstand the high closing pressures in the mould parting line in some applications. For example, a variant of the composition 26MnMoCrNiV6-6-5-4 with reduced amounts of C and Si and increased Ni and Mo, quenched and tempered to 31 - 36 HRC, has been used successfully for large dies with a QT diameter > 400 mm that are used to produce car dashboards and bumpers as well as chairs, rubbish bins and bottle crates. Smaller moulds can also be case-hardened e.g. air-hardening steel X19NiCrMo4. If they are to have a hobbed engraving, it is better to use more annealable steels such as 16MnCr5 or X6CrMo4. Through-hardenable steels (e. g. 90MnCrV8) and maraging steels (e. g. X3NiCoMoTi18-9-5, Figure B.2.23, p. 201) can also be used. The latter are a good choice because they can be given a simple, low-distortion heat treatment after machining and are suitable for welding. In applications with additional corrosive conditions, e. g. PVC processing, stainless martensitic steels (e. g. X39CrMo17 and X90CrMoV18) or steels alloyed with nitrogen under pressure (e. g. X30CrMoN15-1 with 0.35 %N, Chap.B.6, p. 329) are more suitable. Low sulphur contents are necessary to reduce anisotropy in the toughness of hot-formed plastic moulding steels caused by banding. However, this adversely affects the machinability. The addition of small amounts of calcium during deoxidation of the melt results in uniformly distributed spheroidal sulphides and allows cost-effective machining, even for sulphur contents of < 0.005 %. This improves not only the toughness but also the polishability and etchability.

Tools for processing plastics are also surface-finished in some cases. Common methods include hard chromium coatings, nitriding and, above all, PVD coatings. For example, in automotive engineering, they are used on injection moulds for polycarbonate side-windows of cars (CrN) or as a TiN coating on moulds for tail lights made of polymethyl methacrylate. High tool lives are achieved for blow-moulding tools with an arc PVD coating in the form of CrN multilayers and coating thicknesses of between 3 and 6 µm. They out-perform TiN, TiCN and TiAlN coatings owing to their greater bond to the substrate and very good sliding properties with respect to polymer resins.

Nitriding steels such as 14CrMoV6-9 and 31CrMoV9 are used for screws and housings in plastification units such as injection-moulding machines and extruders, if the amount of filler in the polymer is low because the nitrided surface layer has not only a high hardness and a high tempering resistance but also a very low fretting tendency if there is any metal/metal contact of the tools. Preference is given to Al-free steels owing to their greater depth of hardening after nitriding. Duromers and thermoplastics with a high filler content are processed with tools made of secondary-hardening ledeburitic cold-work tool steels (e. g. X153CrMoV12) or PM steels (e. g. X230CrVMo13-4). A higher wear resistance as well as corrosion resistance is obtained with PM steels containing additional Cr (e.g. X190CrVMoW20-4-1; Tab. B.5.1). These steels are produced by hot isostatic pressing and are often applied as a coating to substrates of QT steel. Successful PM steels include the superhard vanadium monocarbide grade X240VCrMo10-5-2 (Tab. B.5.1 and the corrosion-resistant variant X360CrVMo20-10-1. Furthermore, PM technology allows the addition of hard phases (Cr_3C_2, WC/W_2C) to the steel powders to produce MMC coatings with the highest wear resistance (Figure B.4.6, p. 265).

B.5.3 Hot-work tools

Hot-work tools are used at workpiece temperatures of between ≈ 400 and 1200°C, as the typical processing temperatures in Tab. B.5.3 show. The sur-

Table B.5.3 Forming temperatures: Typical process temperatures during processing

Tool material	Casting temperature		Forging temperature
		[°C]	
Zn alloy	400		-
Mg alloy	600		-
Al alloy	700		500
Cu alloy	1000		800
glass	-	$1100^{1)}$	-
steel	-		1200

[1] softening temperature

face temperature of the tools approaches that of the workpiece as the contact time increases and the relative cooling time decreases. The contact time ranges from milliseconds for forging dies and wire rolls to minutes for wire extruders. Creep-resistant and high-strength QT steels, with or without significant secondary hardening, are used depending on the temperature conditions (Tab. B.5.4). At very high temperatures, creep-resistant and oxidation-resistant austenitic steels and nickel-base alloys can also be used because they have a lower tendency to undergo diffusion-dependent creep owing to their denser atomic packing (Figure B.5.11).

Table B.5.4 Hot-work tool steels (classification and examples): 1. with slight, 2. with significant secondary hardness, quenched and tempered to 40 - 50 HRC ($R_m = 1300 - 1700$ MPa), 3. austenitic with 2.1 % Ti and 0.3 % Al, solution-annealed and precipitation hardened to $R_m \approx 1000$ MPa

	HWS	Example application
1	56NiCrMoV7	forging dies, die holders
		and rams of extrusion presses
2	X40CrMoV5-1	pressure die casting and
		extrusion tools for light metals, tools for forging machines
	X32CrMoV3-3	as above, but for copper alloys
3	X6NiCrTi26-15	inner sleeves of containers
		for copper alloys, die blocks
		for open-die forging

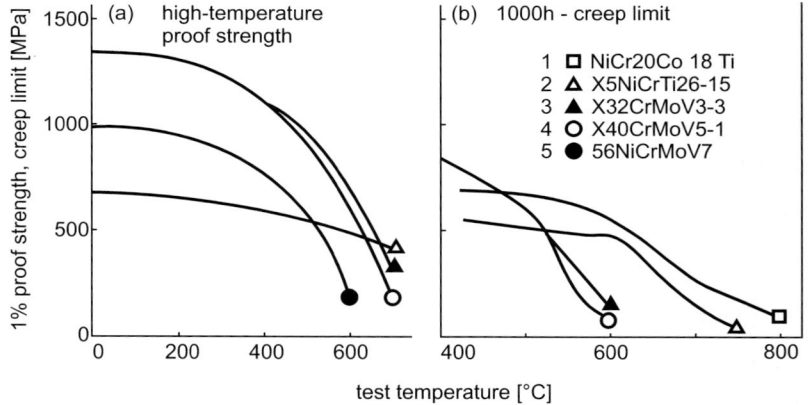

Fig. B.5.11 Tool materials for high-temperature applications: 1 = precipitation-hardened nickel-base alloy; 2 = precipitation-hardened creep-resistant austenitic steel; 3 to 5 = quenched and tempered hot-work tool steels. (a) Proof strength in the hot tensile test. (b) Creep limit at 1000 h measured in a creep rupture test (from H. Berns).

B.5.3.1 Properties

The high-temperature proof stress of quenched and tempered hot-work tool steels increases with the content of Mo and V. At the same time, a certain degree of hot embrittlement becomes apparent. The creep limit lies significantly lower, which indicates thermally activated creep. Above 550°C, the austenitic steel out-performs the best QT steel (Figure B.5.11).

Thermal fatigue caused by temperature changes leads to thermal shock cracks or heat checks on the tool's surface, usually with a netlike pattern. This is due to plane stresses induced by thermal expansion and contraction of the surface. They lead to microplastic deformations that initiate cracks at non-metallic inclusions and segregation carbides. The crack-propagation velocity initially increases because the temperature decreasing towards the core reduces the size of the plastic zone in front of the crack tip, thus lowering the resistance to crack-propagation. At the same time, the temperature difference decreases so that the stress is reduced (simplified: $\sigma = E \cdot \varepsilon = E \cdot \alpha \cdot \Delta T$) and the crack-propagation velocity then drops again (Figure B.5.12). The work-

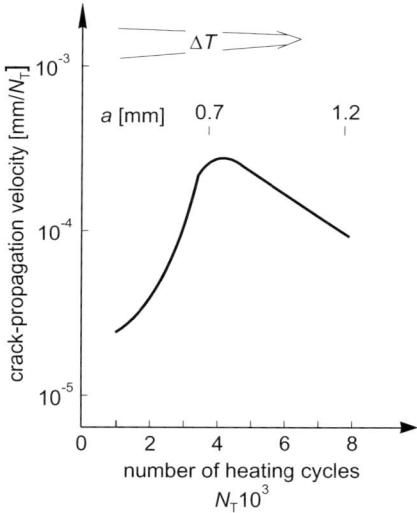

Fig. B.5.12 Thermal fatigue of hot-work tool steel X40CrMoV5-1: The results measured on a test stand show that the crack depth a increases with the number of cycles, but the temperature amplitude ΔT at the crack tip decreases with the crack depth (from H. Berns, F. Wendl).

piece material can now penetrate these thermal shock cracks, thus patterning the surface of the workpiece and the parts 'stick', to the tool. These cracks can be driven even deeper by the wedge effect and may ultimately fracture the tool. The onset of cracking can be delayed by a good cleanness and high ductility as well as a high hot proof stress or creep limit. The higher the fracture toughness, the deeper the crack can penetrate before the tool eventually fractures. This parameter, as well as the impact energy of unnotched specimens, illustrates the influence of the production method on the toughness of high-alloy hot-work tool steels (Figure B.5.13).

ESR lowers the amount of sulphides, and treatment of the melt with calcium spheroidises them. HIP and HIPW are powder metallurgy methods and thus allow production of a macrosegregation-free steel. The thermomechanically treated condition involves diffusion-annealing to decrease microsegregations and to spheroidise the sulphides, followed by hot working to a temperature of < 850°C to avoid the formation of a carbide network on the grain boundaries (Figure B.5.14).

There is comparatively little information available on chemical and tribological loading. The wear of forging dies decreases with the amount of alloying elements that form secondary carbides and it thus increases the creep resistance by precipitation hardening. In addition to surface oxidation by atmospheric oxygen, the chemicals in the workpiece material (glass and aluminium melts) can also corrode the tool.

B.5.3.2 Applications

Primary forming of very viscous glass melts is carried out in compression moulds of e.g. X23CrNi17, quenched and tempered to ≈ 950 MPa. The high chromium content of this steel provides a higher resistance to high-temperature corrosion. Casting moulds and pressure casting dies for copper alloys are made of X32CrMoV3-3 or a variant with 3 % Co, quenched and tempered to 1100 - 1400 MPa. In contrast, steels such as X40CrMoV5-1, quenched and tempered to 1450 - 1600 MPa, are used for aluminium alloys. Alloying variants of this steel are aimed at increasing the toughness. Lowering the C content to ≈ 0.35 % reduces the carbide content, and lower amounts of

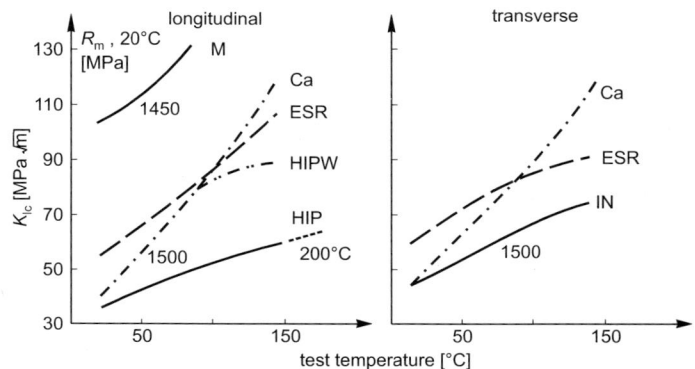

Fig. B.5.13 Influence of the production method on the fracture toughness K_{Ic} of hot-work tool steel X40CrMoV5-1: Quenched and tempered to $R_\text{m} =$ 1450 or 1500 MPa. IN = typical ingot casting, Ca = melt treated with calcium, ESR = electroslag remelting, HIP = hot-isostatic pressing, HIPW = HIP + hot-working, M = thermomechanical hot working (from F. Wendl).

trace elements reduce temper embrittlement. Recently developed steels have a reduced silicon content of $\approx 0.3\,\%$. This delays precipitation of carbides during tempering, thus slightly raising the secondary hardness; however, it also means faster carbide dissolution during hardening. It is therefore recommended that the austenitising temperature is reduced to 1000°C to counteract grain growth. Zinc alloys can also be processed with low-alloy hot-work tool steels. Expensive, vacuum-remelted maraging steels, such as X2NiCoMoTi12-8-8 (Figure B.2.23, p.201), are used for particular die inserts and cores for

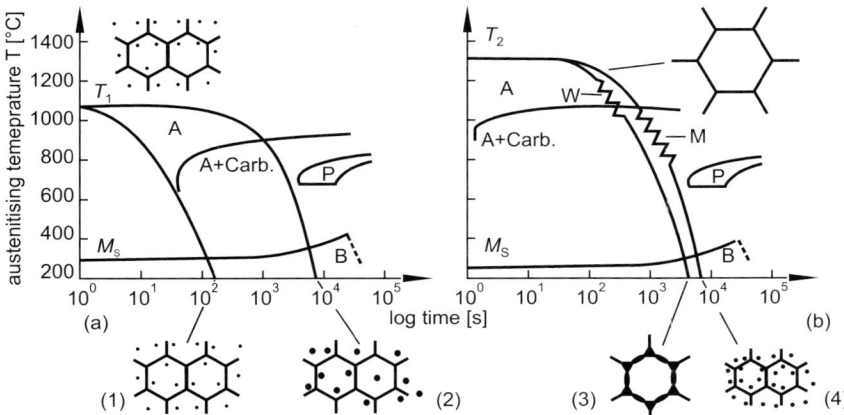

Fig. B.5.14 Precipitation of alloy carbides in austenite (schematic of X40CrMoV5-1): (a) The hardening temperature $T_1 = 1060°C$ has been selected so that approx. 1/4 of the carbides in the initial soft-annealed state are not dissolved in order to avoid grain growth. During cooling of a thin cross-section, the size and distribution of these annealed carbides is preserved (1). Cooling of a thick cross-section leads to the growth of the annealed carbides in the austenite/carbide field (A+Carb.); however, this does not affect their dispersed distribution (2). The austenite subsequently transforms to bainite/martensite. (b) For an initial forging temperature of 1180°C and particularly at a diffusion annealing temperature of $T_2 = 1280°C$, the annealed carbides are dissolved and the grains are coarse. The onset of carbide reprecipitation on cooling is shifted to a higher temperature and shorter time. Both thin and thick cross-sections pass through the A+Carb. field after forging (W). Owing to the lack of annealed carbide nuclei, the grain boundaries act as nucleating sites because they provide the necessary interfacial energy. A brittle carbide network (3) forms before the coarse-grained austenite transforms to bainite/martensite. However, if the material is subjected to thermomechanical hot working (M, cf. Figure B.2.1, p. 166) in the A+Carb. field, nucleation takes place within the grains. This results in finely dispersed carbides. The fine-grained austenite subsequently transforms to bainite/martensite. Thermomechanical forging avoids embrittlement of the grain boundaries by the alloy carbides MC, M_7C_3 and $M_{23}C_6$ (4) (from H. Berns, F. Wendl).

processing of Al-alloys. They are greatly superior to conventional secondary hardenable grades with respect to their hot hardness and toughness. This and their excellent cleanness delays cracking under thermal shock conditions and has even increased the service life by up to 300 % in one case.

The contact time is a key factor when choosing a tool steel for a forming application. Suitable grades for hot rolling include casting steels with $> 1\,\%$ C or chill-casting grades with $> 3\,\%$ C because the wear resistance is of particular importance in this case. High-speed finish rolls in wire lines are sometimes equipped with rings of sintered hard metals. Hammer dies used for forging flat blades are also heated only a little so that surface-hardenable cold-work tool steels with a low hardness penetration and a low tempering resistance are used. More creep-resistant steels, such as 55NiCrMoV6, are used for thicker workpieces, and 56NiCrMoV7, quenched and tempered to 1250 - 1450 MPa, is used for large workpiece cross-sections. Deep engravings, in particular, as used in e. g. forging of crankshafts, requires separating agents whose high vapour pressure facilitates immediate ejection of the workpiece. This also delays tool softening. Preheating of the forging die and many other tools is also important to avoid brittle fracture of already cracked engravings. Mechanically or hydraulically driven forging machines operate with a longer contact time than hammers. Therefore, hot extrusion, e. g. of bolts or deep bowl-shaped objects, require high-alloy hot-work tool steels (e. g. X32CrMoV3-3) or high-speed types of steels with hardnesses of about 45 or 60 HRC, respectively. The wear resistance of dies increases with the content of elements that increase the creep resistance. Practical tests have shown that molybdenum and vanadium are more effective than chromium by a factor of five and twenty, respectively. The saddles of hydraulic open-die forging presses can be made of nickel-base alloys, such as NiCr20Co18Ti, that are strengthened by precipitation of the γ'phase (Fig. B.5.11).

Steel X32CrMoV3-3, with strengths of 1500 to 1650 MPa, is used for forging brass fittings on an eccentric press. Extrusion dies for processing copper alloys are also made of this steel. Steel X15CrCoMo10-10-5 can be used for simple profiles, and a hard cobalt alloy for wire extrusion. The inner sleeves of the container, which usually consist of three shrunk-fit parts, are subjected to long contact times. Copper alloys can be successfully processed with tools made of steel X5NiCrTi26-15, precipitation-hardened by the γ' phase to ≈ 1000 MPa. Extrusion of light metal alloys is often carried out using tools made of 5 % CrMoV steels, e.g. X40CrMoV5-1 with a strength of between 1400 and 1600 MPa, optionally alloyed with 0.01 % Nb. Niobium carbides precipitated during tempering shift the decline of the curve to a higher temperature and keeps the size of the Cr carbides low. This increases the thermal shock resistance.

Tools for cutting hot strip or billets are made of X32CrMoV3-3 with an optional hard-facing of a creep-resistant weld deposit. The thermal loading of deburring tools depends on the residence time after die-forging. Red-hot burrs can be processed with e.g. X32CrMoV3-3, which can be slightly carburised during hardening to increase the wear resistance. Optional nitriding can also increase the service life, similarly to many other hot-work tools.

The development of hybrid technologies and duplex layers based on a combination of plasma nitriding (PN) and vapour deposition enables thin films to be applied to hot-work tools that have a comparatively low substrate hardness. The nitrided layer provides sufficient support underneath the thin film, whose high hardness and dense homogeneous structure provide a higher wear resistance. In practice, both arc PVD and PACVD coatings are applied to PN layers. A secondary hardened X38CrMoV5-1 substrate with a PN receives a TiN followed by a TiCN layer in a continuous vacuum process. The top layer is a TiC coating, which has been successful on precision forging dies at temperatures of up to approx. 1000°C on account of its high-temperature stability. Trials have also been carried out with PVD and PACVD coatings on X40CrMoV5-1 used for inserts and cores of moulds for pressure die casting of Al alloys. The best results were obtained with PVD coatings of CrN monolayers owing to their lower tendency to start melting. Because PVD cannot coat bores properly, PACVD technology is still being developed to overcome this drawback.

Industrial semi-hot forging (600 - 900°C) is becoming increasingly popular. This requires steels that have a creep resistance and wear resistance similar to high-speed steels (HSS) but with the toughness of hot-work tool steels. Steels such as X55CrMoWVCo4-2-1-1 or PM steel X56CrMoWV5-3-2-1 have matrix compositions similar to those of high-speed steels; however, they are low-carbide and thus tougher than HSS. Completely different microstructural concepts are achievable using powder metallurgy. Mixing a carbide-free HWS powder (X38CrMoV5-1) with 15 % coarsely powdered HSS (HS6-5-4) produces a double dispersion microstructure with dispersed carbide clusters. They have a significantly improved resistance to abrasion and metal/metal wear. The fracture toughness K_{Ic} remains at the high value of a pure HWS because any existing crack has to work its way through a tough matrix over long distances and because the carbide clusters are surrounded by a tough austenitic shell arising from a diffusion reaction between the HWS and the HSS.

B.5.4 Tools for machining applications

Fe-base tools are generally made of HSS because their blades are subjected to heating and wear. During continuous and discontinuous cutting, they come into high-pressure contact with the surface of the workpiece. Increasingly higher cutting speeds require an increasingly higher hot hardness, wear resistance, oxidation resistance and increasingly lower adhesion tendency. At carbon contents between 0.8 and 1.4 %, high contents of tungsten, molybdenum and vanadium lead to the formation of primary and eutectic carbides in a secondary-hardening matrix. Because tungsten and molybdenum are homologous elements and have a similar effect, they can be exchanged in a ratio of 2:1

(Tab. B.5.5) on account of their atomic weights (Tab. A.1.4, p. 15). There is no appreciable increase in the cutting efficiency above 18 to 20 % (W + 2 Mo). This initially led to the following basic compositions: tungsten steels HS18-0-1 with ≈ 0.75 % C and HS12-1-2 with ≈ 0.95 % C, molybdenum steel HS2-9-1 with ≈ 0.8 % C and tungsten-molybdenum steel HS6-5-2 with ≈ 0.9 % C. Additional carbon and vanadium increase their carbide contents, and their creep strength is improved by adding cobalt. The large fluctuations in alloying costs shifts the market demand from one basic composition to another.

Table B.5.5 High-speed steels (EN ISO 4957): Common high-speed steels with annealed/service hardnesses and fabrication

W-Mo-V-Co	C content [%]	Annealed hardness [HB]	Service hardness [HRC]	Fabrication
HS6-5-2	0.87	240 - 300	60 - 64	IM, PM
HS6-5-3	1.2	240 - 300	62 - 65	IM, PM
HS6-5-2-5	0.92	240 - 300	62 - 67	IM
HS10-4-3-10	1.3	240 - 300	60 - 64	IM
HS12-1-4-5	1.4	240 - 300	61 - 65	IM
HS18-1-2-5	0.8	240 - 300	63 - 66	IM
HS2-10-1-8	1.1	240 - 300	67 - 69	IM
HS6-5-4	1.3	< 280	64 - 66	PM
HS4-3-8	2.5	< 300	58 - 66	PM
HS6-7-6-10	2.3	< 340	60 - 69	PM

IM = ingot metallurgy, PM = powder metallurgy

B.5.4.1 Properties

During solidification (Figure A.1.9 b, p. 17), M_6C, M_2C and MC carbides precipitate as a net around the metal cells in the advancing front (cf. Figure B.5.3 a, p. 277). The growth of M_6C is promoted by tungsten, silicon and nitrogen. M_2C carbides are promoted by molybdenum, carbon, vanadium and rapid solidification. The precipitation of MC increases with the content of V as well as Nb and Ti. The solidus temperature of tungsten steels is slightly higher than that of molybdenum steels, and only slightly above 1200°C in segregations. This leads to a narrow temperature range for hot forming of these steels, which are difficult to forge owing to their hot strength and carbide content. During soft annealing, the fine lamellar M_2C carbides decompose to smaller M_6C and MC carbides compared to directly precipitated M_6C. Moreover, the M_6C carbide contains ≈ 35 % iron compared to ≈ 5 % in M_2C. This lowers the alloying costs with respect to the carbide volume.

During hardening, the austenitising temperatures must be kept close to the solidus temperature to dissolve the secondary alloy carbides (Figure B.5.15 a). This process is facilitated by chromium and by carbide refinement due to the decomposition of M_2C. Overheating or overtiming leads to coarsening, blocky carbides and ultimately to partial melting. In this case, cooling produces networks of ledeburitic casting structures that adversely affect the toughness. Molybdenum steels are more susceptible to this than tungsten steels. The hardenability of HSSs is extremely high, so that blasting with compressed nitrogen in a vacuum furnace leads to through-hardening, even in thick cross-sections. Austenitizing in a hot bath is usually followed by cooling in a warm bath at $\approx 500°C$ to avoid precipitation of grain boundary carbides (Figure B.5.15 b). Subsequently, slender tools can be straightened in the austenitic state while hot, if hardening distortion has occurred. The martensitic transformation will start during further air cooling. Oil-hardening of thick cross sections may cause carburisation of the surface layer. According to Figure A.1.9 b (p. 17), this lowers the solidus temperature, thus increasing the risk of partial melting. The amount of retained austenite increases as the hardening temperature increases (Figure B.5.15 a). During tempering, precipitation hardening of the martensite by MC and M_2C carbides becomes more pronounced the higher the hardening temperature and with it the content of dissolved molybdenum, chromium and vanadium in the matrix (Figure B.5.15 c). According to H. Fischmeister et al., they are responsible for the 'matrix potential' i.e. for the creep resistance of the matrix. Even after ≈ 3 volume-% carbides have precipitated during tempering, there is still more than twice that amount available for precipitation at the service temperature. Precipitation of carbides

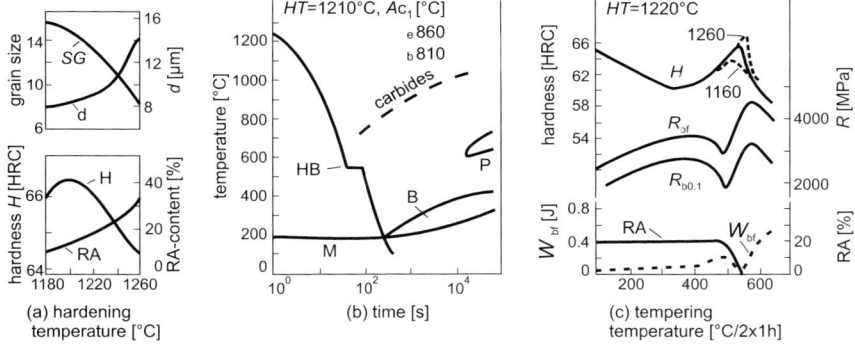

Fig. B.5.15 Heat treatment and properties of high-speed steel HS6-5-2: (a) Influence of the hardening temperature on the hardness H, the proportion of retained austenite RA, the Snyder-Graff grain size SG (number of grains on a line 125 µm long) and the mean grain diameter d. (b) Continuous TTT diagram according to DIN 17350 with a schematic cooling curve for hardening in a hot bath HB (hardening temperature $HT = 1210°C$, cf. Figure A.1.9 b). (c) Influence of tempering on the proportion of retained austenite RA, the plastic work of bending to fracture W_{bf}, the 0.1 bending limit $R_{\text{b0.1}}$, the bending strength R_{bf} and the hardness H (a and c from E. Haberling).

in retained austenite during tempering raises its M_f temperature so that martensite is formed on cooling, which is then precipitation-hardened during a second tempering step. Core segregations and cobalt-rich HSS may still contain some retained austenite after the first tempering treatment. This only decomposes during a second tempering treatment, so that three tempering steps have to be carried out in these cases.

The relationship between the alloy composition and the properties can be summarised as follows: dissolved *carbon* promotes the hardening capacity. It performs the following functions when precipitated as carbides: primary and eutectic carbides - wear resistance; soluble secondary carbides - hardenability; temper carbides - precipitation hardening and creep resistance. The carbon content in steel HSC6-5-2 is increased to 1 %.

Tungsten forms M_6C carbides, hardly contributes to precipitation hardening and is used for solid-solution strengthening of the matrix. The addition of *molybdenum* leads to the precipitation of acicular M_2C carbides that increase the overheating sensitivity and adversely affect the grindability. This is why their decomposition to M_6C and MC carbides is particularly important. Molybdenum increases the matrix potential and contributes to precipitation hardening during tempering.

Vanadium forms MC carbides and dissolves in M_2C. The MC carbide is harder than the other types (Figure. A.4.16, p. 103), which has a positive effect on the wear resistance. An increase in the vanadium content increases the content of MC carbides if carbon is also added in approximately stoichiometric quantities (example: HS6-5-3 with $\approx 1.22\,\%$ C). The use of *niobium* to produce MC carbides leads to a poorly soluble primary carbide so that the amount of wear-reducing carbides can be increased independently of the matrix potential. The higher precipitation temperature of primary carbides compared to eutectic carbides may result in undesirable coarse MC, so that alloying with Nb and Ti are usually restricted to PM methods. A vanadium content of at least 1 % is necessary to achieve a sufficient matrix potential for precipitation hardening during tempering.

Chromium increases the solubility of secondary carbides and thus the hardenability. It participates in precipitation hardening. HSSs therefore contain $\approx 4\,\%$ chromium.

Cobalt increases the solidus temperature and thus the possible hardening temperature. It refines the precipitates during precipitation hardening and delays their growth without being included. Furthermore, it promotes the formation of clusters (initial stage of precipitation) as well as solid-solution strengthening of the matrix. Cobalt is added in amounts of up to $\approx 10\,\%$.

Sulphur is added in amounts of 0.06 to 0.15 % to improve the machinability (example: HS6-5-2 S). The addition of 0.05 % titanium produces finely dispersed titanium carbonitrides in the melt that contribute to the nucleation of spheroidal sulphides. Their hardness can be regulated via the Mn/S ratio. The toughness of S-alloyed HSSs can be improved by influencing the sulphide shape.

Aluminium lowers the carbon solubility in austenite and is thus used in more recently developed steels. In HS6-5-2 it increases the proportion of M_2C, but lowers its stability. As a consequence, the eutectic network microstructure can be broken up more easily during forging on the one hand, and it promotes the equilibrium decomposition of M_2C into MC and M_6C on the other. Al also reduces the solubility of carbon in martensite and thus increases the potential for precipitation of secondary carbides. This increases the hot hardness, wear resistance and the thermal shock resistance.

High-speed steels have a higher bending strength compared to carbide-rich cold-work tool steels (Figure B.5.6, p. 281). The primary and eutectic carbides and the carbide bands are, on average, somewhat smaller than in ledeburitic chromium steels (Figure B.5.3 c). Under fatigue conditions, the number of stress cycles to crack initiation is inversely proportional to the carbide size. The rotary bending test thus indicates a higher service life for HSS compared to ledeburitic chromium steel (Figure B.5.7 a). Production by powder metallurgy or spray-forming with subsequent hot forming refines the carbides and helps to increase the toughness. However, the wear resistance generally decreases. Nevertheless, beneficial effects include improved grindability and spark erosivity. The lack of macrosegregations improves the hot formability and lowers the anisotropy of size changes. PM steels with a greater overall carbide surface area can be hardened at a lower temperature. Additional vanadium carbide can be added to increase the wear resistance of PM steels compared to conventional steels (example: HS6-7-6-10 PM with 2.3 % C). Such compositions are similar to those of PM-CWS that have a high vanadium content and contain the monocarbide; those with higher Mo and W contents approach the range of PM-HSSs. A further development is based on the use of niobium to form MC carbides with a view to reducing the amounts of tungsten and molybdenum. PM steel HS2-3-2 with 3.2 % Nb and 1.3 % C has a cutting efficiency similar to HS6-5-2, but it is less sensitive to overheating. Owing to their good combination of strength and toughness, high-speed steels, produced conventionally and by powder metallurgy, are also used to make cold-work tools. Because toughness is more important than creep resistance for this application, the hardening temperature is lowered by 50 - 80°C.

B.5.4.2 Applications

Tab. B.5.5 gives an overview of common high-speed steels. The longest service life for continuous cutting is obtained with carbide- and cobalt-rich steels such as HS10-4-3-10 or HS12-1-4-5. In contrast, for discontinuous cutting applications and a higher workpiece strength, tougher steels, such as HS6-5-2-5, are more popular. This also applies to milling cutters. The carbide size and segregations increase and the degree of plastic deformation decreases as the workpiece cross-section increases. There may be hardening cracks in core segregations (Figure B.5.16). These defects can be avoided and the toughness improved by using the ESR process or by assembling the tool from a tougher body and HSS elements (e. g. roughing cutters or large broaching tools). PM steels are suitable for delicate tools, e. g. tap drills, on account of their good toughness. They can also be used as precision cutting punches or cold-heading punches because of their high yield strength.

HSS have a lower wear resistance and hot hardness than hard metals, cutting ceramics, cubic boron nitride (CBN) and polycrystalline diamond (PCD). Nevertheless, their good machinability in the annealed state (hardness 240-300 HB), their comparatively good toughness in the hardened and tempered state as well as their less costly manufacture offer decisive advantages over the above-mentioned cutting materials with a high hard-phase content that can only be produced from powder. High-speed steels protected by thin films have a comparable or even superior performance to cutting materials containing hard phases. Hard phases deposited by PVD have proven to be effective coatings (Tab. B.5.2). The high temperature of the secondary hardening maximum means that PVD coatings as well as PACVD coatings can be applied well below the tempering temperature (Figure B.5.9, p. 285 cold-work tool steel). In view of the increasing thermal load due to the high cutting speeds, the usual TiN coatings are being increasingly replaced by more complex coating systems. Multilayer TiAlN coatings, some with gradients, are

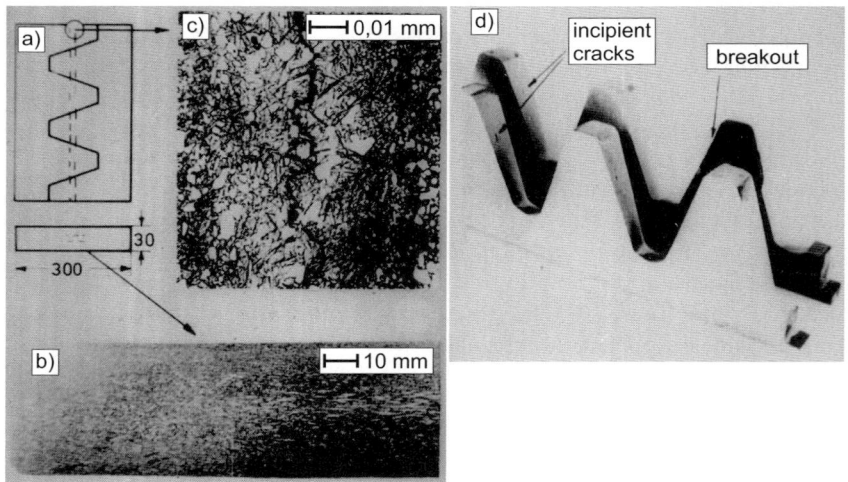

Fig. B.5.16 **Hardening cracks due to segregations**: High-speed steel HS18-1-2-5. (a) Sawing line of a soft-annealed flat bar to produce two toothed profiles. (b) Macroetching of the bar's midline cross-section reveals visible core segregation. (c) Core microstructure with a hardening crack after hardening at 1250°C/warm bath and 540°C/air. (d) Position of core cracks at half the tooth height in the region of the core segregation (from H. Berns).

proving successful on account of their excellent fracture toughness, greater hardness and thermal stability. If the Al concentration in the top layer is high, an Al_2O_3 layer forms on the surface during cutting that provides protection against oxidation. Alternatively, an Al_2O_3 layer can be applied as a final or intermediate layer. Nanocrystalline multilayer coatings with a thickness of less than 3 µm are suitable for high-performance machining of QT and tool steels (even in the hardened state), cast iron as well as titanium and nickel alloys. For example, a TiAlN coating increased the number of drilled holes in GJL by a factor of 2.5 compared to TiCN coatings at different cutting speeds (15 and 33 m/min). The high hardness of TiAlN coatings is required for machining of cylinder blocks made of GJV (Chap. B.1, p. 160). The desirable vermicular shape of the graphites in GJV requires the addition of titanium. As superhard TiC, it leads to heavy abrasive wear of high-speed steel coated only with TiN. Multilayer nanocrystalline diamond coatings have proven successful for machining WC/W_2C and CrB_2 high-Si aluminium alloys, titanium alloys, carbon- and glass-fibre-reinforced plastics as well as laminates (Tab. B.5.2, p. 284). For example, the cutter life for milling AlSi17 was increased by a factor of 5 compared to a TiCN coating.

During machining of calcium-treated steels with coated tools, a chemical reaction takes place between the non-metallic inclusions in the workpiece material and the coating of the cutting material that has a wear-reducing effect. Calcium aluminates enveloped by a sulphide shell (MnS) in the steel react with the titanium in the cutter coating to produce a lubricating and protective layer that reduces abrasive wear of the cutting material.

Charging rack of heat-resistant austenitic cast steel
(SCHMOLZ + BICKENBACH Guss GmbH & Co. KG, Krefeld, Germany)

B.6 Chemically resistant materials

The matrix of chemically resistant ferrous materials consists of ferrite, austenite, martensite or a mixture of these phases. They can be divided into three groups based on the precipitation state of the carbon:

- steels containing dissolved carbon or only a small amount of carbides
- steels containing fine secondary carbides
- steels and cast irons containing coarse eutectic precipitates (carbide, graphite)

B.6.1 General information

The chemical resistance of the iron matrix (Chap. A.4, p. 109) must be transferred to multiphase materials by applying a suitable alloying concept. The matrix properties are of particular importance.

B.6.1.1 Alloying concept

Chemically resistant materials should withstand the corrosive influence of weathering, chemical substances or hot conditions for as long as possible. This resistance is based on surface films that form as a result of chemical attack and which considerably reduce further corrosion. Wet corrosion leads to the formation of a thin passivating layer (Chapt. A.4, p.111) of metal oxides, e.g. M_2O_3 which also contains a certain amount of metal hydroxides. Their structure changes from crystalline to densely amorphous as the chromium content increases. Chromium contents above $\approx 12\,\%$ have a passivating effect on the matrix. Such steels are termed 'stainless'. During high-temperature corrosion, chromium also plays a decisive role in the barrier effect provided by a thicker surface film of scale by forming spinels $FeCr_2O_4$ and mixed oxides of type $(Cr,Fe)_2O_3$ (Chapt. A.4, p. 117). Chromium is thus the key alloying element in heat-resistant steels (Figure B.6.1); aluminium, silicon and titanium are also important. At high carbon contents, chromium is bound as chromium-rich carbides and its passivating effect on the matrix is lost. This is why nickel is used in grey cast iron, although its protective surface film does not quite reach the level of protection afforded by chromium. Copper increases the resistance of the matrix; however, it also promotes precipitation of graphite. Aluminium can only rarely be used in sufficient quantities because it reacts with air and the mould material. Silicon improves the scaling resistance of the matrix. Graphite, on the other hand, burns out so that a coherent lamellar structure has to be spheroidised. This phase is, however, resistant to wet corrosion. In wear- and chemically resistant materials, chromium enrichment of the carbides is compensated either by increasing the chromium content or it is shifted to MC by the addition of V, Nb, Ti so that chromium remains in the matrix. The carbides are usually more resistant than the matrix owing to

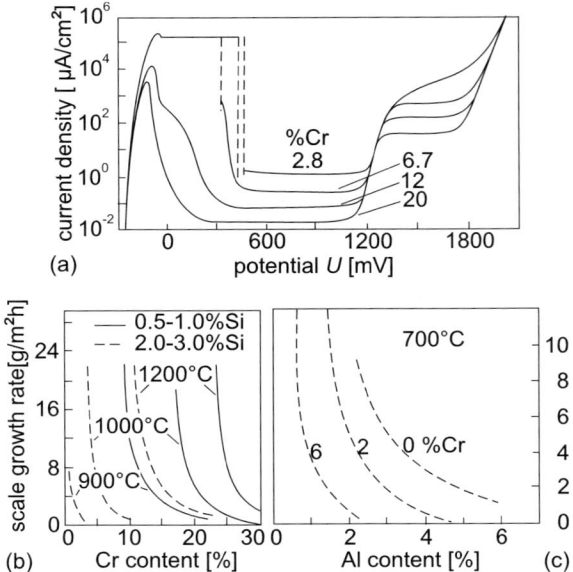

Fig. B.6.1 Influence of the alloy composition on the formation of corrosion-inhibiting surface films: (a) Influence of chromium on the passive current density and broadening of the passive range with respect to wet corrosion in 10 % aqueous sulphuric acid at 25°C (from R. Olivier, cf. Figure A.4.19). (b) and (c): Influence of Cr, Si and Al on the scale growth rate during high-temperature corrosion in air over a period of 120 h (from E. Houdremont and G. Bandel).

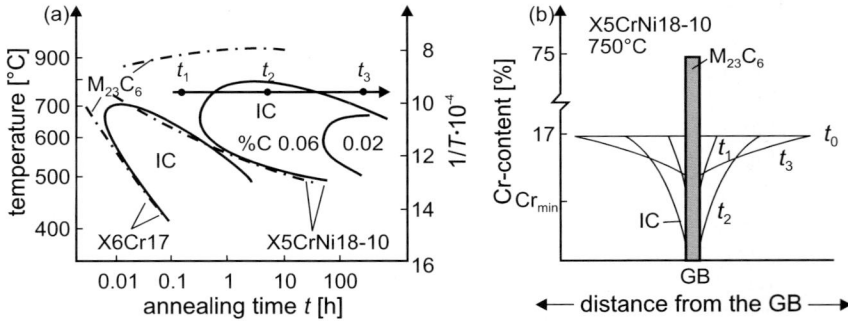

Fig. B.6.2 Intergranular corrosion (IC, schematic): (a) Susceptibility develops more quickly on heating ferritic steel X6Cr17 compared to austenitic steel X5CrNi18-10. It can be delayed by lowering the C content to 0.02 %. (b) Precipitation of chromium-rich carbide lowers the chromium content in the vicinity of the grain boundary (GB). After an annealing time of t_2, the content lies below the minimum value Cr_{min} required for passivation. During holding, the chromium deficit is compensated by post-diffusion of Cr from the interior of the grains until the steel loses its susceptibility to intergranular corrosion at t_3.

their enrichment with Cr and Ti. In contrast, W, Mo or Nb carbides may have insufficient chemical resistance, particularly at elevated temperatures, so that the hard phases decompose before they can contribute to the wear resistance.

In multiphase materials containing chromium-rich carbides or spheroidal graphite, the chemical stability depends on the matrix. A uniform surface film can only develop if the effective elements are evenly distributed in the microstructure, i.e. dissolved. Chromium leads to ferrite, nickel to austenite (Figure A.1.10, p. 18 and A.2.1, p. 21) and both in combination to austenite or martensite. The Schaeffler diagram in Figure A.2.24 shows the alloying limits for the solution-annealed, quenched state. As a consequence of their differing solubilities (Tab.A.1.4, p. 21), mixed microstructures exhibit a slight enrichment of Cr, Si, Mo ... in the ferrite and a slightly increased content of Ni, Mn, Cu ... in austenite/martensite. Dissolved carbon and nitrogen lead to reprecipitation of chromium-rich phases on heating during manufacturing or whilst in service so that the matrix is depleted in chromium, which has an adverse effect on its chemical resistance and toughness. Most stainless steels thus contain only a little C and N, but even in this case, heating in the temperature range in which chromium diffusion takes place leads to local depletion of chromium and susceptibility to corrosion. This undesirable affinity between the protective element and carbon does not occur in nickel-alloyed cast iron with an austenitic matrix because precipitation of C does not withdraw Ni from the matrix so that some of the mechanisms governing localised corrosion are not active.

B.6.1.2 Matrix properties

Owing to their different atomic arrangements, a chemically resistant bcc matrix has other service and manufacturing properties compared to an fcc matrix. This will be illustrated using a few examples.

Corrosion resistance: The smaller octahedral interstices (Figure A.1.3, p. 8) in the bcc lattice and the ≈ 100 times higher diffusion coefficient (Figure A.1.8, p 16) increase the carbide precipitation tendency in ferrite. This adversely affects the microstructural homogeneity of stainless steels and thus its resistance to uniform corrosion. Susceptibility to intergranular corrosion (IC) due to $M_{23}C_6$ precipitated along the grain boundaries (Figure B.6.2) occurs more quickly in a ferritic matrix, sometimes even during quenching.

On the other hand, this matrix is not susceptible to anodic stress corrosion cracking, unlike e.g. austenitic steel X5CrNi18-10, unless, however, the Ni content is more than twice as high, as is the case in some cast irons. Martensitic and unstable austenitic matrices can be damaged by cathodic stress corrosion cracking.

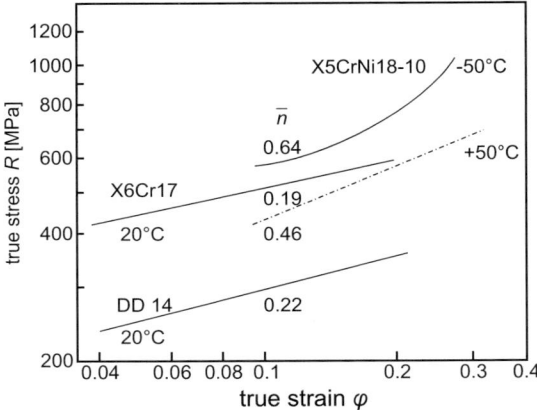

Fig. B.6.3 Comparison of flow curves: Compared to unalloyed deep-drawing steel DD14, the flow stress R in stainless ferritic steel X6Cr17 is higher owing to solid-solution hardening by chromium. Although austenitic stainless steel X5CrNi18-10 has a lower yield strength than X6Cr17, it undergoes greater strengthening, which is expressed by a higher mean strain hardening exponent $\bar{n}$. Lowering of the working temperature increases R and $\bar{n}$ due to higher proportions of deformation-induced martensite (from W. Küppers et al.).

Table B.6.1 Comparison of the properties of a ferritic and an austenitic steel: test temperature 20°C; α from 20 to 500°C,*) market price for 2 mm-thick sheets.

Microstructure	Ferrite	Austenite
Steel	X6Cr17	X5CrNi18-10
proof strength $R_{p0.2}$ [MPa]	≥ 270	≥ 185
tensile strength R_m [MPa]	450-600	500-700
elongation at fracture A [%]	≥ 20	≥ 50
yield ratio $R_{p0.2}/R_m$	≈ 0.75	≈ 0.45
strain hardening exponent $\bar{n}$	0.2	0.45
normal anisotropy r_m	1.0-1.2	0.8-1.2
specific electr. resistance ρ [$\Omega \cdot$m]	0.6	0.75
thermal conductivity λ [W/m·°C]	25	15
coeff. of therm expansion α [m/m°C]	$12 \cdot 10^{-6}$	$18 \cdot 10^{-6}$
costs [€/kg]*)	1.3	2.0

Heat resistance: Ferritic and austenitic matrices with the same chromium contents have a comparable resistance to scaling in air. The higher thermal expansion of the austenitic matrix adversely affects the adhesion of scale during thermal cycling. Nickel has an unfavourable effect in S-containing gases because low-melting nickel sulphides can form. In contrast, Ni is advantageous in C-containing gases because it increases the carbon activity in austenite and lowers its solubility. An increase in the carbon activity in the gas initially leads to carbide formation. If the activity is higher than 1, the carbides in the steel are graphitised. The resulting damaged surface layer consists of ultrafine steel particles surrounded by graphite and are easily abraded. This type of corrosion is known as metal dusting. It is usually caused by damage to the protective oxide layer in colder sections of the equipment with a higher C activity (Figure B.3.10 a, p. 235).

Strength: Ferrite usually exhibits a higher yield strength than austenite (Tab.B.6.1) because it has only two slip directions in each slip plane (Figure A.1.4, p. 8). Austenite undergoes a higher degree of strengthening and thus has a higher tensile strength (Figure B.6.3).

Lattice defects in martensitic matrices also have a strengthening effect (Chapt. A.4, p. 95). At elevated temperatures, the resistance of the more closely packed austenite to diffusion-controlled creep is reflected by a higher hot strength and creep resistance.

Toughness: A bcc matrix exhibits cold embrittlement below T_T (Figure A.4.5, p. 84), whereas a stable austenitic CrNi matrix does not. This embrittlement is intensified by the tendency of a transformation-free ferritic matrix to undergo secondary recrystallisation with grain coarsening, which increases T_T. In contrast, the transformation of a martensitic matrix leads to grain refinement. Therefore quenching and tempering of QT grades lowers T_T. A chromium-rich ferritic matrix decomposes into α' and α at temperatures of up to $\approx 550°C$ and this is associated with so-called 475°C embrittlement. At even higher temperatures, embrittlement is caused by precipitation of the chromium-rich σ phase or the χ and Laves phases in molybdenum-rich steels (Tab A.2.1, p. 52, Figure B.6.4).

Physical properties: A stable ferrite-free austenitic matrix is not magnetisable and has a higher electrical resistance than ferrite (Chapt. A.4, p. 119 and p. 121). Its thermal conductivity is significantly less than that of a ferritic matrix. An austenitic matrix has a higher thermal expansion as a consequence of its closer atomic packing, (Tab.B.6.1).

Cold workability: Ferritic chromium steels require higher metal-working forces than unalloyed deep-drawing steels, although they both undergo a similarly low degree of work-hardening (Figure B.6.3). An austenitic matrix has a lower yield ratio than a ferritic matrix (Tab.B.6.1), it starts to yield at an earlier stage and undergoes a higher degree of work-hardening. Decreasing temperatures and an increasing degree of deformation can lead to flow curve gradients of up to $n = 1.1$ in places, with a mean gradient of up to $\bar{n} = 0.7$. This corresponds to a high uniform elongation. The highest ductility is obtained via the TRIP (transformation-induced plasticity) effect resulting from the transformation of small amounts of austenite to martensite (Figure B.6.5). By adjusting the metal-working temperature (metal-working heat, preheated tools) to the steel composition, the austenite stability can often be sufficiently controlled to allow high degrees of plastic deformation. However, this requires high metal-working forces that increase the load on the tools.

Weldability: The higher diffusion rate in a ferritic matrix leads to grain coarsening, carbide precipitation on the grain boundaries and thus to embrittlement and intergranular corrosion. An austenitic matrix is usually less affected by this, but it does tend to undergo hot cracking owing to its 50 % higher thermal contraction. These cracks occur on the solidus line when the dendrite framework with its film of liquid residual melt is subjected to shrinkage stresses during cooling and thus tears. Primary ferritic solidification (Figure B.6.6, examples a-c) is associated with less shrinkage so that the likelihood of hot cracking is very low. However, in an austenitic steel such as X5CrNi18-10 (example c), the subsequent δ/γ transformation is usually incomplete and leaves a small fraction of δ-ferrite in the weld metal.

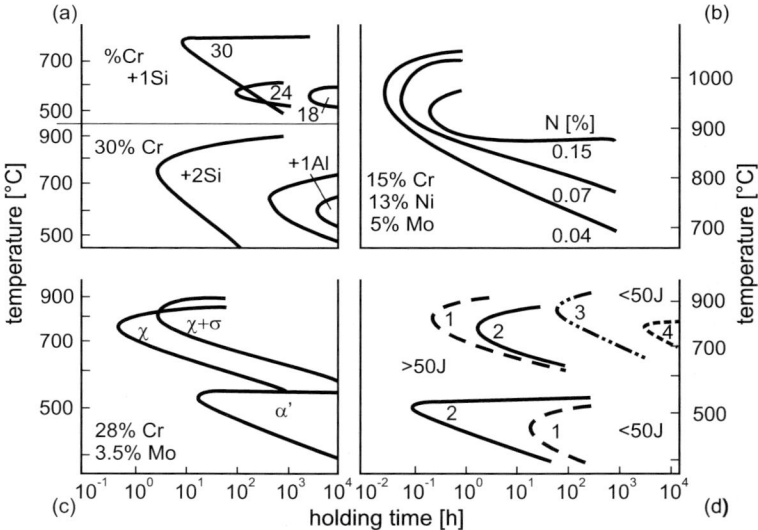

Fig. B.6.4 Precipitation of intermetallic phases: (a) Influence of Cr, Si and Al on the onset of precipitation of the σ phase in ferritic chromium steels (from K. Bungardt et al.). (b) Delay of χ-precipitation by nitrogen (from H. Thier). (c) Onset of χ- and σ-phase precipitation as well as 475°C-embrittlement by the α' phase (from H. Kiesheyer). (d) Precipitation-induced embrittlement; limiting curves for a notch impact energy of KV=50 J at 20°C, 1 = X2CrNiMoN22-5-3, 2 = X1CrNiMoNb28-4-2, 3 = X2CrNiMoN17-13-5, 4 = X5CrNiMo17-12-2 (from R. Oppenheim).

B.6.1 General information

In a martensitic matrix, a higher carbon content may cause cold cracking because tensile stresses develop in the hardened zone on further cooling and result in tears along the weld seam (Figure B.7.5, p. 355). These stresses can be reduced by preheating and their development below T_T may be avoided. Problematic regions with respect to the σ phase, α' decomposition, grain coarsening, hot and cold cracking depend on the alloy composition and can be approximated in the Schaeffler diagram (Figure B.6.7). Susceptibility to intergranular corrosion must also be taken into account (Figure B.6.2). In

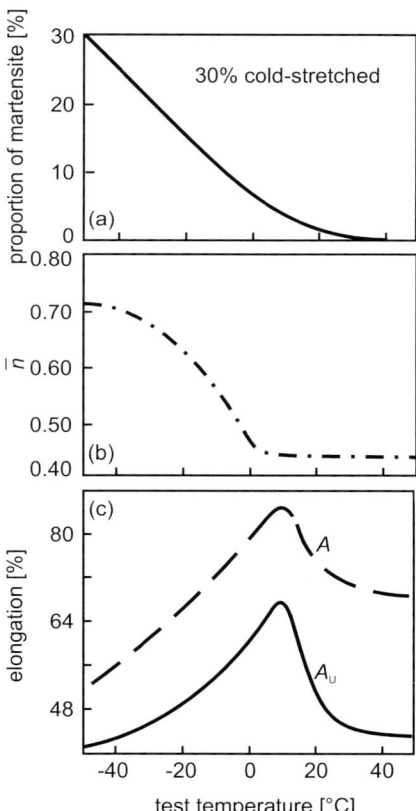

Fig. B.6.5 Deformation-induced austenite transformation: Steel X5CrNi18-10 (from W. Küppers): (a) Cooling alone does not induce transformation of austenite, whereas lowering of the stretching temperature in the tensile test leads to deformation-induced martensite (cf. Figure B.8.3, p. 375). (b) The mean strain hardening exponent $\bar{n}$ increases with the proportion of martensite. (c) Uniform elongation A_u and elongation at fracture A reach a maximum for 3 - 6 % deformation-induced martensite.

B.6 Chemically resistant materials

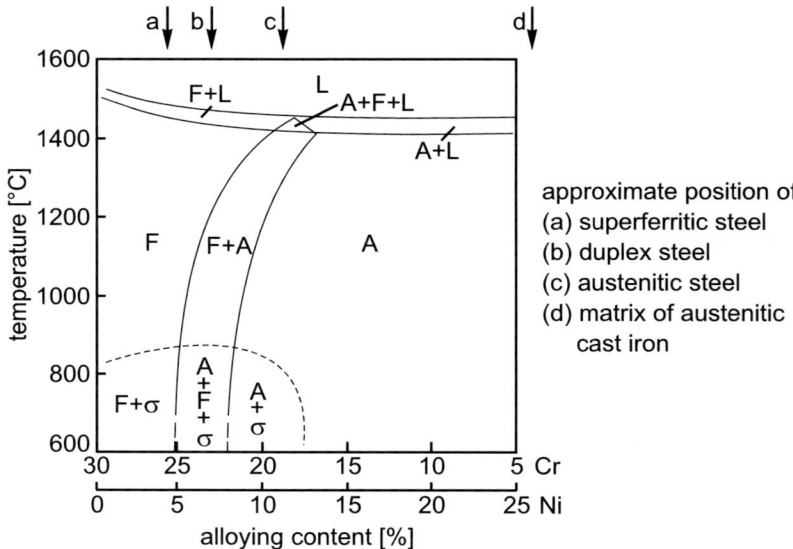

Fig. B.6.6 Isoplethal section through the Fe-Cr-Ni phase diagram at 70 % iron: (from P. Schafmeister and R. Ergang).

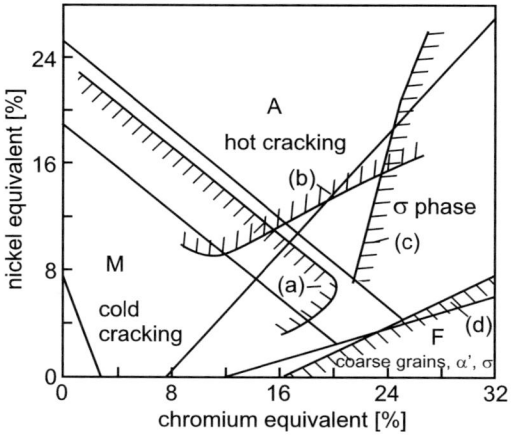

Fig. B.6.7 Problematic regions for chemically resistant steels: (from A. L. Schaeffler, cf. Figure A.2.24, p. 47). (a) Cold cracking in martensitic chromium steels after hot forming, welding or hardening owing to a high hardenability. (b) Hot cracking for predominantly primary austenitic solidification due to the greater thermal shrinkage of austenite. (c) Embrittlement caused by precipitation of the σ phase in δ-ferrite (rapid) and in austenite (slow, cf. Figure B.6.4 d). (d) Embrittlement of ferritic steels due to grain coarsening, σ phase and α' decomposition at ≈ 475°C.

cast iron, the high cooling rate shifts grey solidification towards white so that graphitising annealing is recommended. Modification of the weld filler may reduce some of the problems in the welded material, but it has very little effect on the heat-affected zone (HAZ).

Machinability: As a consequence of their low C content, most stainless steels are so ductile that they tend to produce continuous chips, which can be avoided by chip-breaking precipitates. The addition of 0.15 to 0.35 % S and $< 2\,\%$ Mn leads to the precipitation of MnS inclusions from the melt. However, they dissolve even in weak acids to leave craters that are difficult to passivate and which act as the starting point for further corrosion. This drawback can be remedied by binding sulphur with Ti to form TiS/Ti_2S inclusions that are more corrosion-resistant, although they do not improve the machinability quite as efficiently as MnS. A further development uses hexagonal titanium carbosulphides $Ti_4C_2S_2$, which are also corrosion-resistant and cause less flank wear than MnS. Ti-containing inclusions are elongated to bands far less than MnS during hot working owing to their greater hardness, and they thus improve the cold headability.

B.6.2 Stainless steels

The main objective for all chemically resistant materials is to increase the resistance to uniform corrosion by means of a suitable alloy composition. This is crucially dependent on the chemical composition as well as the concentration, temperature and flow of the corrosive aqueous medium. Resistance diagrams provide a rough indication of the suitability of a particular material. Some diagrams for acid corrosion are given in Figure B.6.8 For example, for the steel shown in (a), a corrosion rate of $1\,\text{g/m}^2\text{h}$ can be expected at 40°C in 20 % sulphuric acid, which corresponds to an approximate decrease in thickness of 1.1 mm/year (Chapt. A.4, p. 112). This value exceeds the commonly used resistance limit of 0.3 $\text{g/m}^2\text{h}$. The selected alloy composition of a stainless steel is also governed by two types of localised corrosive attack: pitting and intergranular corrosion (Figures B.6.9 and B.6.2). The first is caused by mechanical or chemical damage of the Cr-based passivating layer; the second is due to a local drop in the amount of Cr below that required for passivation. Chloride ions are very common (cleaners, deicing salt, seawater, etc.) along with the resulting pitting corrosion due to the removal of oxygen atoms from the passivating layer on stainless steels by chlorine. The resistance to this type of corrosion is improved by increasing the chromium content and by adding molybdenum and nitrogen according to a formula known as the pitting resistance equivalent number (*PREN*).

$$PREN = \%\,\text{Cr} + 3,3 \cdot \%\,\text{Mo} + a \cdot \%\,\text{N} \qquad (B.6.1)$$

Values published for a lie between 10 and 30. The prefix 'super' is often used for grades with PREN $>$ 40, e.g. superaustenitic steel.

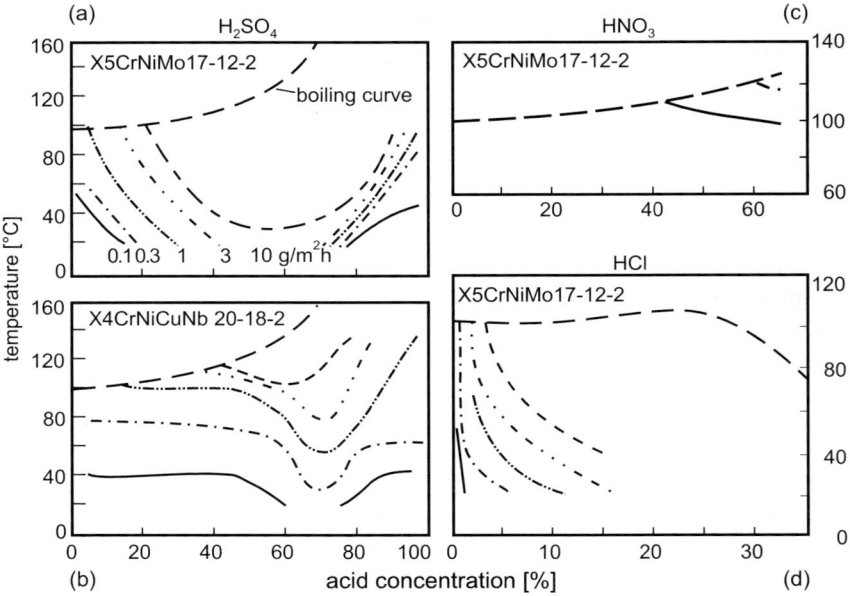

Fig. B.6.8 Resistance diagrams for uniform corrosion in aqueous acids: (data from Böhler AG). (a, b, d) In non-oxidising acids, such as H_2SO_4 and HCl, corrosion usually occurs in the active range and is retarded by Cu. (c) Oxidising acids, e.g. HNO_3 (< 65 %), passivate the steel.

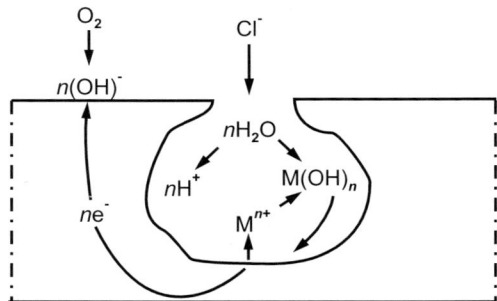

Fig. B.6.9 Pitting corrosion of stainless steels: Localised damage of the passivating layer, frequently caused by chloride ions, is followed by active metal dissolution (M) and pitting. An increase in acidity within the pit as a result of hydrolysis (Eq. A.4.15, p. 113) is particularly noticeable because e.g. Cr forms poorly soluble hydroxides. The pH value and breakdown potential drop. There is a large surface area for the cathodic half-reaction (Chap. A.4, p. 109) versus a small anodic surface area of the pit wall with a high dissolution rate.

Common fabrication steps such as welding, stress-relief annealing etc., may lead to intergranular corrosion. There are four ways of avoiding this: 1. Steel melts with a particularly low carbon content (ELC = extra low carbon). 2. Binding of the carbon as the more stable MC carbides. This is achieved by alloying (stabilising) with hyperstoichiometric amounts of $Ti \leq 7 \cdot \% C$ or $Nb \leq 12 \cdot \% C$ to suppress the formation of chromium-rich grain-boundary carbides. Titanium is more cost-effective and lighter. However, the lighter titanium carbides and carbonitrides rise to the surface during ingot casting and tend to deposit on the mould wall, which can have an adverse effect on the surface quality of flat products. Above 1150°C, they dissolve more quickly than niobium carbides, e.g. during welding. Because $M_{23}C_6$ precipitates faster than MC, the microstructure may become susceptible to intergranular corrosion during subsequent reheating in the critical range, e.g. during multi-pass welding. A hyperstoichiometric stabilisation ratio of M/C helps to suppress this knife-line attack in a narrow strip along the HAZ. 3. Increasing the chromium content. 4. Subsequent stabilising annealing. However, for 3 and 4, the embrittlement range of other precipitates, such as the σ or χ phase, must be avoided. The large number of stainless steels can be conveniently classified according to their microstructure, PREN and resistance to intergranular corrosion.

B.6.2.1 Properties

Stainless steels for general applications are specified in EN 10088. A selection of pressure vessel steels for forgings is given in EN 10222-5. Cold-heading and cold-extrusion steels are covered by EN 10263-5 and strip for springs by EN 10151. Steel castings are specified in EN 10283 and 10213 as well as SEW 410.

We differentiate between ferritic, austenitic and duplex steels; the microstructure of the latter consists of approximately equal proportions of ferrite and austenite. The C content is usually less than 0.1 % to increase the weldability and minimise localised corrosion. It can be more than 1 % in martensitic steels.

(a) Ferritic steels

The basic grade is X6Cr13 (Tab.B.6.2). In Figure B.6.10, based on Figure A.1.10 b, the austenitic phase field is extended by C, N and Mn so that at least partial transformation takes place. This effect is desirable because the austenite fraction slows down ferrite grain growth during processing, which has a favourable effect on the transition temperature T_T. During rapid cooling, these austenite grains transform into martensite so that stabilising annealing at 750 to 850°C becomes necessary to obtain a ferritic structure in which the low contents of C and N are present as fine carbide/nitride precipitates. Interstitial elements are bound with e.g. Ti to reduce susceptibility to intergranular corrosion. Because chromium carbides do not precipitate in

Ti-stabilised steels during annealing, the chromium content can be slightly reduced (X6CrNiTi12). ELC grades include X2CrNi12 and stabilised X2CrTi12. The addition of < 1 % Ni and < 1.5 % Mn slightly re-extends the austenite field that was reduced by the low C content (Figure B.6.10), which can be employed to generate the aforementioned useful fraction of austenite during hot processing that transforms into tough martensite on cooling. This method produces a corrosion-resistant, dual-phase microstructure with a higher strength (Figure A.2.2b, p. 22 and Figure B.2.4, p. 169), whose transition temperature is lowered by Ni.

Increasing the chromium content to 17 % increases the corrosion resistance (X6Cr17, X3CrTi17 or Nb-stabilised as X3CrNb17). Mo improves the resistance to pitting corrosion (X6CrMo17-1, X2CrMoTi17-1). Austenite stabilisation dwindles at 17 % Cr, which is counteracted by adding 1.5 % Ni (X6CrNi17-1). The addition of 0.15 to 0.35 % S improves the machinability (X6CrMoS17, X2CrMoTiS18-2), whereby the high CrMo content of the latter grade is designed to increase the pitting corrosion resistance. An increase up to superferrite is found in steels X1CrNiMoNb28-4-2 and X2CrMoTi29-4. The low C content and stabilising with Nb produces fewer and rounder MC carbides compared to stabilising with Ti, which, together with Ni leads to a reduction in T_T (Figure B.6.11 a). As discussed in Chapt. B.2 (p. 182),

Table B.6.2 Mechanical properties of some ferritic (I), austenitic (II) and ferritic-austenitic (III) stainless steels.

	Name	$R_{p0.2}$ [1] [MPa]	R_m [MPa]	A [1)2)] [%]
I	X2CrNi12	260	450-600	20
	X6Cr13	230	400-630	20
	X6Cr17	240	400-630	20
	X6CrMoS17	250	430-630	20
	X6CrMo17-1	280	440-660	18
II	X5CrNi18-10	190	500-700	45
	X6CrNiTi18-10	190	500-700	40
	X8CrNiS18-9	190	500-750	35
	X5CrNiMo17-12-2	200	500-700	40
	X2CrNiMoN17-11-2	280	580-800	40
	X2CrNiMoN17-13-5	280	580-800	35
	X1CrNiSi18-15-4	210	530-730	40
	X3CrNiCu18-9-4	175	450-650	45
III	X2CrNiMoN22-5-3	450	650-880	25
	X3CrNiMoN27-5-2	460	620-880	20
	X2CrNiMoCuN25-6-3	500	700-900	25

[1] minimum values, thickness < 160 mm, [2] longitudinal specimens

aluminium contributes to passivation by forming a surface film. The chromium effect is assisted by adding up to $\approx 2\,\%$ Al in ferritic steels (X2CrAlTi18-2).

Owing to the high diffusion velocities in a bcc lattice (Figure A.1.8, p. 16), low-transformation ferritic steels are prone to grain coarsening and thus require hot working to refine the grains and lower T_T. This is why they are not generally used for steel castings, except for wear-resistant grades according to SEW 410 with a ferritic matrix and a high amount of eutectic carbides (GX70Cr29, GX120CrMo29-2), which border white cast irons (Chapt. B.4, p. 261). Enough Cr is added to ensure that the matrix still contains sufficient for passivation after the chromium-rich $M_{23}C_6$ carbides have precipitated. These crack-prone casting materials are slowly cooled after stress-relief annealing.

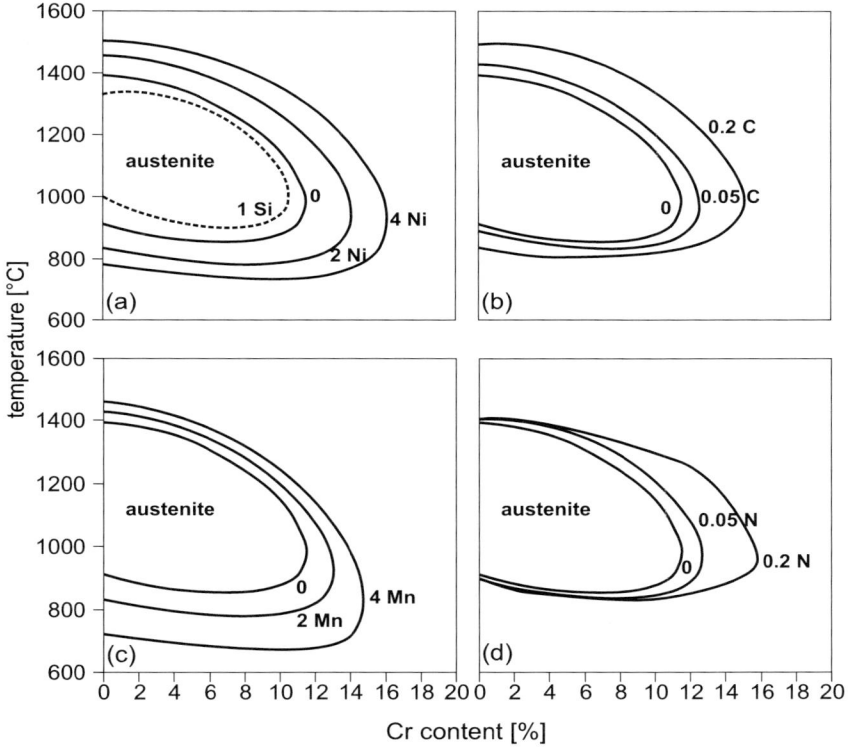

Fig. B.6.10 Constitution of chromium steels: The austenite field is cropped by chromium and this is intensified by silicon, whereas it is expanded by nickel, manganese, carbon and nitrogen (alloying contents in %, calculated with ThermoCalc).

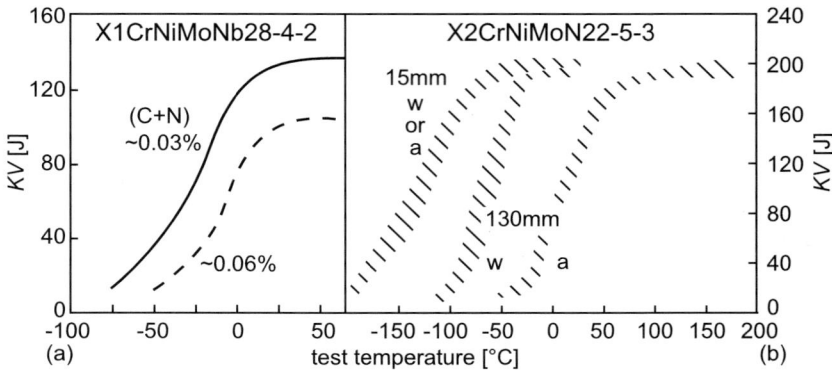

Fig. B.6.11 Low-temperature toughness of stainless steels: (a) The transition temperature of the ferritic steel, solution-annealed at 1050°C/water, increases with the (C+N) content (from R. Oppenheim, KV = ISO-V notch impact energy). (b) The duplex steel, after solution-annealing at 1060°C and cooling in water (w) or air (a), exhibits a lower transition temperature than ferritic steels because propagation of cleavage cracks in the ferrite grains, which arise at low temperatures, is hindered by the tough austenite grains. Slower cooling, e.g. in thick cross-sections, leads to precipitation-induced embrittlement (from P. Gümpel, G. Chlibec).

(b) Austenitic steels

According to the Schaeffler diagram (Figure A.2.24, p. 47), an austenitic microstructure requires 17 to 18 % Cr for the lowest possible nickel content, i.e. the lowest costs (basic grade X5CrNi18-10, Tab.B.6.2). Increasing the chromium equivalent to improve the resistance to pitting corrosion requires a higher nickel equivalent to avoid excessive ferrite. Molybdenum has a factor of 1.4 in the chromium equivalent and a factor of 3.3 in the PREN, which means it saves nickel with respect to the pitting corrosion resistance (X5CrNiMo17-12-2). The use of stabilisation to adjust the resistance to intergranular corrosion is associated with an increase in the 0.2 proof strength (e.g. X6CrNiNb18-10), but with a decrease for ELC grades (X2CrNi19-11). This is countered by adding up to 0.22 % N (X2CrNiN18-10). Such solid-solution hardening increases the yield point, whereby the susceptibility to intergranular corrosion due to nitrogen is much less than that due to carbon. Therefore, even in the basic grade X5CrNi18-10, the C content is often kept at the lower limit and a little N is added to compensate. Nitrogen also replaces nickel in the equivalent and delays precipitation of intermetallic phases so that high chromium and molybdenum contents can be used without increasing the susceptibility to embrittlement (X2CrNiMoN17-13-5, Figure B.6.4 b). The resistance to pitting corrosion increases as well (Figure B.6.12). High nickel contents lower the solubility of carbon in austenite, which is why only an ELC grade is suitable

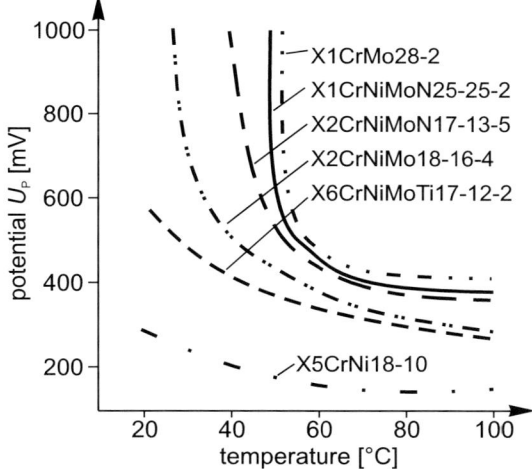

Fig. B.6.12 Pitting corrosion of stainless steels in artificial seawater: ((3 % NaCl in aerated water, potentiokinetic measurement at 20 mV/h, resting potential $U_R \approx 400$ mV). The curves show the breakdown potential for pitting corrosion U_P. The material is resistant if $U_P > U_R$. The condition $U_P = U_R$ gives the critical temperature above which a steel becomes susceptible to pitting corrosion (from P. Gümpel, N. Arlt).

(X1CrNi25-21). Heat treatment of austenitic steels usually involves solution annealing at 1000 to 1150°C followed by quenching in water.

Tough austenitic steels tend to form continuous chips during machining. This can be countered by adding 0.15 to 0.35 % S. Alloying with ≤ 2 % Mn results in the formation of manganese sulphides during solidification that contribute to short chips, lower cutting forces and reduced tool wear. Because the weldability of components made of free-cutting steels is not of prime importance, up to 0.1 % C is allowed in order to reduce the amount of Ni (X8CrNiS18-9).

Silicon improves the resistance to highly concentrated nitric acid (fuming nitric acid, > 65 %). Its ferrite-stabilising effect is compensated by increasing the Ni content (X1CrNiSi18-15-4). The resistance to non-oxidising acids (e.g. sulphuric acid) can be improved by alloying with copper, which also has an austenite-stabilising effect and saves Ni (X3CrNiCu19-9-2). The initially active metal dissolution produces a deposit of more noble copper, thus resulting in anodic polarisation and passivation of the steel. The cold-headability can be improved by adding 3 to 4 % Cu to stabilise the austenite against deformation-induced martensite, which leads to excessive strengthening (Figure B.6.3 and B.6.5). Copper overcomes this problem with a lower flow stress than Mn or N and is cheaper than Ni. Pressed parts usually remain non-magnetic.

324   B.6 Chemically resistant materials

In contrast, the costs of spring steel are reduced by replacing Ni by about twice the amount of Mn and also by up to 0.15 % C and up to 0.25 % N because weldability is not required (X12CrMnNiN17-7-5). This meagre austenite stabilisation results in a high degree of strengthening during cold rolling, along with strengthening by the martensite fraction, and thus to spring-hardness, i.e. a high elastic limit. In continuation of the low-corrosion lightweight steels (Chapt. B.2, p. 182), the soluble Al content can be increased by C to such an extent that passivation without Cr occurs (GX100MnAlSi30-10). This intensifies the aforementioned problems with Al burnout.

(c) Ferritic-austenitic steels

Whereas in austenitic steels an increase in the PREN by Cr and Mo must be 'bought' with more expensive Ni, it is also possible with less Ni if ferrite fractions are acceptable (Figure A.2.24, p. 47). So-called duplex steels contain almost equal proportions of ferrite and austenite (Figure A.2.2 c and A.2.1 b, p. 21 f.). The ferrite/austenite phase boundaries act as two-dimensional lattice defects (Figure A.2.4, p. 23) and increase the 0.2 proof strength to 450 - 550 MPa, i.e. significantly higher than that of the individual phases (Tab.B.6.2). The pitting resistance of the basic grade X2CrNiN23-4 is increased by Mo (X2CrNiMoN22-5-3) or Cr (X3CrNiMoN27-5-2). Similar to austenitic steels, up to 2.5 % copper is added if passivation is inadequate (X2CrNiMoCuN25-6-3). Tungsten reinforces the effect of Mo, so that superduplex steel X2CrNiMoCuWN25-7-4 with 0.25 % N, 0.8 % W and 0.8 % Cu reaches PREN > 40.

Nitrogen is added in amounts of up to 0.3 %. It promotes the formation and solid-solution strengthening of austenite and delays the precipitation of embrittling intermetallic phases. The heat treatment consists of solution-annealing at 1000 to 1150°C and accelerated cooling, usually in water, to prevent precipitates. As the solution temperature rises, the precipitates dissolve first and the austenite is stabilised. Furthermore, the equilibrium shifts towards a higher proportion of ferrite (Figure B.6.6). The good resistance to pitting, crevice and stress-corrosion cracking must be weighed against the drawbacks of susceptibility to σ-, α'- and cold-embrittlement (Figure B.6.4 d - steel 1 and B.6.11 b). The possibility of α' embrittlement can be completely excluded by limiting the long-term service temperature to ≤ 280°C. The small amount of carbon in the steel dissolves in the austenite thus almost doubling its C content. This increases the risk of intergranular corrosion, in spite of it being an ELC steel. However, the high chromium content usually prevents depletion of chromium to below the critical value for intergranular corrosion. Furthermore, nitrogen delays the precipitation of $M_{23}C_6$, so that the steel is usually resistant to intergranular corrosion after welding. The high diffusion rate of nitrogen promotes the formation of austenite during welding after primary ferritic solidification, so that a duplex microstructure can also arise in the weld seam in spite of the high cooling rate.

The wear resistance is improved by using castable grades with a higher carbon content (GX40CrNiMo27-5), which results in the precipitation of eutectic $M_{23}C_6$ carbides. The austenite content is low and increases to about a third on addition of nitrogen and dissolution of the precipitates as the solution-annealing temperature rises. The cracking susceptibility of carbidic grades often precludes water quenching, thus requiring the use of forced air, possibly combined with oil cooling through the α' field. The alloy content precipitated in the carbides lowers the PREN of the matrix.

(d) Martensitic steels

Cropping of the austenite field by chromium (Figure A.1.10 b, p. 18) can be overcome by adding austenite-stabilising elements, such as C or Ni (Figure B.6.10), to produce a hardenable stainless steel (Tab.B.6.3). Small amounts of δ-ferrite are suppressed in the basic grade X20Cr13 by a little N from the air and Ni from scrap. The hardenability increases with the C content (X30Cr13, X39Cr13, X46Cr13), although in the latter steel, not all the secondary carbides dissolve at a hardening temperature of about 1000°C (cf. Figure B.6.13 b). At higher hardening temperatures, retained austenite does not contribute to hardening unless it can be transformed by sub-zero cooling or cold working (shot peening, surface rolling). Owing to the deeper hardness penetration due to chromium, the steels can be hardened in air if they are not too thick. An increase in the Cr content and the addition of $\approx 0.7\,\%$ Mo

Table B.6.3 Mechanical properties of some martensitic stainless steels: I quenched and tempered, II hardened, III soft-martensitic, IV precipitation-hardened

	Name	$R_{p0.2}$ [1] [MPa]	R_m [MPa]	A [1)2)] [%]	H [3] [HRC]
	X20Cr13	600	800-950	12	45
I	X30Cr13	650	850-1000	10	
	X17CrNi16-2	700	900-1050	12	
	X39CrMo17-1	550	750-950	12	52
	X46Cr13				56
II	X50CrMoV15				57.5
	X70CrMo15				59
	X30CrMoN15-1 [4]				59.5
	X105CrMo17				59.5
III	X3CrNiMo13-4	620	780-980	15	
	X4CrNiMo16-5-1	700	900-1100	16	
IV	X5CrNiCuNb16-4	1000	1070-1270	10	

[1] minimum values, thickness < 60, some < 160 mm, [2] longitudinal specimens, [3] hardened with low tempering, approximate mean value, [4] with 0.35 % N

improve the pitting corrosion resistance of knife steel X50CrMoV15 in the dishwasher. In alloys with 0.15 % V, fine carbides remain after austenitising that slow down grain growth and make the steel less sensitive to overheating. The non-dissolved secondary $M_{23}C_6$ carbides improve the sharpness retention of the blade. Tempering at $< 200°C$ keeps the hardness of martensitic chromium steels high. If elevated tempering temperatures are used, account must be taken of the fact that chromium participates in the precipitation of temper carbides above 400°C. This depletes the surrounding matrix of this element so that the narrow region along the phase boundary to the carbide becomes susceptible to corrosion (Figure B.6.14).

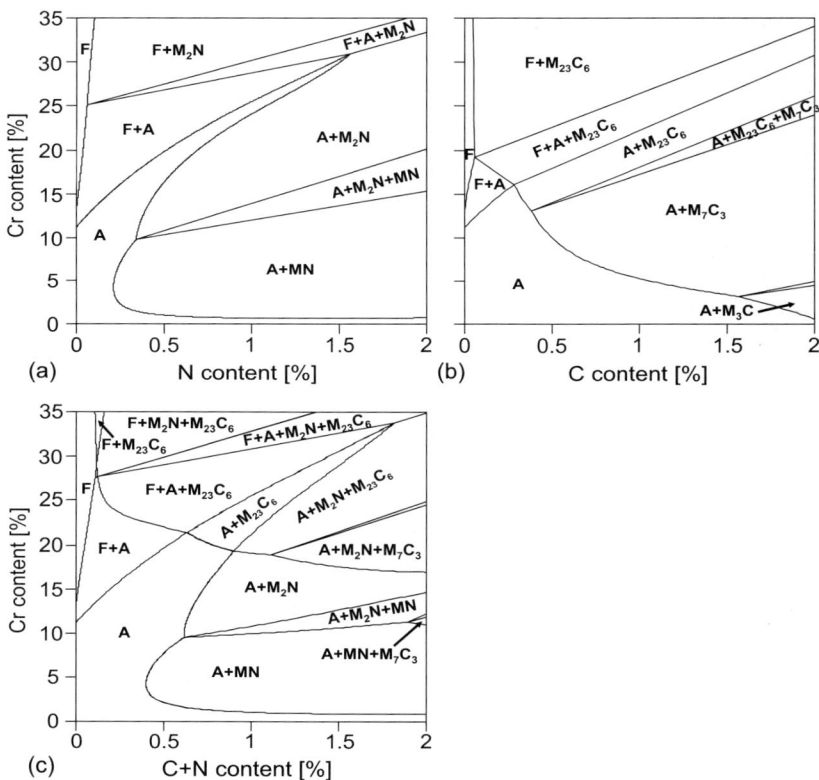

Fig. B.6.13 Influence of carbon and nitrogen on the isothermal (1050°C) phase diagram of chromium steels; calculated with ThermoCalc: (a) Fe-Cr-N system, (b) Fe-Cr-C system, (c) Fe-Cr-C-N system for C/N = 0.85, which corresponds to a mole ratio of approximately 1. For 13 to 17 % Cr, the maximum interstitial solubility of austenite increases in the order C, N, C+N.

Chromium depletion starts to reverse above 600°C, similar to intergranular corrosion along the grain boundaries (Figure B.6.2), and localised corrosion disappears again. Martensitic stainless steels should therefore be tempered at well over 600°C during QT treatment so that the carbon content is almost completely precipitated as $M_{23}C_6$ and the matrix chromium content decreases - uniformly in this case. This is why steels intended for QT treatment contain not only 13 but even 16 - 17 % Cr. δ-Ferrite can be avoided by alloying with Ni (X17CrNi16-2) or by increasing the C content (X39CrMo17-1); the latter steel is also used in a low-tempered state with a high hardness. A further increase in the carbon content raises the amount of undissolved secondary carbides of type $M_{23}C_6$. Steels X65Cr14 and X70CrMo15 also contain coarse eutectic carbides in segregation bands whose proportion increases up to steel X105CrMo17 and gradually converts into M_7C_3. The enrichment of Cr and Mo in the carbides reduces the matrix content of these elements. The fraction of hard carbides increases the steel hardness and the resistance to grooving wear.

Austenitisation of chromium steels using nickel instead of carbon (Figure B.6.10 a) produces soft martensitic steels such as X3CrNiMo13-4 or, with a higher corrosion resistance, X4CrNiMo16-5-1. After hardening from $\approx$ 1000°C, they do not exhibit noteworthy corrosion susceptibility during tempering between 400 and 600°C (Figure B.6.14). However, their high Ni content already induces the start of the reverse transformation to austenite just below 500°C. Raising the tempering temperature increases the fraction of austenite - an increasing proportion of which transforms into new martensite on cooling.

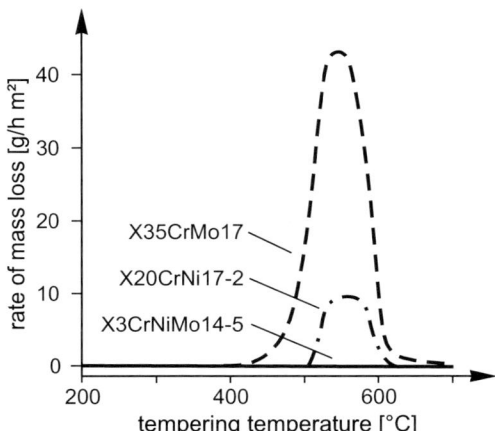

Fig. B.6.14 Corrosion resistance of martensitic chromium steels as a function of the tempering temperature: Tempering time 2 h, tested in boiling 20 % acetic acid, test duration 24 h (from N. Arlt et al.) .

Tempering at 550 to 600°C can produce a fine distribution of austenite in the tempered martensite along with a little new martensite to give an excellent combination of yield point and notch impact energy. The austenite does not even transform at -196°C and the low-temperature toughness remains high (Figure B.6.15).

An excessive tempering temperature and the resulting new martensite may lead to the uptake of hydrogen during service. A few hundredths of a percent of N contribute to avoiding δ-ferrite after hardening and also supports secondary hardening during tempering. Like high-strength maraging steels (Figure B.2.23, p. 201), the strength of soft martensitic stainless steels can also be increased by precipitation hardening during tempering. One method involves precipitating very fine copper-based crystals from the martensite (X5CrNiCuNb16-4 with $\approx 4\%$ Cu); another involves precipitating intermetallic phases of Al with Fe and Ni (X7CrNiAl17-7 with $\approx 1.2\%$ Al). These steels are also known as 17/4 PH and 17/7 PH (precipitation hardening), respectively. A more recent development is based on precipitation hardening with beryllium and titanium. Ageing of maraging steel X1CrNiMoBeTi13-8-1 with 0.25% Be and 0.2% Ti at 470°C gives e.g. $R_{p0.2} = 1900$ MPa, $A = 3\%$, HV = 600. Applications include stainless steel springs.

Martensitic-ferritic steels are obtained if there is just enough austenite stabilisation to enter the two-phase field between austenite and ferrite at the hardening temperature (Figure A.1.10 c, p. 18 and B.6.10). This indeed occurs on going from X20Cr13 to X12Cr13, which is suitable as a cold-heading grade owing to its lower C content and lower hardness. If it is to be machined, it is alloyed with 0.15 and 0.35% sulphur and is designated X12CrS13. X14CrMoS17 with $\approx 0.7\%$ Mo has a higher corrosion resistance and a microstructure consisting of approximately equal proportions of tempered martensite and ferrite.

(e) Steels with a high nitrogen content

The highest degree of solid-solution hardening of iron is obtained with C and N atoms dissolved in the interstices (Figure A.4.12 a, p. 94). Because the solubility of C and N is low in ferrite, this effect is generally exploited in austenite where it is higher, i.e. in austenitic and martensitic steels. The high chromium content in stainless grades limits the solubility of carbon at the solution-annealing or hardening temperature, thus promoting the formation of carbides. In the isothermal Fe-Cr-C phase diagram (Figure B.6.13 b), the phase boundary line between austenite and austenite + carbide has a hyperbolic shape, i.e. the C solubility in austenite decreases with increasing Cr in the steel. The reverse is true for N (Figure B.6.13 a). This tendency is also maintained when both C and N are added, e.g. at a C/N ratio of 0.85 (Figure B.6.13 c). It is thus possible to dissolve more Cr (corrosion resistance) and interstitial elements (hardness increase) compared to standard steels. This is limited by the increasing fraction of retained austenite, some of which can

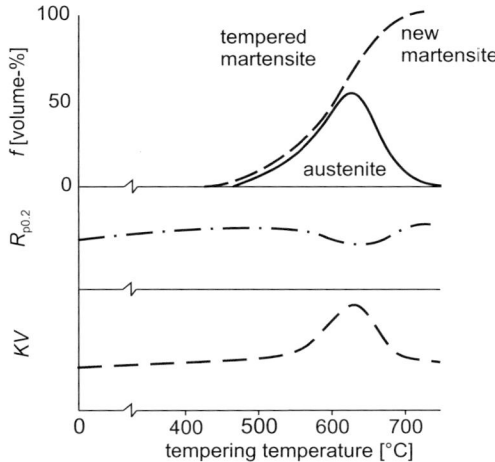

Fig. B.6.15 Tempering of soft martensitic steels: Microstructural phase fraction f, proof strength $R_{p0.2}$ and notch impact energy KV depending on the tempering temperature (schematic from P. Brezina).

only be decomposed by sub-zero cooling. However, steel X30CrMoN15-1 with 0.35 % N has to be produced by pressure metallurgy because the Cr,Mo content is insufficient to keep enough nitrogen in solution during primary ferritic solidification in air. Nevertheless, the effort is worthwhile because a hardness of up to 60 HRC can be obtained, which is significantly higher than that of low-carbide standard steels, and because the dissolved nitrogen considerably increases the PREN. Under rolling contact conditions, the service life of particle-contaminated bearings made of roller bearing steel X30CrMoN15-1 exceeds that of the standard steel 100Cr6 by e.g. one order of magnitude.

Mössbauer spectroscopy provides information on the nearest neighborhood of alloy atoms with respect to iron atoms in bcc steels. The results show that - in a steel with 15 % Cr and 0.6 % C - both elements tend to form clusters. These clusters comprise a few hundred atoms and contain a greater proportion of alloying atoms compared to regions between clusters. This situation arises in austenite, with sizes several orders of magnitude smaller than microsegregations. However, the addition of 0.6 % N leads to a more homogeneous distribution of the Cr atoms and approaches short-range ordering. This is increased even further by alloying with 0.3 % each of C and N. Short-range atomic ordering, which increases in the order C, N, C + N, raises the solubility of interstitial elements (Figure B.6.13) and stabilises the austenite, which is expressed by a higher fraction of retained austenite.

Exploitation of short-range ordering to increase the solubility and austenite stability is also useful in austenitic steels. Nickel can be excluded as an austenite former because it lowers the solubility of the solid-solution hardeners C and N in austenite, whereas manganese increases it. A steel with 18 % each of Cr and Mn does not have a homogeneous austenite phase field with C, but it does with N (Figure B.6.16).

At atmospheric pressure, $\approx 0.55\,\%\,\mathrm{N}$ dissolve in the melt (X8CrMnN18-18). Without the use of pressure metallurgy, up to $1\,\%\,\mathrm{C}+\mathrm{N}$ can be added in combination (X35CrMnN18-18 with $0.6\,\%\,\mathrm{N}$), which leads to $R_{p0.2} \approx 600\,\mathrm{MPa}$ in the solution-annealed and quenched state. Despite the high degree of cold-work hardening to a true fracture stress of $R_\mathrm{f} > 2500\,\mathrm{MPa}$, values of $\approx 70\,\%$ elongation at fracture are reached. This ductile behaviour is attributed to a high concentration of free electrons (n_e) in the austenite, which can be determined by electron spin resonance measurements of paramagnetic materials.

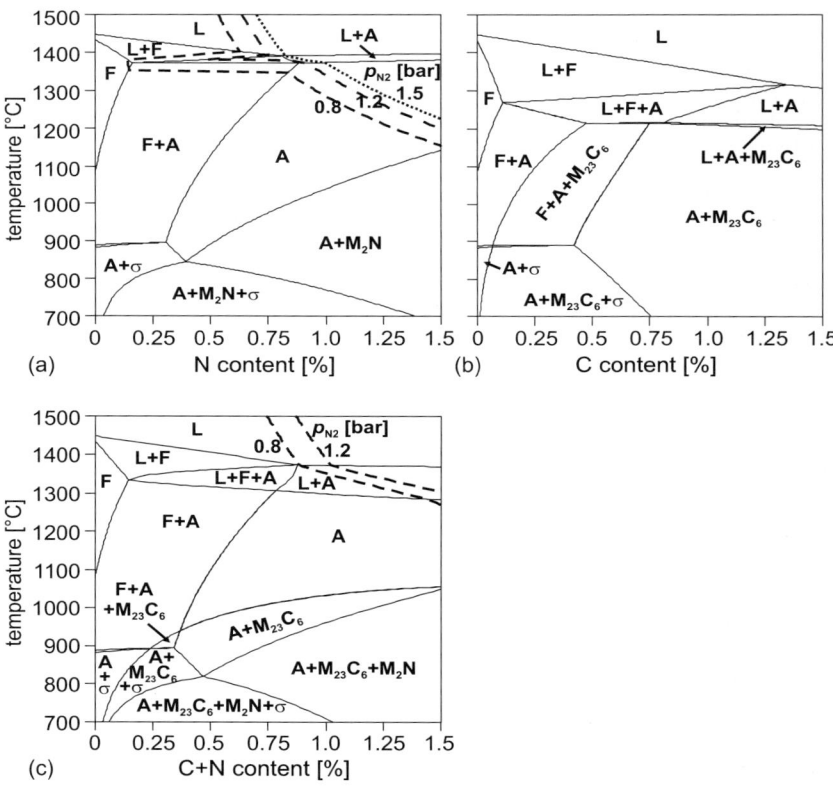

Fig. B.6.16 Phase diagrams for iron with 18 % Cr and 18 % Mn: (a) alloyed with nitrogen, (b) alloyed with carbon and (c) alloyed with both for C/N = 0.6 (calculated with ThermoCalc). The nitrogen solubility is limited by the equilibrium partial pressure p_{N_2}.

In binary iron alloys, n_e is increased by elements with a higher atomic number than Fe (e.g. Ni or Co) and decreased by those with a lower atomic number (e.g. Cr or Mn). The aforementioned CrMnCN steel is thus reliant on the combination of interstitial elements to promote metallic ductile bonding by means of free electrons.

As the measured values in Figure B.6.17 show, C has little effect, whereas n_e is significantly increased by N, although it drops again above $\approx 0.5\,\%\,N$. The highest values are obtained with a combination of C + N at a lower Cr content. The n_e maximum of N-alloyed steels is shifted to a higher interstitial concentration by alloying with C+N, which means that metallic ductile behaviour promoted by free electrons is accompanied by a higher degree of solid-solution strengthening. This produces a high-strength and nonetheless tough austenitic steel.

B.6.2.2 Applications

The main objectives in using stainless steels are to keep uniform corrosion low and to avoid localised corrosion. Uniform corrosion occurs in acids and strong bases. The shape of the cathodic partial current density/potential curve within the steel's passive range is important (Figure B.6.18). At a low pH and in

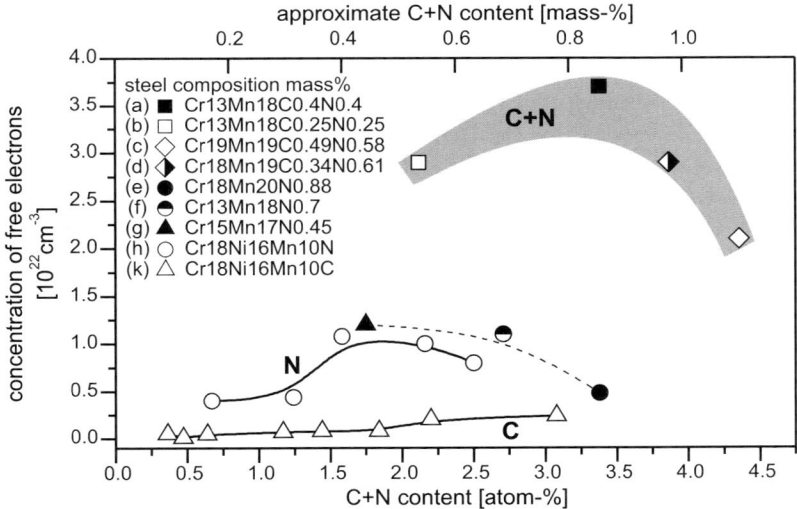

Fig. B.6.17 Concentration of free electrons n_e in the fcc lattice of austenitic stainless steels: the concentration increases in the order C, N and C+N (from V.G. Gavriljuk et al.).

non-oxidising acids (sulphuric or phosphoric acid), acid corrosion (case a) occurs. In oxidising media with a higher redox potential U_red there is metastable or stable passivation (cases b and c, respectively). In a strongly oxidising acid (e.g. fuming nitric acid), the resting potential U_R may lie within the transpassive range (case d). In neutral, optionally aerated water, the activation potential U_act is lower compared to acidic conditions, indeed it lies below the associated redox potentials $U_\text{red} = 0\,\text{V}$ (H^+ reduction) and $U_\text{red} = 0.41\,\text{V}$ (O_2 reduction, Tab.A.4.3). Stainless steels are resistant to uniform corrosion under these conditions. Nevertheless, there is still a risk of localised corrosive attack by impurities in the water. They can be enriched in encrustations and crevices by alternating wet/dry cycles. Uniform corrosion and localised corrosion are affected by the concentration, temperature and flow rate of the corrosive medium. The resistance increases with the surface quality (pickled or peened, ground, brushed, polished).

There are three application fields for stainless steels:

a) chemical equipment with uniform corrosion and localised corrosion,
b) human environments without significant uniform corrosion but with possible localised corrosion due to climatic conditions, foods and cleaning agents,
c) mechanical and vehicle engineering that have particular requirements with respect to strength and design weight.

(a) Chemical-engineering equipment

Cr, Ni, Mo and Cu lower the passivating current density I_Pa and thus reduce the rate of uniform corrosion in the active range. They also facilitate passivation (Figure B.6.18 b, c). This is the reason why the chemical industry uses austenitic steels X1NiCrMoCu25-20-5 and X1NiCrMoCu31-27-4 for containers and pipelines in contact with sulphuric and phosphoric acid and ferritic-austenitic steel GX2CrNiMoCuN26-6-3-3 for high-strength pump components and fittings. These materials undergo passivation in the presence of oxidising additives and exhibit good resistance to impurities such as chloride and fluoride ions. Passivation is hampered in hydrochloric acid, and high-alloy steels can only be expected to exhibit a sufficiently low rate of corrosion at low concentrations and temperatures (Figure B.6.8 d).

The potential of boiling 65 % HNO_3 lies within the transition region between passive and transpassive (Huey test). Nitric acid is passivating up to this point. Austenitic steel X2CrNi19-11 with low contents of C, Si, B, Mo, P and S, exhibits good resistance in the Huey test. Molybdenum-free duplex steel X2CrNiN23-4 improves the 0.2 proof strength. Highly concentrated nitric acid leads to corrosion in the transpassive range. Alloying with silicon (X1CrNiSi18-15-4) has a favourable effect in this case. Uniform corrosion and stress corrosion cracking occur in strong bases such as NaOH and KOH. Their rates decrease as the nickel content increases.

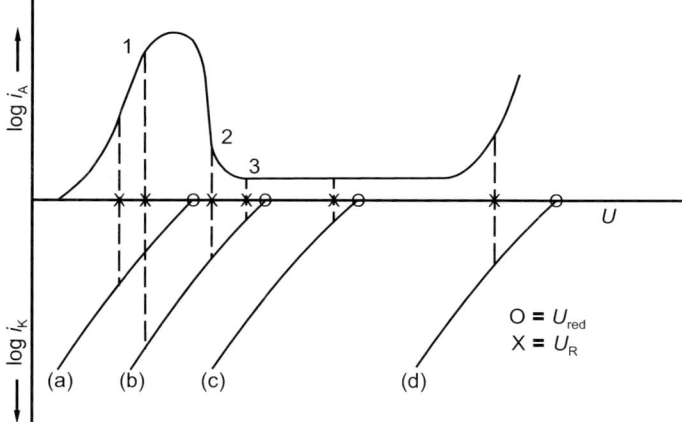

Fig. B.6.18 Position of the cathodic versus the anodic partial current density/potential curve (schematic, cf. Figure A.4.19, p. 112): (a) The free corrosion or resting potential U_R lies within the active range. (b) Of the three possible values of U_R, only 1 and 3 can be permanently established. This metastable passivity means that a passivated surface is not repassivated after intermediate activation. (c) U_R lies within the passive range. (d) U_R lies within the transpassive range.

In addition to the aforementioned aggressive substances, chemical engineers are confronted with a large number of organic and inorganic substances that have to be processed (crushed, sieved, mixed or separated). This involves a combination of corrosion and wear, thus necessitating e.g. hard martensitic steels or carbide- and boride-containing hard-facing alloys based on iron, cobalt or nickel (Chapt. B.4, p. 264). Equipment used in chemical process engineering includes pressure vessels, distillation columns, absorbers, heat-exchangers, heating coils, measuring instruments, etc. that are made from a wide range of steels in the form of flat products, bars, pipes, castings and forgings.

Not only uniform corrosion, but also pitting corrosion, in particular, due to concentrated chloride solutions has to be controlled using steels with a high PREN (Eq. B.6.1, p. 317). Examples include bleaching equipment in the pulp industry, where e.g. steels X2CrNiMoN17-13-3 and X2CrNiMoN22-5-3 are used along with titanium in acidic environments. Similar conditions and steels are found in the textile and leather industries as well as in facilities used for salt production and seawater desalination. In the latter case, temperatures up to 180°C and elevated pressures necessitate steels with the highest PREN and yield point (X1NiCrMoCuN25-20-7). Superferrite X1CrNiMoNb28-4-2 can be used as an alternative to austenitic and ferritic-austenitic steels e.g. in the final heating stage of heat-exchangers. High nickel equivalents in an austenitic steel or a ferritic microstructure improve the resistance to stress corrosion cracking.

Steel X2CrNi12 is used in plant engineering for a completely different reason: it reduces maintenance costs as a construction steel for important rust-prone plant sections. It can replace grades such as S355 or even weather-resistant steels such as S355WP (Chapt. B.1, p. 128), which do not have sufficient corrosion resistance in industrial atmospheres. Maintenance and repair costs are even higher in offshore engineering equipment. Readily weldable molybdenum-alloyed steels, such as e.g. X2CrNiMo17-12-2, are used owing to the chloride in seawater. Wire guy lines can be made of stable austenitic steels with a high resistance to crevice corrosion, e.g. X3CrNiMnMoNbN23-17-5-3, which can be cold-drawn to a strength of ≈ 1500 MPa.

The conditions in the lower part of the scrubber tower of flue gas desulphurisation (FGD) plants are extremely aggressive: temperatures of $\approx 120°C$, pH values as low as ≈ 1 and the presence of chloride ions. This necessitates the use of nickel-base alloys such as NiCr22Mo9Nb (Inconel 625) and GNiCr20Mo15. The concentration and temperature decrease towards the top of the tower so that steels X1NiCrMoCu25-20-5 and X2CrNiMoN17-13-3 can be used. The SO_2 in the flue gas reacts with water and lime in the air to produce gypsum $CaSO_4$. The flow of solid particles in the gypsum suspension causes erosion corrosion in this aggressive environment. Under these operating conditions, ferritic-austenitic steels are generally used, e.g. X2CrNiMoN22-5-3 for pipelines and castable grades such as GX2CrNiMoCuN26-6-3-3 for pumps and valves. Surface-hardened rings of X40Cr13 are fitted to briquetting rolls.

(b) Human environments

Ferritic and austenitic steels, e.g. X6Cr17 and X5CrNi18-10, are used for interior fittings such as panelling, lifts, escalators, doors, countertops, lockers etc. In contrast, outdoor facades and roofs are made of molybdenum steels such as X5CrNiMo17-12-2. Stabilised grades are used for welded structures. In this case, niobium is usually preferred over titanium because it imparts a better surface quality on visible polished parts. Chimney linings of X5CrNiMo17-12-2 are used to prevent damp brickwork caused by condensation in the upper part of the flue. Modern boilers are made of e.g. X3CrTi17 and X6CrNiMoTi17-12-2.

Stainless steels are commonly used in the food-processing industry because they can be easily cleaned and liberate practically no metal ions into the products. The foods and beverages themselves exert only a low corrosive load, in contrast to chloride-containing cleaners or salt added to sausage products and fish brine. Therefore, not only X6Cr17 and X5CrNi18-10 are widely used, but also molybdenum-containing grades such as X6CrMo17-1 and X5CrNiMo17-12-2 along with their stabilised variants. Examples include tanks for milk, wine and water as well as bakery and medical equipment. Beer barrels are made of aluminium alloys as well as steel X5CrNi18-10, or even the weight-saving high-strength ferritic-martensitic steel X5CrNiMoTi15-2. Domestic goods, such as pots, bowls, jugs, basins and cutlery, are made of cold-rolled strip and sheets of X6Cr17 and X5CrNi18-10. Cheaper cutlery made of X6Cr13 is not usually dishwasher-resistant. This also applies to blades of molybdenum-free martensitic steels such as X46Cr13. Consequently, steels X50CrMoV15 or X39CrMo17-1 are used instead; the former is harder and stays sharp for longer, the latter is more corrosion-resistant. The highest hardness and dishwasher resistance was obtained with forged and with stamped blades in a trial series made of steel X30CrMoN15-1 with 0.35 % N. This material is also used for medical instruments as well as for cutting and chopping tools in food processing, and is supplemented by the carbide steel X90CrMoV18, which has a higher wear resistance. Nickel-free austenitic steels with a high nitrogen content, e.g. X5CrMnMoN16-14-3, are used for spectacle frames and armbands to avoid nickel-induced skin allergies.

(c) Mechanical and vehicle engineering

There are a number of material standards that regulate various non-corroding machine elements. Springs are made of cold-worked and optionally artificially aged steels such as X10CrNi18-8 and X7CrNiAl17-7. Roller bearings are made of e.g. X108CrMo17. Steels alloyed with nitrogen under pressure, e.g. X30CrMoN15-1 with 0.35 % N, are more resistant under pitting conditions. Cold-worked screws are made of ferritic steel X6Cr17 and ferritic-martensitic steel X12Cr13 as well as a series of austenitic CrNi and CrNiMo steels. Shafts and other machined parts are often made of free-cutting stainless steels containing MnS or TiS inclusions. Soft martensitic steels are generally used for hydropower equipment. One example is a heavy Pelton wheel of GX4CrNiMo13-4, tempered at $> 600°C$ to $R_{p0.2} > 520\,\text{MPa}$. For pump impellers and compressor rotors, a 0.2 proof strength of $> 800\,\text{MPa}$ is generally obtained by tempering at $< 600°C$.

Steel X5CrNi18-10 is sometimes used for the outer panelling of rail vehicles and X2CrNi12 for rust-prone structural components. The latter steel considerably increases the service life of the shipping hold of rail and commercial vehicles used for transporting bulk materials. Intergranular corrosion can be avoided by stabilising with Ti, and the yield point is increased by the martensitic fractions obtained by alloying with Ni and Mn (e.g. X2CrNiTi12 with $R_\text{e} > 380\,\text{MPa}$). Tanks of road tankers are usually made of steel X6CrNiMoTi17-12-2. Panelling of commuter rail cars has been manufactured almost completely stainless. The high-strength ferritic-martensitic steel X5CrNiMoTi15-2 was used to save weight. The service life of buses has been increased by underbodies of X5CrNi18-10 and wheel houses of X2CrNi12.

Steel X8CrMnN18-18 with 0.55 % N is not susceptible to chloride-induced anodic stress corrosion cracking. This is also observed for steels with $> 22\,\%$ nickel and nickel-base alloys. The common feature of the aforementioned alloys is a high austenite stability, even under cold-working conditions. This differentiates them from standard steels, e.g. X5CrNi18-10, in which areas of martensite can form in the plastic zone in front of the crack tip (Figure B.6.3 and B.6.5). This microstructural component is prone to hydrogen-induced cathodic stress corrosion cracking. Hydrogen is known to escape from growing cracks. This leads to the conclusion that acid corrosion is taking place in the crack, which is driven by hydrolysis (Eq. A.4.15, p. 113). Stress corrosion cracking is a problem in civil engineering applications, in particular, where highly loaded fastening elements are subjected to crevice and pitting corrosion as a result of wet/dry cycles with growth of deposits and concentration of chemicals to result in localised surplus stress.

High-strength machine components can be made of soft martensitic steels with $R_{p0.2} > 800$ MPa or the precipitation-hardened variants 17/4 PH and 17/7 PH with $R_{p0.2} > 1000$ MPa (EN 10088). The Schaeffler diagram (Figure A.2.24, p. 47) shows that the CrMo content of martensitic steels and thus also their corrosion-resistance are limited by retained austenite or δ-ferrite. Duplex steels do not have this drawback; however, they only reach $R_{p0.2} > 530$ MPa. Austenitic steels remelted under compressed N_2, such as X5CrMnMoN16-14-3 with 0.9 % N, reach $R_{p0.2} > 550$ MPa and also exhibit a high resistance to pitting corrosion. The 0.2 proof strength can be more than doubled by cold-working. Steel X5MnCrNiN23-21-2 with the highest CrMn content takes up 0.85 % N without remelting under pressure and is used for non-magnetisable drill collars. As part of the drill string they enclose a magnetic field-dependent positioning and guidance system for accessing crude oil/natural gas reservoirs. They are subjected to high mechanical and chemical loads that are intensified by a depth-dependent temperature increase. Potential applications of the newly developed steel X35CrMnMoN18-18 with 0.6 % N include pumps as well as tools for processing minerals that are exposed to corrosive conditions. Non-magnetisable roller bearings can be adequately hardened by cold working the surface layer. Cold-expanded retaining rings promise a higher 0.2 proof strength than the previously used X8CrMnN18-18 with 0.55 % N. They are shrunk onto the ends of generator shafts where they hold down the ends of the winding against centrifugal force.

B.6.3 Heat-resistant steels

Heat-resistant rolling and forging steels (EN 10095) are intended for operating temperatures above 550°C (see wustite in Chapt. A.4, p. 115). EN 10295 regulates steel castings.

B.6.3.1 Properties

Heat-resistant steels are similar to stainless grades in many ways, so that much of that discussed in Chapt.B.6.1 applies here. However, their microstructure approaches equilibrium during high-temperature service. Solution-annealing and quenching as for stainless superferrites and austenites only brings benefits for processing, at best. Carbon and nitrogen can precipitate during service, depending on the temperature, and are not suitable for stabilising austenite. The nickel content of austenitic steels must be high enough to avoid δ-ferrite. Martensitic steels are not commonly used because the change in volume associated with the phase transformation causes delamination of the surface scale under thermal cycling conditions. Increasing the nickel content lowers the higher thermal expansion of austenitic steels (Figure A.4.23, p. 120). Alloying with cerium/lanthanum improves scale adhesion.

Ferritic rolling steels are alloyed with Cr, Al, Si, whereas castable steels have only Cr and Si because aluminium can cause oxide skins and inclusions in the casting (Tab.B.6.4). Ferritic steels are prone to grain coarsening and to σ-, 475°C- and cold-embrittlement.

Table B.6.4 Heat-resistant steels classified according to their limiting temperature T_{max} with respect to scaling resistance in air for 120 h with four intermediate cooling periods in accordance with EN 10095. The mass loss was ≤ 1 g/m²h at T_{max} and ≤ 2 g/m²h at $T_{max} + 50°C$

T_{max} [°C]	ferritic steels rolled	ferritic steels cast	austenitic steels rolled	austenitic steels cast
850	X10CrAlSi13		X8CrNiTi18-10	
		GX40CrSi13		
900		GX40CrSi17		GX25CrNiSi18-9
950				GX25CrNiSi20-14
1000	X10CrAlSi18		X15CrNiSi20-12	
1100				GX40CrNiSi25-20
1150	X10CrAlSi25		X15CrNiSi25-21	
		GX40CrSi28		GX40NiCrSi35-26
1100		GX40CrNiSi27-4	X15CrNiSi25-4	
		ferritic-austenitic steels		

Austenitic rolling and castable steels based on Cr, Ni (Si) embrittle more slowly and to a lower degree overall, and are more suitable for processing by cold working and welding. They also have a higher creep strength than ferritic steels (Figure B.6.19). Coarse as-cast grains are favourable in this case, whereas δ-ferrite is not.

Chromium is bound by inward diffusion of C, N and S so that it cannot protect the matrix. These processes are faster in ferritic than in austenitic steels. *Nickel* lowers the solubility of C and N, which slows down carburisation (Figure B.6.20).

The formation of low-melting nickel sulphides lowers the resistance of austenitic steels to reducing S-containing gases. Steels with 15 to 20 % Cr and more than 30 % Ni exhibit very little, if any, embrittlement by the σ phase.

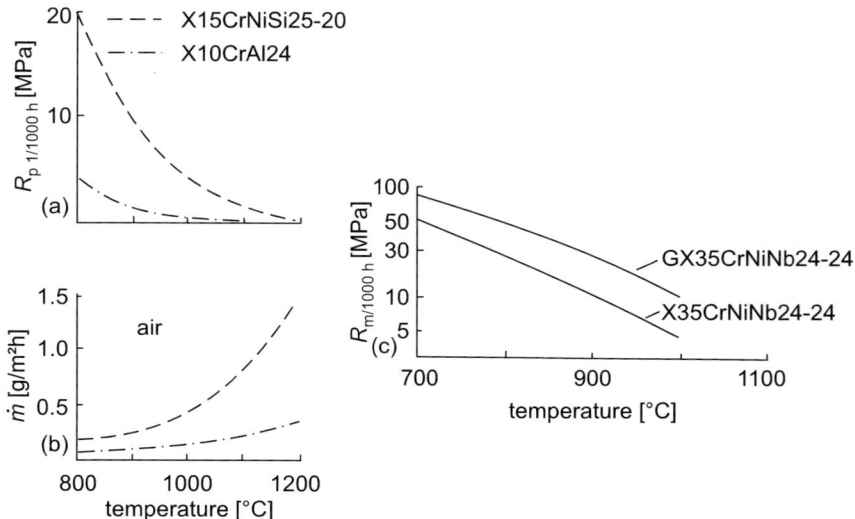

Fig. B.6.19 Scale growth rate and creep behaviour of heat-resistant steels:
(a) Comparison of the 1000 h-1 % creep limit of a ferritic and an austenitic steel.
(b) Comparison of their scale growth rates $\dot{m}$. (c) Comparison of the 1000 h creep rupture strength of a cast and a forged steel. (a and c from W. Steinkusch).

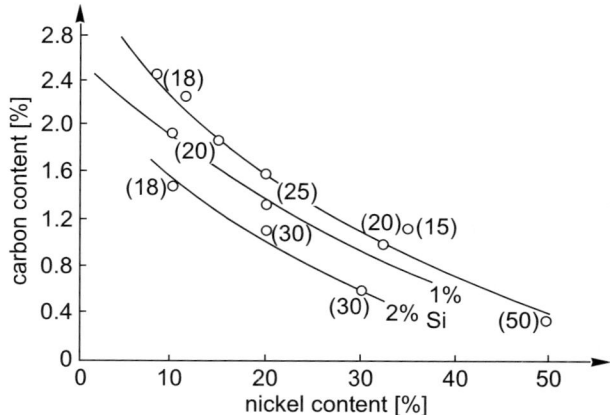

Fig. B.6.20 Carburisation of heat-resistant austenitic CrNi(Si) steels: Carbon content in steel after annealing for 200 h at 1000°C in carbon granules; chromium content of the steels in brackets (from U. W. van den Bruck and C. M. Schillmöller).

Aluminium improves the durability in air and S-containing gases. However, this effect is lost in N-containing atmospheres due to the formation of Al nitrides. N_2 gas dissociates above approx. 800°C to such an extent that technical levels of nitriding are measurable (Chapt. B.3, p. 239). In Ni-rich steels, Al can cause precipitation of γ'-phase Ni_3Al (Tab. A.2.1), whose precipitation-hardening effect may be undesirable. Al delays the precipitation of the σ phase in ferritic steels (Figure B.6.4 a) and accelerates 475°C-embrittlement. Silicon increases the oxidation resistance and assists nickel in slowing down carburisation and nitriding (Figure B.6.20). It has an opposite effect to that of Al in ferritic steels with respect to the σ phase and 475°C-embrittlement. *Niobium* binds incoming carbon in austenitic steels during carburising so that less chromium is withdrawn from the matrix.

B.6.3.2 Applications

Heat-resistant steels are primarily used for structural components that are exposed to hot air and hot process gases. Not only is their resistance to high-temperature corrosion important, but also their creep resistance e.g. for pressure-retaining components. Under discontinuous operating conditions, wet corrosion due to condensates may also occur during the cold phase.

Heat-resistant steels are widely used for the construction and operation of *industrial furnaces* for heat-treating metals and firing of cement clinker and clay ceramics. Housings, doors, valves, transport rollers, conveyor belts as well as burners, mounts, walking beams and fans are made of rolling or casting steel whose composition is tailored to the service temperature, gas composition and the required creep resistance. The lower alloying costs and higher resistance to S-containing combustion gases of ferritic steels must be weighed against the higher creep strength, lower tendency to carburise and to σ- and α'-embrittlement as well as the improved weldability of austenitic steels. The higher carbon content of casting alloys increases their wear resistance.

Combustion grates in lignite-fired power stations or waste incineration plants consist of parallel rows of grate bars between which the incoming combustion air is blown. The rows are arranged in a cascade and the ends of the bars rub against each other because every second row moves backwards and forwards to transport combustion material down the slope. The grate bars are cast from steels GX40CrSi17 or GX40CrSi23. The wear resistance of the latter steel is increased by carbon contents of up to 0.7 %. Sulphur from the lignite dissolves in the $M_{23}C_6$ precipitates thus withdrawing chromium from the matrix and lowering its scaling resistance. In addition, the sodium-containing ashes produce low-melting areas of scale that promote wear. Under waste incineration conditions, the metal is attacked not only by sulphur and chlorine, but also by metallic aluminium originating from packaging waste. The aluminothermal reaction with the scale in rubbing contact may even cause partial melting of the steel surface.

Facilities in the *petrochemical industry* are also reliant on heat-resistant steels. For example, the internal fittings in furnaces used to generate hydrogen by thermal decomposition of e.g. steam/natural gas mixtures are made of centrifugally cast pipes of steel GX30CrNiSiNb24-24 or a corresponding alloy with 30 % each of Cr and Ni. Some of them allow a service temperature of ≥ 925°C, which is 30 or even 50°C higher than that allowed by steel GX40CrNiSi25-20, and increases the efficiency of the process. The latter steel has also been used in pyrolysis furnaces to produce ethylene by hydrocarbon cracking at temperatures of 900 to 1000°C. The process temperature could be increased to more than 1000°C by using steel GX50CrNi30-30. This grade contains ≈ 2.5 % Si, which delays carburisation and facilitates the formation of new scale during burnout of coked deposits. In the coal gasification process, a raw gas consisting of H_2, CH_4 and CO, along with a little CO_2 and H_2S, is generated from pulverised coal and steam heated by nuclear process heat to 750 - 950°C and 40 bar. Although there is still no prospect of cost-coverage, much effort went into the development of materials for the immersion heater, which transfers heat from the high-temperature reactor helium to the process gas. To this end, the internal oxidation of steel X10NiCrAlTi32-20 was suppressed by limiting the amount of Si, Al and Ti, and the adhesion and healing of the protective surface film was improved by increasing the chromium content to 25 % and alloying with cerium.

In the USA, the housing of an automobile *catalytic converter* and the upstream exhaust pipe are exposed to temperatures of up to 750°C and are thus usually made of ferritic steel X6CrTi12. Higher exhaust gas temperatures are reached in Europe because of the higher driving speeds, so that austenitic steels such as X6CrNiTi18-10 or X15CrNiSi20-12 are often used which, however, exhibit a higher thermal expansion. This drawback was overcome by developing ferritic steel X2CrNbTi18, which, compared to X6CrTi12, contains higher amounts of Cr and free Ti, is scale-resistant up to 950°C and more creep-resistant due to the addition of ≈ 0.7 % Nb. Inside the catalyst housing sits a honeycomb-shaped ceramic body coated with an extremely thin Pt/Rh film on a porous intermediate layer (wash coat). Alternatively, corrugated foils made of a steel with 20 % Cr and 5 % Al, which is also used for electrical resistance heating, is wound around a porous body. Oxidation of the surface of a steel containing > 4 % Al produces a very resistant outer layer of Al_2O_3 that is very similar to the ceramic support. Although Al contents > 5 % would increase the durability of the surface film, they make steel processing more difficult because they form increasingly embrittling Fe-Al or Cr-Al precipitates. Attempts were made to achieve a film thickness $< 50\,\mu m$ in a single work step using melt-spinning and also suppress embrittlement by means of a high cooling rate.

The increasing alloy content of heat-resistant austenitic steels ultimately leads to Ni-base alloys such as NiCr20Ti, GNiCr28W or GNiCr50Nb, that take full advantage of the lower solubility of nickel for interstitial C and N atoms from carburising and nitriding gases. C and N are bound as precipitates in the surface layer by carbide- and nitride-forming elements (Cr, W, Nb, Ti), thus preventing them from penetrating even deeper.

B.6.4 Cast iron

General atmospheric corrosion of unalloyed cast irons with flake or spheroidal graphite is comparable to that of unalloyed steels. During high-temperature corrosion of cast iron, graphitisation of the cementite in pearlite (Chapt. A.4, p. 115) already occurs below the wustite temperature (Chapt. A.2, p. 51). It starts at about 450°C and is associated with a loss of strength and a volume increase. This does not occur in unalloyed steel. Although the addition of small amounts of Cr, Cu, Sn can increase the pearlite stability to about 550°C, GJL already begins to show signs of oxygen penetration along the matrix/graphite lamella boundary which, according to the Richardson-Ellingham diagram, leads to the oxidation of iron in air below ≈ 700°C. This internal oxidation is accelerated by intermediate cooling with parting of the boundary interfaces. Above this temperature, graphite starts to burn out preferentially. Under cyclic temperature conditions, the onset of the ferrite/austenite transformation causes repeated volume changes in the substrate that promote detachment of the scale. If account is also taken of the fact that the creep strength decreases as the temperature increases, an approximate limiting temperature for unalloyed ferrous materials can be obtained: GJL = 500°C, steel = 550°C, GJS (with $\approx 2.5\,\%$ Si) = 600°C.

There are two ways of increasing the chemical resistance of grey cast iron: alloying with Si or with Ni. Alloying with $\approx 5\,\%$ Si produces a ferritic cast iron with an increased scale resistance and $\approx 15\,\%$ Si leads to exceptional acid resistance. Alloying with Ni produces an austenitic cast iron with a high resistance to wet and/or high-temperature corrosion. The designation of these highly alloyed grades is made up of GJ for cast iron, L or S for lamellar or spheroidal graphite, F or A for a ferritic or austenitic matrix, X for high-alloy with the alloying elements and contents in mass-%. If chemical loading is accompanied by wear, carbide-rich white cast iron or steels are used.

B.6.4.1 Ferritic cast iron

An increase in the Si content in cast iron with spheroidal graphite from the typical value of 2.5 % Si to 4 - 6 % significantly increases the scaling resistance. The ferrite/austenite transformation temperature also increases to over 800°C, thus allowing a service temperature of ≤ 800°C. Addition of 1 % Mo increases the creep strength. To prevent expansion during service, a low-pearlite microstructure must be obtained e.g. by annealing. The high Si content and the size of the ferritic as-cast grains increases T_T to above room temperature in the notched impact test. Even elongation at fracture is usually less than 10 % (cf. No. 1 in Tab.B.6.5).

Table B.6.5 Cast iron with enhanced chemical resistance: No. 1 with a ferritic matrix, Nos. 2 to 8 with an austenitic matrix, in accordance with EN 13835.

Nr.	Name	C [%]	$R_{p0.2}$ [MPa]	R_m [MPa]	A [%]	ISO-V [J]
1	GJSF-XSiMo5-1	< 3.8	360	450	4	-
2	GJLA-XNiCuCr15-6-2	< 3.0	-	170	-	-
3	GJSA-XNiCr20-2	< 3.0	210	370	7	13
4	GJSA-XNiCrNb20-2	< 3.0	210	370	7	13
5	GJSA-XNiMn23-4	< 2.6	210	440	25	24
6	GJSA-XNiMn13-7	< 3.0	210	390	15	16
7	GJSA-XNi35	< 2.4	210	370	20	-
8	GJSA-XNiSiCr35-5-2	< 2.0	200	370	10	-

An important field of application of SiMo cast irons is exhaust manifolds and turbine casings in turbochargers of combustion engines (Figure B.6.21). The appropriate EN standard is being drafted. These materials are also used with vermicular graphite that interrupts internal oxidation similar to spheroidal graphite, but which has a slightly lower stiffness, higher thermal conductivity and thus contributes to relieving stress during thermal cycling.

Cast iron alloys containing 13 - 17 % Si and 1 - 0.3 % C with resistance to wet corrosion were already being developed in the 1920s. In the Fe-Si system, the austenite field is cropped at 1.7 % Si (Tab. A.1.4, p. 15) and there is a eutectic at 21 % Si and 1200°C that is shifted to a lower Si content by C. According to ASTM 518, a cast iron alloy contains 14.5 % Si and 0.85 % C. Thermodynamic calculations indicate that this composition corresponds to the eutectic, and that ferrite and graphite precipitate from the melt and SiC can precipitate on further cooling. Alloying with 4 % Cr or 3.5 % Mo is optional. With a hardness of 520 HB, a tensile strength of only 110 MPa indicates a high degree of brittleness. One of the advantages of these materials is their good resistance to acids owing to the formation of a surface film of Si. Evidence for this is provided by some literature values and a comparison with Figure B.6.8: H_2SO_4/70 %/boiling < 1 g/m^{2}h; HCl/30 %/20°C < 1 g/m^{2}h; boiling saturated sodium chloride solution < 0.1 g/m^{2}h; HNO_3/≤65 %/boiling < 0.1 g/m^{2}h. Applications include containers, pipelines, evaporator bowls, pump components, coolers etc. for the production of acids or for pickling equipment. The use of FeSi15 is limited on the one hand by manufacturing difficulties (castability, cracks, machinability) as well as its inadequate mechanical properties and, on the other, by further developments in stainless steels.

B.6.4.2 Austenitic cast iron

These materials are used under conditions of wet and high-temperature corrosion, and are also used for sub-zero applications and non-magnetisable components. Their chemical composition differs accordingly. Approx. 20 % Ni are required to obtain an austenitic matrix. Similar to steels (Figure A.2.24, p. 47), Cr supports this process, but has to be limited to < 3 % because it forms carbides (Tab. B.6.5, No. 3, 4 and 8). Ferromagnetic carbides are undesirable in non-magnetisable applications. Chromium has a beneficial effect on the imperviousness, strength and chemical resistance of castings. If some of the Ni is replaced by Cu, flake graphite forms that improves the already high resistance to wet corrosion (No. 2). The other grades are usually produced with spheroidal graphite to provide greater ductility and/or scaling resistance. The addition of 4 % Mn increases the austenite stability so that applications at sub-zero temperatures are possible (No. 5). The partial replacement of Ni by 7 % Mn produces a cost-effective non-magnetisable alloy (No. 6) and avoids, like Cu, the ferromagnetic austenite field at high Ni contents (Figure A.4.22, p. 118). The scaling resistance can be increased by alloying with 5 % Si; however, the Ni content has to be increased to > 30 % to suppress long-term embrittlement (No. 8). Nickel and silicon shift the graphite eutectic to a lower C content so that austenitic cast irons contain only < 3 % C. Both elements lower the solubility of C in the matrix, thus limiting its effects on the austenite stability

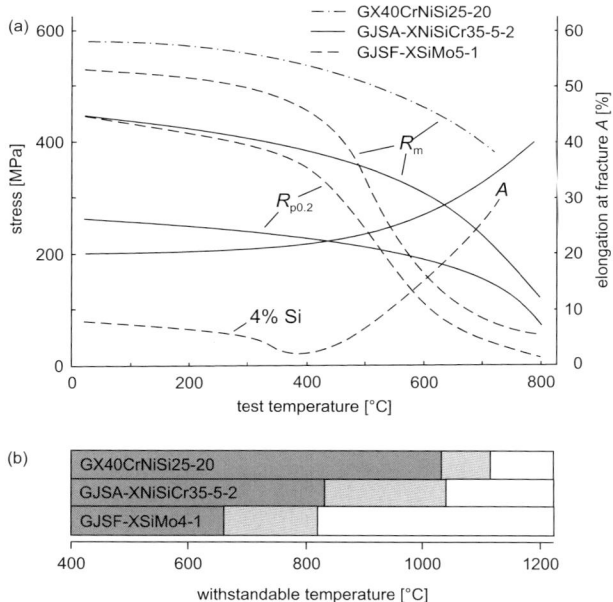

Fig. B.6.21 Comparison of heat-resistant ferrous materials: (a) Mechanical properties at elevated temperatures (from a review by K.Röhrig). (b) Withstandable temperature for exhaust manifolds and turbocharger housings (from W. Kallen and K. Röhrig, light grey = scatter band).

and yield point. The addition of 0.12 to 0.2 % Nb improves the weldability (No. 4). Approximately 1 % Mo increases the creep strength of No. 3. A low thermal expansion is obtained with 35 % Ni (No. 7, Figure A.4.23, p. 120).

For applications at elevated temperatures (Figure B.6.21), annealing between 875 and 900°C is recommended by EN 13835 (except for Cu-alloyed grades) to stabilise the microstructure and thus prevent a size change due to carbide decomposition during service. Silicides dissolve at temperatures above 1000°C. More carbon remains dissolved after accelerated cooling, thus increasing the strength values. This increase disappears again on stess-relief annealing between 625 and 650°C, however, size changes induced by residual stress do not occur on machining. The relief of residual stresses is also important to avoid stress corrosion cracking, e.g. in warm tropical seawater. Alloy No. 3 in Tab. B.6.5 can be used for a wide range of applications that require resistance to wet corrosion and heat: pumps, valves, compressors, bushes, non-magnetisable housings, turbocharger casings and exhaust manifolds that require a higher service temperature than that allowed by ferritic cast iron. Under wet corrosion conditions, GJLA-XNiCuCr15-6-3 has a many times higher resistance to pitting corrosion than the austenitic standard steel X5CrNi18-10, although its resistance to uniform corrosion is lower. Principal applications of the other alloys are: No. 2 - bases, dilute acids, seawater; No. 4 - as for No. 3, but with better weldability; No. 5 - cryoengineering to -196°C; No. 6 - non-magnetisable castings, e.g. for electrical installations; No. 7 - dimensionally stable machine tools, instruments; No. 8 - high-temperature resistant applications, e.g. housing components for gas turbines.

B.6.4.3 White cast iron / carbide-rich steels

According to EN 10295, GX130CrSi29 with $\approx 2\,\%$ Si is classified as a heat-resistant casting steel with a ferritic matrix containing eutectic $M_{23}C_6$ carbides that impart a high wear resistance. This is advantageous for friction pairs e.g. in pusher furnaces or between grate bars in firing systems. From the microstructural point of view, this material borders on hypoeutectic white cast irons. This also applies to GX160CrSi18, which contains eutectic M_7C_3 carbides in a hardenable matrix, although this can hardly be used to advantage on account of the tempering effect at elevated service temperatures. Embedding the carbides in a hard martensitic matrix is recommended under conditions with wear and wet corrosion (Chapt. B.4, p. 261). Although a wear-resistant white cast iron such as GX260CrMo27-2 is hardenable, it is not sufficiently corrosion-resistant, in spite of the high Cr content, because too much Cr is withdrawn from the matrix by M_7C_3 carbides. The obvious solution is to increase the Cr content, but this may lead to δ-ferrite that lowers the matrix hardness. One way of overcoming this is to add V, Nb, Ti, which form primary MC carbides. This reduces the amount of subsequently precipitated chromium carbide and the matrix chromium content increases as in e.g. GX145CrNbMoTi15-6-2. This effect is exploited in mineral processing and toolmaking. The triplex grade GX150CrNiMo40-6 was specially developed for constructing pumps. Its microstructure consists of a third each of chromium carbide, ferrite and austenite. The duplex matrix has a higher strength to support the carbides compared to a ferritic one, and its chemical durability corresponds to that of a duplex steel. This casting steel also borders on white cast iron.

Gas turbine components made of a creep-resistant martensitic cast steel
(SCHMOLZ + BICKENBACH Guss GmbH & Co. KG, Krefeld, Germany)

B.7 Creep-resistant materials

Creep-resistant materials are used in machines and facilities operated at high temperatures e.g. power engineering equipment. They must be able to withstand the highest possible operating loads at elevated temperatures and also be sufficiently resistant to high-temperature corrosion. In contrast to heat-resistant materials, the mechanical properties of creep-resistant materials are of prime importance.

The influence of thermally activated processes in the matrix increases with the test temperature and duration because they lower the yield point, which reaches higher values for sudden short-term loadings (Figure B.7.1). At low temperatures, loads below the yield stress measured in slow tensile tests are withstood for long periods without plastic deformation. The diffusion of interstitial elements starts above room temperature (Figure A.1.8, p. 16), and time-dependent permanent deformation ε is observed, particularly for stresses above the yield point:

$$\varepsilon = \varepsilon_0 + a \cdot \log t \qquad (B.7.1)$$

This process is known as logarithmic creep. The creep rate continuously decreases with $\dot{\varepsilon} \sim t^{-1}$. The metal atoms also start to diffuse if the temperature rises to $\approx 350°C$. Dislocations are then able to leave their slip planes, which

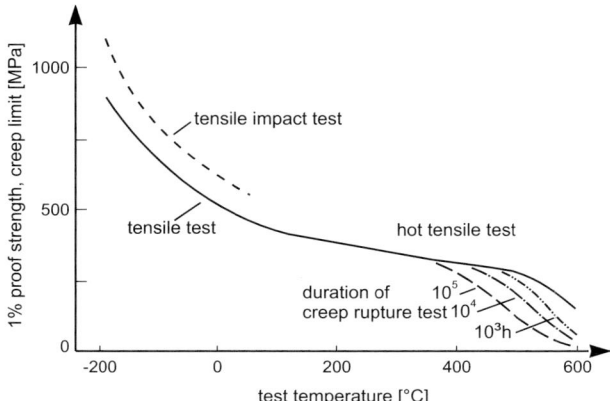

Fig. B.7.1 Influence of the test temperature and duration on the proof stress: Approximate representation of the behaviour of a low-alloy structural steel. Thermal activation of slip processes increases with the test temperature and lowers the proof stress measured in the tensile and hot tensile test. An increase in the strain rate by several orders of magnitude in the tensile impact test shortens the time of exposure to the test temperature and increases the proof stress. Prolonging the test duration in the creep rupture test by several orders of magnitude at temperatures above $\approx 350°C$ causes the creep limit to drop with respect to the high-temperature proof stress as a result of thermally activated creep.

is known as climbing. This non-conservative movement of dislocations lowers their density and they arrange themselves into networks (subgrain boundaries). This recovery process eases slip. As the temperature increases, an increasingly lower level of stress is required to cause measurable creep. Strengthening induced by creep deformation cannot prevent slip occurring, which means that a constant stress below the yield point measured at elevated temperature will eventually lead to rupture after a certain loading time.

Creep curves recorded in a creep rupture test (Figure B.7.2 a) at high temperatures usually exhibit three different stages (Figure B.7.2 b). In the primary stage (transient creep), the steel starts to creep after application of a constant force or stress and follows the relation $\varepsilon \sim t^m$ with $m < 1$. The resulting strengthening lowers the creep rate to a minimum value $\dot{\varepsilon}_{\min}$, which remains virtually constant in the secondary stage because deformation-related strengthening (increase in the dislocation density) and thermally related softening (recovery, decrease in the dislocation density) are virtually balanced (stationary creep). The diameter of the subgrains is inversely proportional to the applied stress. Norton's Creep Law applies at low stresses and long loading times, (Figure B.7.2 e):

$$\dot{\varepsilon}_{\min} \sim \sigma^n \quad (B.7.2)$$

However, at high stresses, the creep rate becomes exponential

$$\dot{\varepsilon}_{\min} \sim e^{\alpha \sigma} \quad (B.7.3)$$

Towards the end of the primary creep stage, voids start to form as a result of diffusion at the grain boundaries. These voids gradually coalesce so that microcracks with a length of one grain diameter start to appear towards the end of the secondary stage. This internal damage as well as external necking increase the creep rate in the tertiary stage and the material ultimately ruptures. There are empirical relationships between the three creep stages:

$$m \sim \dot{\varepsilon}_{\min} \quad (B.7.4)$$

$$\dot{\varepsilon}_{\min} \cdot t_R = const \quad (B.7.5)$$

Desirable for technical applications is a high degree of strengthening after a small primary strain ($\varepsilon_I \downarrow$) to give a lower level of $\dot{\varepsilon}_{\min}$ that is maintained for the longest possible time ($t_{II} \uparrow$). Lattice defects increase the resistance to creep (Figure A.2.4, p. 23). Whereas solid-solution hardening by interstitial elements is particularly effective up to 400°C higher temperatures require substitutional alloying atoms which diffuse more slowly. Their effect is essentially maintained up to the melting temperature. Cold or semi-hot strengthening by dislocations is exploited in e.g. austenitic steels. However, this type of strengthening is only effective up to the recrystallisation temperature. Hardening by grain refinement, as at room temperature, does not occur at high temperatures. On the contrary, grain boundaries act as disturbed lattice regions with faster diffusion rates and cause sliding between the grains. This grain boundary sliding leaves the grains themselves largely unaffected. It is hindered by coarse grains and by precipitation hardening of the grain boundaries. Precipitation hardening is also used to a large extent to strengthen the grains themselves. Because the precipitates can gradually coarsen due to creep loading or even dissolve, thus losing their effectiveness, there is also an upper limiting temperature for this

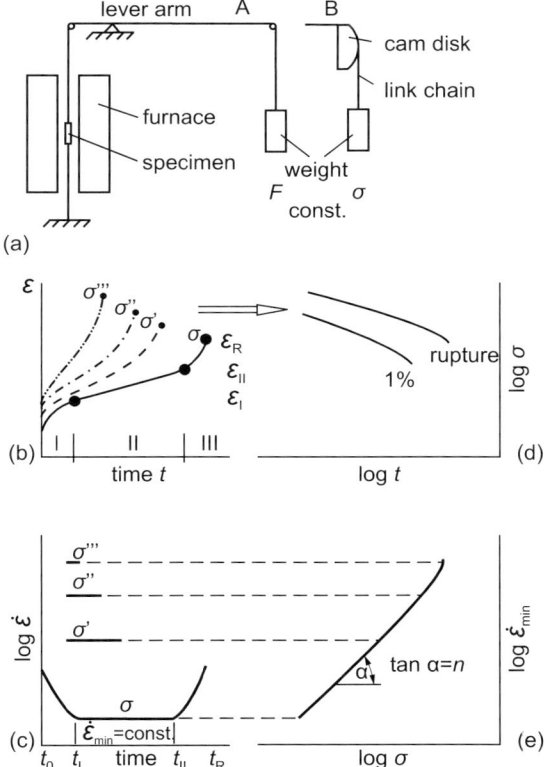

Fig. B.7.2 The creep rupture test and its evaluation: (a) A tensile load is applied to the specimen at a given test temperature with a constant force F (case A) or stress (case B). In case B, the lever arm is shortened by a cam disk as the specimen elongates so that the force follows the reduced cross-section of the specimen. (b) Four creep curves at a constant stress that increases from σ to σ'''. The shape of one curve can be simplified to $\varepsilon \sim t^m$ where $m < 1$ in the primary (I), $m = 1$ in the secondary (II) and $m > 1$ in the tertiary (III) creep stages. (c) The first derivative of the creep curve gives the minimum creep rate $\dot\varepsilon_{\min}$ for each stress and this gives (e) Norton's Creep Law according to Eq. B.7.2, which loses its validity at high stresses. In evaluating (b) to give (d), only the 1 % creep stress curve and the creep rupture curve are taken from the creep curves and they are used to derive the 1 % creep limit and the creep rupture strength for a given test duration.

strengthening mechanism. Owing to the slower diffusion rate in the fcc lattice, austenitic materials exhibit a significantly higher creep resistance at temperatures above 650°C compared to normalised or quenched and tempered bcc materials. In addition to the aforementioned creep by slip, at high temperatures and lower stresses diffusion-controlled creep occurs without movement of the dislocations (Nabarro-Herring creep). If the atoms diffuse preferentially along the grain boundaries, this is known as grain boundary creep (Coble creep).

Cast irons also contain graphite as a microstructural component, which is generally spheroidised to improve the mechanical and chemical properties. In the creep rupture test, creep causes the matrix to detach from the graphite nodules to form relatively large voids at an early stage that bring forward damage in the tertiary creep stage. These materials thus rupture earlier than steels with a comparable matrix.

Four groups of creep-resistant materials emerge from these metallographic conditions: 1. *Low-alloy creep-resistant structural steels* that are used in the normalised or QT state. They are strengthened by carbides and carbonitrides of Cr, Mo, V, W, Nb. 2. *Highly creep-resistant chromium steels* that are quenched and tempered, and strengthened in the same way. Their chromium content of 9 to 12 % imparts a greater scaling resistance. 3. *Highly creep-resistant austenitic Cr-Ni steels*, some of which are used in the semi-hot worked or precipitation-hardened state. 4. *High-alloy cast iron*. Information on creep-resistant and highly creep-resistant materials is given in EN 10216-2 (seamless pipes), 10222-2,-5 (pressure vessels), 10269 (fasteners), 10213 (cast steel), 13835 (cast iron).

B.7.1 Properties

B.7.1.1 Normalised as well as quenched and tempered steels

In unalloyed creep-resistant structural steels, such as P235GH+N in Tab. B.7.1, the creep rate is lowered by dissolved nitrogen. If this content is increased from 0.002 to 0.015, for example, the 10^5 h creep rupture strength at 400°C can double. Accordingly, killing the steel with aluminium to make it more resistant to ageing (Chapt. B.1, p. 130) binds nitrogen as AlN, thus lowering the creep rupture strength. Metals having a favourable effect on the creep-rupture strength include manganese as well as unintended alloying with e. g. chromium and molybdenum that originate from scrap metal added to the melt.

The quenched and tempered steels in Tab. B.7.1 owe their creep rupture strength (Figure B.7.3) to a dispersion of temper carbides (and nitrides) of Cr, Mo, W, V and Nb as well as the dissolved remainder of these elements. Incoherent hard particles (Tab. A.2.1) cannot be cut by dislocations. Diffusion-dependent climbing over such obstacles takes time, which thus has a retarding effect. For a precipitation volume of $< 3\,\%$, the diameter and spacing of the

B.7.1 Properties

Table B.7.1 Creep-resistant materials: The alloying contents not specified in the name are given as percentages in brackets. Minimum yield point or 0.2 proof strength R_e and thermal conductivity λ at room temperature. Mean coefficient of thermal expansion α between 20 and 600°C (cast iron between 20 and 200°C).

Group	Entry	Material	HT	R_e [MPa]	α [µm/m·K]	λ [W/m·K]
creep-resistant structural steel	1	P235GH	N	225	14.5	55
	2	13CrMo4-5	QT	290	14.5	44
	3	10CrMo9-10	QT	280	14.0	35
highly creep-resistant chromium steel	4	X20CrMoV11-1[1)] (0.3 V, 0.6 Ni)	QT	490	12.3	24
	5	X10CrMoVNb9-1 (0.2 V, 0.08 Nb, 0.05 N)		450	–	–
highly creep-resistant austenitic steel	6	X6CrNi18-10 ($\leq$ 0.11 N)	SA	190	18.5	16
	7	X3CrNiMoBN17-13-3 (0.003B, 0.15 N)		260	18.5	16
nickel-base alloy	8	NiCr20TiAl (2.3 Ti, 1.4 Al)	SAA	600	15.2	13
cast iron	9	GJSA-XNiCr20-2	A	210	18.7	12.6

[1)] Previously under the same material number 1.4922 as X20CrMoV12-1,
HT = Heat Treatment, N = normalised
QT = quenched and tempered, SA = solution-annealed
SAA = solution-annealed and artificially aged, A = annealed

particles increase with the tempering temperature as a result of Ostwald ripening (Chapt. A.2, p. 48). An optimum initial precipitate size can be obtained in this way. However, this initial microstructure is not retained in service (Figure B.7.4).

Ostwald ripening continues and is accelerated by creep deformation. Furthermore, the carbides try to attain equilibrium, which results in further dissolution and precipitation of new particles. Simultaneous creep deformation can facilitate nucleation of these new precipitates. Although precipitation on the grain boundaries impedes grain boundary sliding, it does promote the formation of voids and cracks. Slip is localised to precipitation-free grain boundary regions, which lowers the ductility (Figure B.7.4 c). This results in creep embrittlement. All these microstructural changes take place, even if the selected tempering temperature of creep-resistant steels is 150 to 200°C above the service temperature. It seems reasonable to assume that this would anticipate the approach of microstructural equilibrium. However, a coarse approximation

B.7 Creep-resistant materials

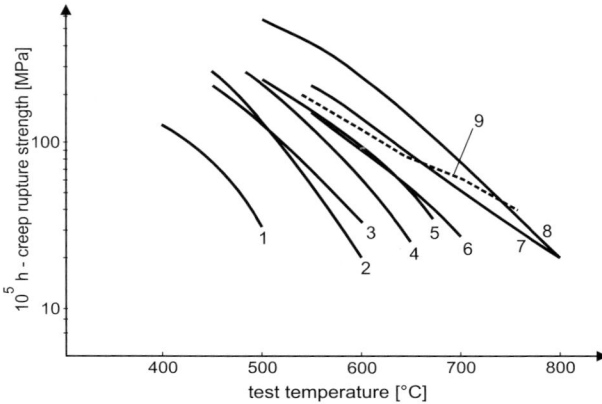

Fig. B.7.3 Creep-rupture strength of creep-resistant materials: see Tab. B.7.1; approximate values of steels according to the respective EN standards for a test duration of 10^5 h and of austenitic cast iron for a test duration of 10^3 h.

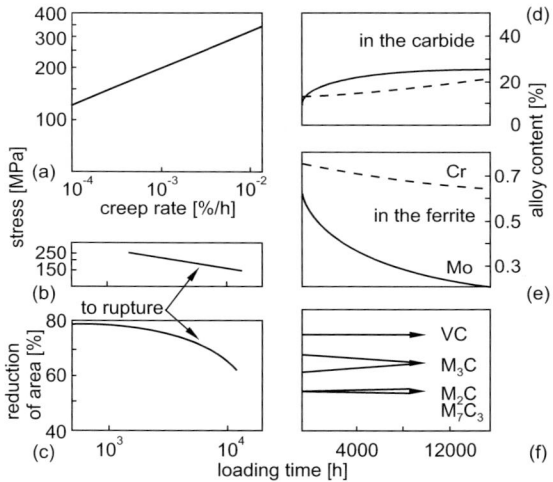

Fig. B.7.4 Microstructure and creep rupture behaviour: at 550°C for an example steel with 0.24 % C; 1.15 % Cr; 0.86 % Mo and 0.28 % V ($\approx$ 21CrMoV5-11), quenched 960°C 15 min/oil and tempered 730°C for 1 h/air to $R_{eL} = 644$, $R_m = 751$ MPa, $Z = 67$ %. (a) Mean minimum creep rate $\dot{\varepsilon}_{min}$; (b) creep rupture curve; (c) reduction of area under creep conditions; (d) and (e) precipitation of Cr and Mo into the carbides; (f) qualitative changes in the amount of individual types of carbides. The content of iron-rich M_3C carbide decreases in favour of the alloy carbides M_2C and M_7C_3 (from D. Horstmann et al.).

according to Eq. A.3.3 (p. 69) indicates that a service time of 10^5 h would have at least a similarly large 'tempering effect' P_A as tempering itself. Creep deformation also has an accelerating effect. The higher tempering temperature used for creep-resistant steels compared to QT steels results in a relatively low proof strength at room temperature.

The cooling rate during hardening decreases as the component's cross-section increases. This has the following implications: in low-alloy steels, martensite converts to a ferritic/bainitic microstructure, which leads to an inhomogeneous distribution of the temper carbides after tempering. Although increasing the alloy content improves the hardness penetration, alloy carbides may already start precipitating during cooling. Slowly cooled steel 10CrMo9-10 contains lamellar M_2C and steel X22CrMoV12-1 contains $M_{23}C_6$ located on the former austenite grain boundaries. These types of precipitates adversely affect the ductility. In steels with 12 % chromium, similar to hot-work tool steels (Figure B.5.14, p. 297), the carbide network on the grain boundaries may become so wide during slow cooling of thick components that its complete dissolution during hardening is prevented, thus causing embrittlement. The low M_s temperature of these steels is also associated with a susceptibility to cold cracking during cooling from the forging or hardening temperature. The high hardenability and low M_s temperature necessitate special precautions during welding of creep-resistant steels with 12 % chromium. These include preheating and control of the interpass temperature (Figure B.7.5).

During reheating of low-alloy creep-resistant steels for stress-relief annealing, intergranular precipitation cracks may occur in the heat-affected zone

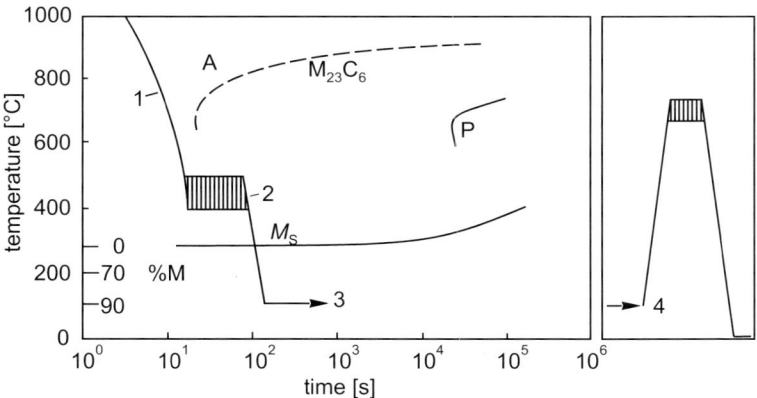

Fig. B.7.5 Temperature profile during welding of steel X20CrMoV11-1:(1) Cooling of the weld bead; (2) interpass temperature between carbide precipitation and growth of martensite; (3) cooling of the weld seam to above 100°C and (4) immediate tempering. If the proportion of retained austenite is high: repeat the tempering step.

(stress-relief or reheat cracking). This is caused by the precipitation of alloy carbides that not only hinder the relief of residual welding stresses (relaxation), but also strengthen the grains as a secondary hardening effect so that sliding mainly occurs along the grain boundaries in low-precipitate zones. The general requirements for these defects are: 1. high residual stresses that occur in welded joints > 70 mm thick; 2. threshold concentrations of the elements Cr, Mo, V, Nb; 3. a high dislocation density after welding that promotes nucleation of precipitates (martensite > bainite > ferrite/pearlite).

Temper embrittlement (Chapt. A.2, p. 48) can occur during slow cooling of heavy shafts from the tempering temperature. Figure B.2.19, p. 197 shows an example of grain-boundary segregation of phosphorus in a bcc matrix. Dissolved carbon and boron atoms displace phosphorus from the grain boundary and reduce temper embrittlement. Nickel increases the activity of carbon, whereas chromium binds it. Molybdenum dissolved in the matrix lowers phosphorus segregation and thus embrittlement.

If the temperature during long-term loading exceeds the wustite temperature (570°C) in the Fe-O system (Chapt. A.4, p. 115), the chromium content required for protection against high-temperature corrosion becomes increasingly important. The limit lay at $\approx$ 12 % for a long time because greater levels necessitated additional carbon in order to avoid δ-ferrite (Figure B.6.10). This phase does not contain the dispersion of fine precipitates that are obtained during tempering of a basic martensitic microstructure, which adversely affects the creep resistance. Because a high tempering temperature leads to precipitation of nearly all the carbon, the amount of carbides also increases with carbon content. This can dilute the Cr and Mo fractions dissolved in the carbides, thus lowering the stability of the precipitates against undesirable coarsening during service. Moreover, excessive amounts of precipitates cause damaging voids at an earlier stage. This prompted a lowering of the C content and enrichment of Mo, W, V, Nb in the smaller amount of carbides, along with a reduction in the Cr content to $\approx$ 9 % and alloying with cobalt to suppress δ-ferrite. Refinement and stabilisation of the precipitates was achieved by partial exchange of C by N, whose solubility in the melt is improved by Cr. In contrast, in more recently developed boron-alloyed steels with 9 to 12 % Cr, the nitrogen content is limited to avoid boron nitride. Boron replaces some of the carbon in $M_{23}C_6$, delays coarsening of these precipitates and thus also coarsening of the dislocation structure. Figure B.7.6 shows that the time-to-rupture increases considerably if the boron content is about 0.01 %. Twice this amount leads to coarse tungsten borides that withdraw W from the matrix and promote void formation. The development of creep cracks in the finely grained zone of welded joints was not observed in boron-alloyed steels. However, a steel containing only 9 % Cr (e.g. X8CrCoWVNbB9-3-3 with 0.2 % V, 0.05 % Nb and 0.01 % B) brings us back to the problem of scaling resistance. For example, tests by Grabke et al. with a 11 % Cr steel exposed to H_2-2.5 % H_2O at 600°C for 100 h showed a thicker Cr-rich scale layer and a lower Cr depletion of the surface layer (to 8 % Cr) compared to a 9 % Cr steel (to 5 % Cr). Any damage to the scale film due to temperature fluctuations or particle-laden gases cannot heal if there is only 5 % Cr in the surface layer. Particularly at the start, scaling is not determined solely by diffusion of chromium within the scale layer but also by its diffusion into the surface layer from deeper regions. This is why further developments focus on 12 % Cr steels with a lower C content and additional cobalt to keep them free of δ-ferrite, e.g. X12CrCoWMoVNbB12-5-2.

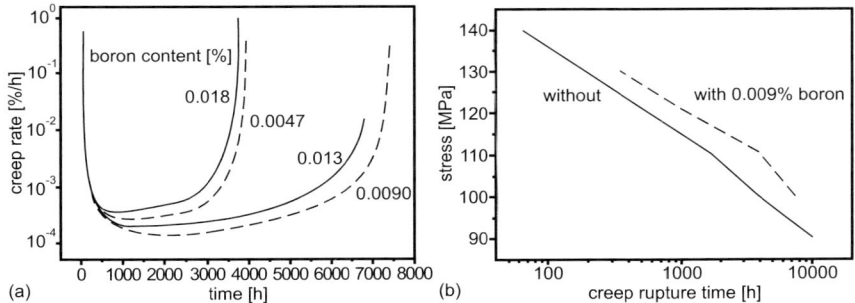

Fig. B.7.6 Influence of boron on the creep behaviour of steel X8CrCoWVNbB9-3-3 at 650°C:from S.K. Albert et al. (a) Creep rate $\dot{\varepsilon}$ at a stress of 100 MPa showing the three creep stages: primary (strengthening), secondary, tertiary (softening, damage); (b) time-to-rupture t_R plotted against the applied stress (compare Figure B.7.2 c, d, p. 351).

B.7.1.2 Austenitic steels

Highly creep-resistant austenitic steels should not contain δ-ferrite, particularly as a grain boundary network, because it has a lower creep resistance than austenite and undergoes embrittlement more quickly. A fully austenitic microstructure can be obtained by alloying with nickel (Figure B.6.6). Carbon and nitrogen do not provide a permanent contribution to the nickel equivalent (Figure A.2.24, p. 47) because they are partially precipitated during service. Lowering the chromium content delays embrittlement of austenite by the σ phase during long-term service. This results in a creep-resistant austenite containing 16 % chromium and 16 % nickel (X7CrNiMoBNb16-16).

This steel contains niobium to avoid sensitisation during service with intergranular corrosion (IC) during stoppages. Compared to titanium, stabilisation (Chapt. B.6, p. 319) with niobium provides a higher gain in the creep-rupture strength by MC precipitates within the grains during service. However, precipitation hardening can be so extensive in the HAZ of welded components that grain boundary cracking occurs during operation. Creep embrittlement due to niobium can also be expected (Figure B.7.7). Avoiding IC

B.7 Creep-resistant materials

susceptibility by means of a low carbon content (ELC) is associated with a loss of creep rupture strength. This can be more than compensated by adding nitrogen (X3CrNiMoBN17-13-3). Nitrogen inhibits the precipitation of intermetallic phases and raises the low room-temperature proof strength of austenitic steels. Above $\approx 0.18\,\%$ nitrogen, these advantages may be cancelled by creep embrittlement due to the precipitation of Cr_2N during service. The solid-solution hardness is increased not only by chromium and nickel, but also by alloying with molybdenum ($<4\,\%$), tungsten ($<4\,\%$) and cobalt ($<20\,\%$). This also shifts recrystallisation to higher temperatures. Molybdenum and tungsten form χ and Laves phases (Tab. A.2.1, p. 52). The latter can be used for precipitation hardening in service. Cobalt increases the solubility of carbon in austenite, which allows higher carbon contents in the steel and results in strengthening by increased carbide precipitation during service (X40CrNiCoNb13-13). Boron ($<0.1\,\%$) is preferentially dissolved at the grain boundaries and thus lowers their energy as well as the tendency to intergranular precipitation. Boron lowers grain boundary sliding by slowing

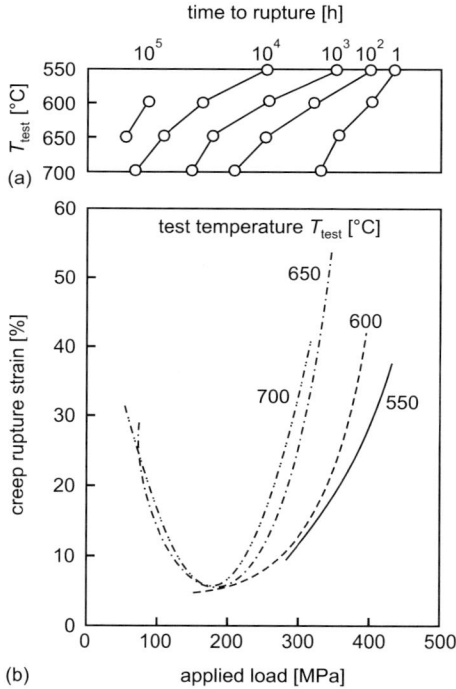

Fig. B.7.7 Creep embrittlement: niobium-stabilised austenitic steel. (a) Test temperature T_{test} and the time-to-rupture, depending on the applied load; (b) minimum of the creep rupture strain (from J.D. Murray).

down diffusion. One drawback is cracking in the HAZ during welding of boron-containing steels due to partial melting of the grain boundaries.

Preliminary cold- and semi-hot working by 5 to 20 % reduces primary creep; ε_i is significantly reduced. Carbides precipitate on the resulting dislocations during service. This also lowers $\dot{\varepsilon}_{min}$ and t_B is longer. Although this advantage is particularly pronounced at a test temperature of 650°C, it disappears again above 700°C. The widespread use of cold work-hardening by prestraining is not always possible, particularly for thicker dimensions. It is difficult to obtain a dispersion of carbides to increase the creep resistance, as in QT steels, owing to the absence of martensite with its high density of dislocations that act as disperse nuclei for the barely coherent (Tab. A.2.1, p. 52) carbides during artificial ageing. As already discussed, inducing lattice defects by cold working is only possible to a limited extent. If the material is not prestrained, the grain boundaries provide suitable precipitation sites during artificial ageing. Although the preferred network arrangement of carbides inhibits grain boundary sliding, it lowers the ductility. This is avoided by starting with the solution-annealed state and letting precipitation take place during service. This affects intragranular regions more strongly owing to creep strain.

In addition to non-cuttable hard carbides, cuttable γ' precipitates of type $Ni_3(Al,Ti)$ can be used for strengthening. Particles with a usual size are not climbed by dislocations, but are cut by dislocation pairs: at room temperature, the first dislocation produces an antiphase boundary in a coherent particle, the second simply restores the original atomic ordering. Their contribution to the yield point remains small. At higher temperatures, the antiphase boundary may be disturbed by diffusion so that the second dislocation must also expend cutting energy. The creep resistance thus increases with the volume fraction of the γ' phase, which in turn grows with the amount of aluminium and titanium. However, this requires a minimum nickel content (e.g. X5NiCrTi26-15 with 2.1 % Ti, 0.3 % Al and $\approx$ 4 vol.-% γ'). Nickel-base alloys allow higher volume fractions of the γ' phase than steels (e. g. NiCr20TiAl according to Tab. B.7.1 with $\approx$ 18 vol.-% γ'). With 5 % each of aluminium and titanium, the γ' phase increases to $\approx$ 60 volume-%. High-temperature materials consisting of only intermetallic phases are currently being developed. γ' Precipitates are produced in steels and nickel alloys during artificial ageing at 700 to 750°C. The higher the service temperature lies above the ageing temperature, the higher is the expected degree of coarsening, which ultimately leads to complete dissolution of the particles. This reduces the creep rupture strength (Figure B.7.3, NiCr20TiAl), and indeed, above 800°C it is below that of a solution-annealed, predominantly solid-solution hardened steel (X3CrNiMoBN17-13-3 or X10NiCrAlTi32-20).

Increasing the nickel content to avoid δ-ferrite is associated with primary austenitic solidification of the weld material (Figure B.6.6, p. 316) along with the resulting susceptibility to hot cracking (Chapt. B.6, p. 314). One way of avoiding this manufacturing defect is to use a weld filler with a lower nickel content that, after dilution, results in primary ferritic solidification and < 6 % δ-ferrite after cooling. This avoids a confined ferrite network around the austenite grains that would have a particularly unfavourable effect on the creep rupture behaviour. However, even small amounts of δ-ferrite accelerate creep embrittlement by precipitation of the σ phase. A second way of avoiding hot cracks is based on reducing the amount of impurities, such as sulphur and phosphorus, that contribute to low-melting films on the grain boundaries. Alloying with 3 to 5 % manganese binds the sulphur in the weld pool and lowers the susceptibility to hot cracking, which is increased by niobium and lowered by molybdenum. Metallurgical methods must be accompanied by welding techniques that help to reduce welding stresses: low heat input, thin electrodes, low amperage, low interpass temperature and narrow weaving.

B.7.1.3 Cast iron

Chapt. B.6, p. 341 discussed the use of cast irons at elevated temperatures from the viewpoint of their resistance to high-temperature corrosion. There are comparatively few studies on their long-term mechanical behaviour. EN 13835, issued in 2003, is limited to the creep rupture strength at 1000 h of some austenitic cast irons with spheroidal graphite based on studies carried out decades earlier. As can be seen in Figure B.7.3, the 10^3 h values lie in the range of the 10^5 h values of austenitic steels. Early damage due to detachment of the graphite nodules significantly limits the long-term service of cast iron in the creep range.

B.7.2 Applications

The development of power engineering equipment is closely associated with the development of creep-resistant steels. The main focus is on the generation of electrical power as well as the extraction and utilisation of fuels.

B.7.2.1 Steam power plants

In a fossil fuel-fired steam power plant, the temperature and pressure of the superheated steam (e.g. 540°C, 250 bar) are limited by the creep rupture strength of steels with a bcc lattice. Steel grades used in the construction of *steam boilers* include: 15Mo3 for boiler tubes, 13CrMo4-5 for headers, 10CrMo9-10 for reheaters, X22CrMoV12-1 for superheaters and superheated steam pipes. Steel 13CrMo4-5 is used in the air-hardened and tempered state and has a microstructure of ferrite and bainite. Isothermal transformation of steel 10CrMo9-10 at $\approx$ 700°C leads to a microstructure of ferrite grains surrounded by M_2C, which increases the creep rupture strength. Important design factors for steam turbines are the creep rupture strength as well as the permissible strain, which is obviously lower in a machine with narrow tolerances compared to a boiler tube.

An increase in the *steam temperature* to 600°C improves the efficiency of the energy conversion process. One way of achieving this is to use austenitic steels. Their fcc lattice structure provides sufficient creep rupture resistance, even at temperatures > 600°C. They have a lower thermal conductivity than bcc steels; however, this means that the turbine must be started more slowly owing to the greater thermal stresses. Further drawbacks are the higher thermal expansion and the higher costs. This is why austenitic steels are only used to a limited extent in steam power plants, e.g. in the boiler section (superheater harp). On account of their better weldability, some titanium-stabilised steels or an ELC grade such as X3CrNiMoBN17-13-3 are also suitable.

The aforementioned highly creep-resistant bcc chromium steels are used in steam turbines. Lowering the chromium content decreases the carbon and carbide contents without formation of δ-ferrite during forging or heat treatment, but still provides sufficient scaling resistance. The carbides contain more alloying elements and are stabler. In addition, increasing the nitrogen content to a value close to the solubility limit of the melt produces stabler nitrides. Partial exchange of vanadium by niobium provides a further increase in the stability of the precipitates after quenching and tempering. However, this requires a higher hardening temperature to dissolve the particles beforehand. Steels used include those with 9 to 11 % Cr, up to 2 % Mo+W, 0.2 to 0.3 % V and e.g. 0.08 % Nb, 0.06 % N or 0.01 % B. One design example is given in Tab. B.7.2.

Table B.7.2 Applications of creep-resistant steels in a steam turbine: Example of a 3-train turbine with an output of ≈ 1000 MW consisting of HP, IP, LP = high-, intermediate- and low-pressure sections with reheating.

	HD	IP	LP
steam admission			
pressure [bar]	270	70	15
temperature [°C]	600	600	400
shaft			
diameter [mm]	850	1100	1600
steel	X12CrMoWVNbN10-1-1		26NiCrMoV14-5
blades	X12CrMoWVNbN10-1-1		X20Cr13
housing			
inner	X12CrMoWVNbN10-1-1		creep-resistant
outer	G17CrMo5-10		plate
bolts	X19CrMoNbVN11-1		

A higher steam temperature increases the temperature levels in the entire system. This means that even the turbine shafts in the low-pressure section can reach temperatures of over 350°C so that their creep rupture properties must be also taken into account. Grain boundary diffusion of elements P, Sn, As, Sb, also known from temper embrittlement, (Figure A.2.19, p. 42) occurs during long-term service. To reduce the associated embrittlement, both the deleterious elements and the supporting manganese and silicon are limited to extremely low levels (clean steels) by careful selection of scrap feedstock and the appropriate ladle metallurgy. This also decreases the amount of non-metallic inclusions. High-purity Ni-Cr-Mo-V steels, e.g. 26NiCrMoV14-5, exhibit a higher toughness, which largely remains during service. The problem of stress corrosion cracking in low-pressure rotors (diameter e.g. 1.6 m) does not appear to be solved by the use of clean steels. However, steam purity, a reduced yield point and design measures to lower the stress help to reduce this problem.

B.7.2.2 Gas turbines

Stationary gas turbines are commonly used as small power sources in the chemical and petrochemical industries. Combined gas/steam concepts use large gas turbines (GT) with more than 300 MW output that are installed upstream of a steam turbine (ST). The gas turbine operates at high temperatures (gas inlet temperature for natural-gas firing up to 1350°C). Their hot exhaust gases are used to produce steam for the steam turbine. This increases the efficiency of $>40\%$ for ST to $>60\%$ for GT plus ST. The internally cooled blades and the hot gas housing are made of nickel-base alloys. The blades in the front rows, which are exposed to the highest temperatures, are made of single crystals (e.g. 250 mm long) or are directionally solidified with an elongation of the grains in the direction of centrifugal force. Both measures aim at eliminating grain boundary sliding. A ceramic coating is used as a thermal barrier. Highly creep-resistant chromium steels are used for the rotor disks. In addition to natural gas and oil, an upstream coal gasification unit is also used as a firing system. Direct injection of coal dust into the gas turbine causes excessive wear by the fly ash. However, many gas turbines used for petrochemical processes operate under such conditions and thus require special protective coatings on the edges of the blades. Jet turbines operate with gas temperatures of up to $\approx 1500°C$. The higher thermal loads on the turbine blades are also met by using high-alloy nickel-base alloys with directional solidification or as single crystals. A further measure is to intensify internal cooling of the blades. The extreme surface temperature requires a high resistance to hot-gas corrosion. This is attained using coatings that also act as thermal barriers. Overall, the higher thermal load in jet turbines must be paid with a shorter service life of the components compared to stationary gas turbines. Developments in materials engineering for jet turbines can only be partially transferred to large stationary turbines owing to the large difference in dimensions. Powder metallurgy was used to develop ODS (oxide dispersion-strengthened) alloys in which very stable oxide particles act as precipitates. For example, Y_2O_3 was ground together with an alloy powder. After hot-isostatic pressing, this mechanically alloyed powder contains a fine dispersion of particles that are scarcely affected by service conditions, unlike the precipitates, which means that the creep resistance is retained. This PM material has not yet gained acceptance over cast blades.

B.7.2.3 Estimation of the service life

The rapid implementation of newly developed steels is hindered by the long and expensive period of testing. These tests must first prove whether the initial microstructural stability is maintained over the design life and also exclude premature softening and embrittlement. Approximate design lives include: $3 \cdot 10^3$ - 10^4 h for civil jet turbines, 10^4 - $3 \cdot 10^4$ h for stationary gas turbines and 10^5 - $3 \cdot 10^5$ h for steam power plants. Extrapolation methods are thus used in an attempt to estimate long-term values from short-term data. One method is based on the Monkman-Grant relationship (Eq. B.7.5). Another includes the temperature dependence of the creep rate.

$$\dot{\varepsilon}_{\min} = A\sigma^n \cdot \exp\left(Q/RT\right) \tag{B.7.6}$$

At a constant stress, $\log \dot{\varepsilon}_{\min} \approx (1/T)$ gives a straight line that allows limited extrapolation to lower temperatures. Longer times-to-rupture (t_R) can then be inferred using Eq. B.7.5. A direct estimation can be made by plotting t_R against the test temperature at a constant stress in an iso-stress diagram (Figure B.7.8). This also indicates e.g. the effect of combined alloying with C+N and V+Nb on the stability of the precipitates and on the time-to-rupture. If values at measured different test temperatures and stresses are available, a plot of $\log \sigma$ against the Larson-Miller parameter

$$P_{\text{LM}} = T(C + \log t_R) \tag{B.7.7}$$

gives a curve that can also be extrapolated to lower stresses to a limited extent (see the tempering parameter P_A from Eq. A.3.3, p. 69). All these and similar empirical methods suffer from their inability to include microstructural changes correctly. Newer extrapolation methods are thus more closely based on physical relationships and include changes in the precipitate size. This is also important in estimating the remaining service life of components that have exceeded their design life of e.g. 10^5 h. Mobile metallographic examinations are used to obtain information on the condition of the microstructure. In some cases, this allowed the service life to be increased to $3.5 \cdot 10^5$ h, i.e. ≈ 40 years. Particularly in bcc steels, experience shows that a higher short-term strength leads to a lower long-term strength, i.e. the finer the precipitates are at the start of creep loading, the faster they coarsen. More recent developments thus

B.7 Creep-resistant materials

aim at minimising the driving force behind microstructural coarsening in advance by lowering the free energy of the material using near-equilibrium precipitates with a particular size (low interfacial energy). Prestages of carbide and nitride precipitates, as shown in Eq. A.2.3, (see p. 50 and Figure B.7.4 f), are avoided for long-term applications. Service life estimations often have to take account of superimposed damage from cyclic thermal and/or cyclic mechanical loads, such as those occurring during start-up and shut-down of a plant. In this case, the number of cycles lies below $5 \cdot 10^4$ and thus within the range of LCF (low-cycle fatigue) of the material. High-frequency vibrations experienced by high-speed components cause HCF (high-cycle fatigue). Predicting the service life of components operated at high temperatures under creep fatigue conditions is one of the most complicated problems facing material engineers, especially because three-dimensional temperature and stress distributions must be taken into account.

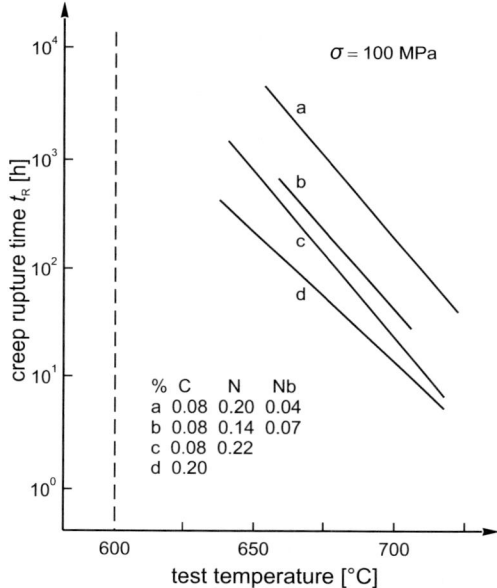

Fig. B.7.8 Iso-stress creep diagram: for steel X20CrMoV12-1 modified by alloying with niobium and nitrogen under pressure and subjected to solution-annealing at 1250°C, thermomechanical working, cooling to room temperature and tempering at 750°C. Tested at a constant stress $\sigma = 100$ MPa (from F. Krafft). Lowering of the test temperature increases t_R as well as the reliability of extrapolating to e.g. 600°C.

B.7.2.4 Petrochemistry

Chapt. B.6, p. 339 discussed the use of heat-resistant steels that have to withstand creep as well. The petrochemical sector uses creep-resistant steels for a variety of applications, such as pipelines, fittings, heat-exchangers and pressure vessels, in which the creep rupture strength is of prime importance. Process gases with a high partial pressure of water vapour may cause a type of hydrogen damage in equipment operated at high temperatures which differs from that discussed in Chapt. A.4, p. 114 (wet corrosion). Pressurised hydrogen dissociates, penetrates the steel and decomposes the iron carbides to produce methane. The CH_4 molecules cannot diffuse out of the steel so that an internal pressure develops in situ that eventually leads to microcracks in the microstructure. The component becomes brittle. One countermeasure is to reinforce the M-C bond in Fe_3C by adding Cr, Mo, V to produce a more stable M_3C so that methane only starts to form at higher temperatures and/or pressures. Low-alloy creep-resistant structural steels thus have an inherently higher resistance to pressurised hydrogen compared to unalloyed steels. A further improvement is obtained by increasing the alloying levels to produce alloy carbides such as M_7C_3, M_2C or MC (Tab. A.2.1, p. 52). According to SEW 590, the creep-resistant steel 20CrMoV13-5 with M_7C_3 can be used at up to $\approx 480°C$ under a hydrogen pressure of ≤ 700 bar. It recommends the highly creep-resistant steels X20CrMoV11-1 and X7CrNiMoBNb16-16 for higher temperatures.

B.7.2.5 Valves

The intake valves in combustion engines are cooled by the incoming mixture. Exhaust valves are subjected to higher thermal loads and reach temperatures between 500 and 1000°C. They also have to withstand high-temperature corrosion due to the combustion process as well as high-cycle fatigue. In addition to this complex load on the valve disk, the cold end of the stem is worn by the tappet. Valve steel X45CrSi9-3, quenched and tempered to $R_m \approx 1200$ MPa, is suitable for intake valves as well as exhaust valves subjected to low loads. The end of the valve stem is inductively hardened to ≈ 56 HRC. Austenitic steels are used under higher loading conditions. Their fatigue strength can be improved by a high yield point, which is achieved with steel X53CrMnNiN21-9 by dissolving not only the carbon but also most of the $\approx 0.4\%$ N. Partial exchange of nickel by manganese increases the solubility of nitrogen in the melt. Alloying with Mo, V, Nb improves the creep resistance (X60CrMnMoVNbN21-10), although Mo and V do not improve the oxidation resistance. In contrast, omitting Ni makes steel more resistant to sulphur in the fuel. A high solution-annealing temperature ($\approx 1150°C$) increases the 0.2 proof strength to more than 500 MPa. The carbide and nitride precipitates obtained after artificial ageing at $\approx 750°C$ slightly increase the 0.2 proof strength. Under service conditions, the amount of precipitates increases and the toughness decreases.

Nickel-base alloys, e.g. NiCr20TiAl, are used under even harsher conditions. Hollow designs filled with sodium are internally cooled by heat transfer using a flow of Na from the disk to the stem. The corrosion and wear resistance of the disk can be improved by hard-facing the seat with a weld deposit of a cobalt-base hard alloy, e.g. CoCr26W12 with 1.6 % C. Non-hardenable, highly creep-resistant disks can be joined by friction welding to a hardenable stem of X45CrSi9-3 in order to facilitate hot forging of the disc and allow inductive hardening of the end of the stem.

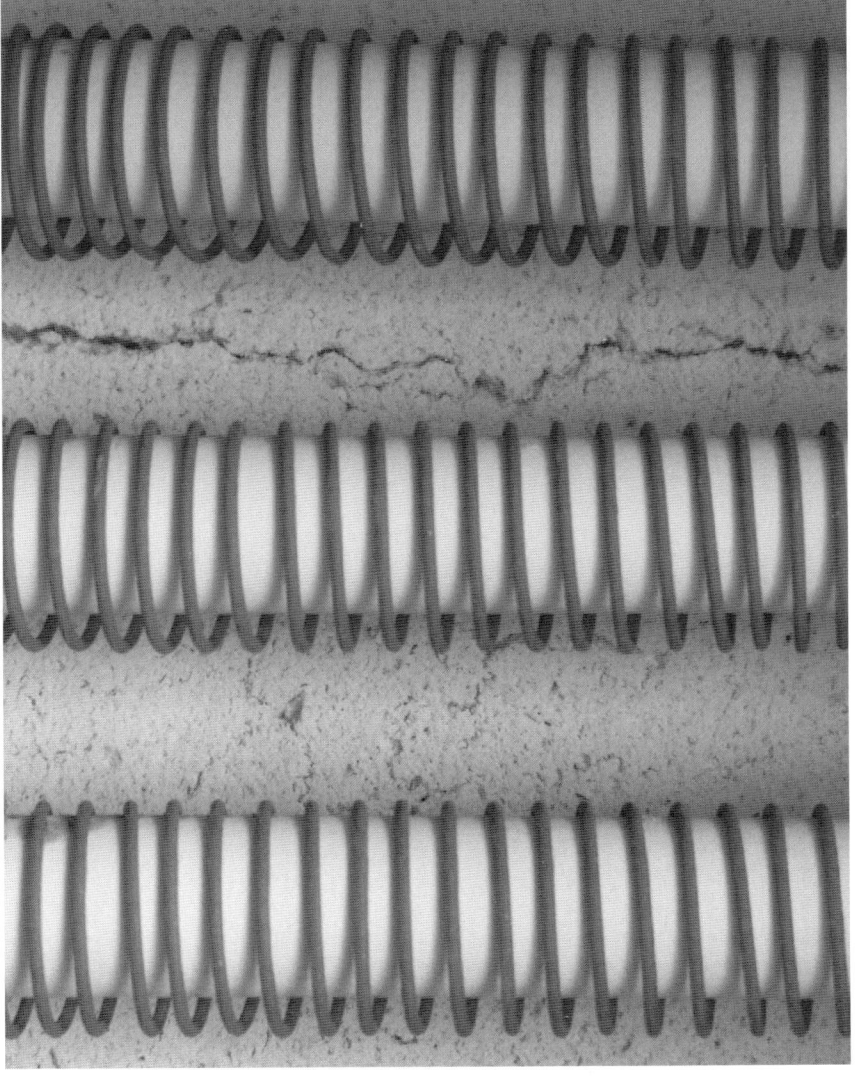

Spiral of electrical resistance heating wire for a muffle furnace
(Ugitech S.A., Ugine, France)

B.8 Functional materials

All the ferrous alloys discussed up to this point are used as structural materials, i.e. as force-transmitting components whose key characteristics are their mechanical properties in addition to their tribological and chemical behaviour. This chapter discusses ferrous materials and related alloys that can perform certain functions on the basis of certain physical effects.

B.8.1 Magnetically soft materials

A narrow hysteresis curve (Figure A.4.21, p. 118) with a small coercive force H_c and a high saturation polarisation J_s indicates that the material responds to even a small magnetic field strength H with a high polarisation J (or magnetic flux density $B = J + \mu_o H$), where $\mu_o = 4\pi 10^{-7} \text{Hm}^{-1}$ (unit H = Henry) is the magnetic permeability of free space. The permeability $\mu = B/H$ is the gradient of the hysteresis loop. Materials with $H_c < 1000\,\text{A/m}$ (Tab. B.8.1) are regarded as magnetically soft.

Their wide range of applications can be roughly divided into two fields: (a) signal conversion, e.g. by a relay in electrical circuits or by magnetic heads in oscillating circuits, (b) energy conversion in electric generators, transformers and motors.

(a) Signal conversion

IEC 60404-8-6 states that the magnetic quality is governed by the maximum coercive force, which lies between 2 and 300 A/m, depending on the material. The minimum values of the magnetic polarisation (in T = Tesla) and permeability (in H/m) are also of interest. They are measured statically in DC relays and in an alternating field up to 100 kHz for oscillating circuits. The standard specifies that static and dynamic measurements are to be carried out with four specimens (ES, LR, SR, SW), that differ in their structure (bars, laminates) or their shape (ring, rod). There are four alloy classes: (A) pure iron, (C) Fe-Si, (E) Fe-Ni, (F) Fe-Co (Tab. B.8.1). The designation of E31-10, for example, is made up as follows: E3 = alloy with 42 - 49 % Ni, 1= round hysteresis loop (not oriented), 10 = minimum permeability divided by 1000. If the hysteresis loop has a rectangular shape due to its texture or to heat treatment in a magnetic field, the core number is 2. Ageing of the magnetic properties during service can be avoided by heating the material at e.g. 100°C.

A lower content of tramp elements as well as fewer non-metallic inclusions and lattice defects increases the mobility of the Bloch walls and the magnetisability of iron. Contributing factors are the highest possible cleanness on melting and annealing for coarse grains and stress-relief. H_c can be lowered by adding 40 - 80 % Ni. Alloying with 25 - 50 % Co is used, above all, to increase J_s. Fields of application include highly sensitive relays, chokes and transformers in power electronics, magnetic cores in data engineering as well

B.8 Functional materials

Table B.8.1 Materials with special magnetic properties: H_c = coercive force, J_s = saturation polarisation for H = 4000 and 5000 A/m, P = core loss at 50 Hz for a polarisation of 1.5 T, $(BH)_{max}$ = static energy product, room temperature. I = magnetically soft relay materials A, E, F (IEC 60404-8-6) and electrical steel sheet M (EN 10106, 10107). II = magnetically hard permanent magnets (IEC 60404-8-1). III = non-magnetic, non-magnetisable steel (SEW 390) and cast iron (EN 13835).

	Designation	Main elements approximate values [%]	H_c [A/m]	J_s [T]	P [W/kg]	$(BH)_{max}$ [kJ/m³]
	A60	Fe	60	1.6		
	E41	Fe-38Ni	3	1.2		
I	F11	Fe-48Co	60	2.25		
	M235-35A¹⁾	Fe with		1.6	2.35	
	M125-35P	< 4.5 Si(+Al)		1.88	1.25	
	AlNiCo 9/5	Fe with Co, Ni	47 · 10³			9
	AlNiCo 72/12	Al, Cu, Nb/Ti	12 · 10⁴			72
II	CrFeCo 35/5	Fe-30Cr-20Co	50 · 10³			35
	RE₂Co₁₇ 220/70	Fe-50Co-25Sm	70 · 10⁴			220
	REFeB380/100²⁾	Fe-30Nd-2B	10 · 10⁵			380
III	X2CrNiMnMoNNb23-17-6-3		≈ 0	< 0.01	$\mu_r ≈ 1.003$	
	GJSA-XNiMn13-7				$\mu_r ≈ 1.02$	

¹⁾ Thickness 0.35 mm, A = non-grain-oriented, P = grain-oriented with high permeability; ²⁾ RE = rare earth

as switching and shielding systems. Large magnets in particle accelerators with piece weights of up to 100 t are made as cast or forged parts of vacuum-degassed iron containing < 0.01 % C. The main objectives are to produce a high core quality and homogeneous properties. Values of e.g. $H_c ≈ 50$ A/m are obtained in large cast magnetic cores.

(b) Energy conversion

The key property for applications in alternating magnetic fields is a low core loss P, in addition to the aforementioned properties. The core loss comprises the hysteresis loss P_H in a constant field plus the eddy-current loss P_E, which in turn is the sum of the classical loss P_{EC} and the abnormal loss P_A, thus $P = P_H + P_{EC} + P_A$. Silicon reduces the loss, but leads to difficulties in cold-processing above levels of ≈ 3.5 %. Aluminium also has a favourable effect on P, so that cold-rolled sheets contain e.g. < 4.5 % (Si + Al). Given that

P_E increases quadratically with the material thickness (Figure B.8.1 a), these losses can be minimised by passing the magnetic flux through thin sheets of electrical steel that are separated by insulating layers and stacked to produce a laminated core. Such cores are used in motors, generators and transformers. However, there is a limit to reducing the sheet thickness because undesirable internal oxidation of the surface layer during annealing becomes more important and this surface damage increases P_H. The direction of the magnetic flux rotates in motors and generators, thus necessitating electrical steel sheets with the greatest possible isotropy. Because texturing cannot be avoided during cold-rolling and annealing (Figure A.2.6, p. 25), the manufacturing objective is to produce the most favourable texture i.e. a cube-on-face. In this case, the preferred orientation of the {100} plane (Figure A.1.4, p. 8) of the grains is parallel to the plane of the sheet. The direction of the <100> cube edge, however, varies from grain to grain. The optimum grain size is ≈ 0.1 mm. EN 10106 regulates this non-grain-oriented electrical steel sheet. To take grade M250-35A as an example: M signifies electrical sheet and strip, 250 is the hundredfold of the maximum permitted core loss at 1.5 T and 50 Hz in W/kg and 35 is the hundredfold of the nominal thickness in mm. A signifies non-grain-oriented, finally annealed. Materials that have not been finally annealed are assigned the code letter K (EN 10341) e.g. M260-50K.

The flux direction does not rotate in transformers, and anisotropic sheet properties are thus desirable. In this case, the manufacturing objective is to produce a Goss texture in which the preferred orientation of the diagonal plane {110} of the bcc cube is parallel to the plane of the sheet. Alignment of the

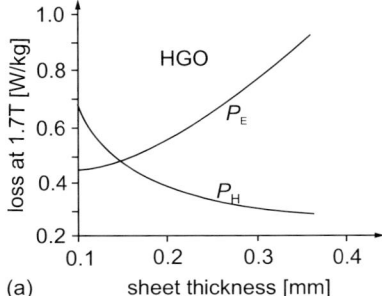
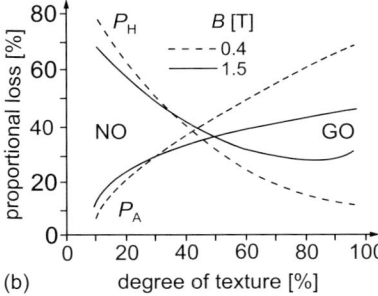

Fig. B.8.1 Contributions to core losses in electrical steel sheet:(a) Increase in the eddy-current loss P_E with the sheet thickness, increase in the hysteresis loss P_H with decreasing sheet thickness due to internal oxidation and the resulting loss of Bloch wall mobility (HGO = high-permeability grain-oriented from F. Bölling et al.). (b) Large P_H fraction, but small influence of the magnetic flux density in non-grain-oriented (NO) sheet. Large fraction of abnormal loss P_A in grain-oriented (GO) sheet, which decreases with increasing flux density (from A.J. Moses and W.A. Pluta).

individual grains in the direction of the cube edge means that this grain-oriented electrical steel sheet has a preferred orientation in service. The orientation accuracy of the texture increases with the degree of cold reduction; however, the tendency to incomplete secondary recrystallisation increases with decreasing sheet thickness s. This leaves fine-grained zones without Goss texture, and P_H increases (Figure B.8.1 b).

The orientation accuracy also increases with the grain size d. Modern grades differentiate between RGO (regular grain-oriented, $d < 5$ mm) and HGO (high-permeability grain-oriented, $d < 20$ mm). Silicon leads to a transformation-free ferritic steel and thus makes it easier to obtain a coarsely grained microstructure. Because the Bloch wall spacing l is proportional to d, but P_A increases with the l/s ratio, this spacing has to be reduced in coarse-grained HGO grades by laser treatment. This involves passing a narrow laser beam at short intervals over the sheet perpendicular to the rolling direction. The thus-induced residual stresses shorten the domains (domain refinement). Scratching the sheets has the same effect. The designation according to EN 10107 differs from the above one only in the last letter: in M150-35S, S signifies regular grain-oriented products (RGO), finally annealed. In M90-27P, the last letter signifies grain-oriented products with a high permeability (HGO), finally annealed. EN 10303 includes specifications for non-oriented (NO) and grain-oriented (GO) strip with a nominal thickness down to 0.05 mm and frequencies from 400 to 1000 Hz. Figure B.8.2 compares both products at the mains frequency. A reduction in the core loss is of great economic importance in power engineering applications. Developments are aimed

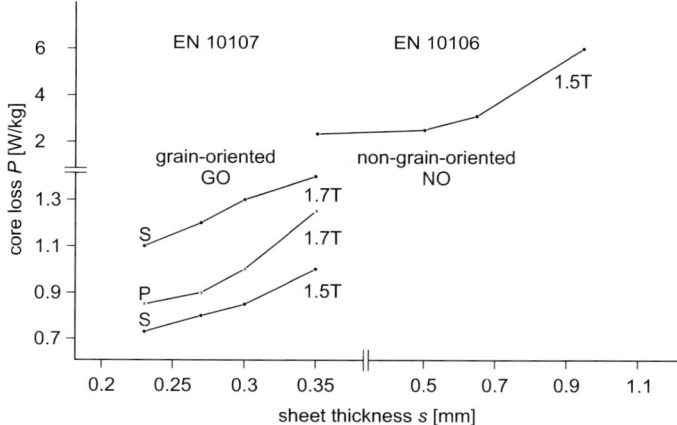

Fig. B.8.2 Core loss (EN 10107 and 10106): The upper limit for the lowest loss grades of finally annealed electrical steel sheets at a polarisation of 1.5 or 1.7 T and a frequency of 50 Hz. S = standard products, P = products with high permeability, with and without domain refinement.

at improving the surface layer and secondary recrystallisation in order to decrease the optimum sheet thickness and the core loss P (Figure B.8.1). One avenue of research is focusing on the rapid quenching of iron-boron-silicon melts by spraying them onto a rotating copper chill (melt spinning). The material does not have enough time to crystallise and thus solidifies as thin amorphous, i.e. glassy ribbons. The thickness of these metallic glass ribbons is about one order of magnitude below that of electrical steel sheet, and they thus exhibit very low eddy-current losses. The lack of crystalline anisotropy and grain boundaries contribute to the excellent soft magnetic properties of these metallic glasses. The low saturation polarisation as well as problems relating to manufacturing and costs are hindering rapid implementation of these materials. A slightly slower solidification rate produces microcrystalline ribbons. This method allows silicon contents of up to 6.5 % without a loss of cold processability. Ferritic chromium steels are used in corrosive environments (Chapt. B.6, p. 319). Their C content of $< 0.03\,\%$ is bound by Ti/Nb to increase the mobility of the Bloch walls. Because the saturation magnetisation of iron decreases by approx. 1.4 % with every 1 % Cr, steels with 13 % Cr are preferred and those with 18 or even 28 % Cr are only used if necessitated by aggressive media. MnS, TiS or $Ti_4C_2S_2$ precipitates act as chip breakers and improve the machinability.

B.8.2 Magnetically hard materials

This group of materials is characterised by a very wide hysteresis loop with coercive forces H_c that are usually $> 10000\,\mathrm{A/m}$. Advantageous are a high saturation polarisation J_s as well as the highest possible value of the maximum energy product $(BH)_{\max}$. The latter corresponds to the area of the largest rectangle that can be constructed beneath the demagnetisation curve in the second quadrant of the hysteresis loop and is given as a static energy product in kJ/m^3 (Tab. B.8.1). The highest possible thermal stability of the magnetic properties within the service temperature range is desirable. This requires a stable microstructure that does not age.

Of the large number of available materials, only a few iron-containing examples will be described here. Cobalt increases J_s in these materials. Its austenite-stabilising influence is compensated by adding Cr, Mo, W, V and Al. Silicon has a similar effect, but it lowers H_c. In hardenable steels with $\approx 1\,\%$ C, the mobility of the Bloch walls is limited during tempering by fine carbide precipitates. These steels are rarely used nowadays owing to their low values of H_c and $(BH)_{\max}$ along with their inadequate microstructural stability. In iron-containing AlNiCo and CrCoMo alloys there is a miscibility gap that is used to induce spinodal decomposition. After solution-annealing in the ferrite field and rapid cooling, the solid solution decomposes at temperatures of 500 to 650°C without nucleation according to $\alpha \rightarrow \alpha + \alpha^{'}$. The coherent α-Fe-Co-(Ni) particles in the Al- and CrMo-rich $\alpha^{'}$ matrix can be given a preferred orientation by heating in a magnetic field. FeCoVCr alloys transform after solution-annealing in the austenite field to a lamellar $\alpha + \gamma$ microstructure during tempering at $\approx 600°C$. A preferred orientation can be obtained by cold rolling. Both $\alpha + \alpha'$ and $\alpha + \gamma$ microstructures aim to avoid Bloch walls by embedding separate α domains in a weakly magnetisable matrix.

Magnetically hard oxides are also used in addition to these alloys. Examples include oxides with barium or strontium and boron phases with rare earth metals, such as neodymium. The hard oxidic ferrites of type $BaFe_{12}O_{19}$ or $SrFe_{12}O_{19}$ and boron-containing $Nd_2Fe_{14}B$ attain high coercive forces and $(BH)_{max}$ values. The selection of materials must take account not only of the magnetic characteristics but also the manufacturing feasibility and the costs. Metallic materials with lower alloying levels can be produced cost-effectively by casting and metal working. Anisotropic properties with a pronounced preferred orientation can be obtained by directional or monocrystalline solidification. Some high-alloy grades as well as those containing oxides and boron have to be manufactured by powder metallurgy or by melt spinning (e.g. NdFeB). Another method is to embed a magnetically hard powder in a polymer matrix (P magnets). IEC 60404-8-1 specifies code numbers for the individual groups of magnetically hard materials (Tab. B.8.2). A second appended number indicates isotropic (0) or anisotropic (1) properties and a third serves as a grade number, e.g. R1-1-3 for anisotropic, cast AlNiCo44/5, where the number in front of the slash is $(BH)_{max}$ in kJ/m^3 and the number behind is a tenth of the coercive force in kA/m.

Table B.8.2 Code numbers of magnetically hard materials: Alloys (R) as well as ceramics (S) in a ceramic or polymer matrix (U) (IEC 60404-8-1). RE = rare earth metal.

alloys (R)		bonded materials (U)
AlNiCo	R1	U1
CrFeCo	R6	
FeCoVCr	R3	
RECo	R5	U2
REFeB	R7	U3
ceramics (S)		
ferrites	S1	U4

Permanently magnetic materials are used in statically or dynamically loaded magnetic circuits. The first group includes magnetic closures in furniture and refrigerator doors, magnetic clamps in fixing plates, couplings and contactless bearings in mechanical engineering applications as well as permanent magnets in measuring instruments. The second group includes applications in DC motors and generators as well as loudspeakers. Rare earth magnets impart a high torque to e.g. a synchronous motor in a mechatronic power steering unit with a compact design.

B.8.3 Non-magnetisable materials

This group comprises structural materials that are not affected by a magnetic field and are not heated by eddy currents in strong alternating magnetic fields. Predominantly stable austenitic steels without δ-ferrite are used because they do not undergo a martensitic transformation during service, not even as a result of cold working. The non-magnetisability is measured by the relative permeability $\mu_r = \mu/\mu_o$, whose value is < 1.01 for most grades. SEW 390 (steels) and SEW 395 (cast steel) provide further information.

Cost-effective grades include manganese steels of type X40MnCr18, which were used for cold-expanded retaining rings. In the meantime, however, they have been replaced by stainless steels such as X8CrMnN18-18 with 0.55 % N that have proven more resistant to stress corrosion cracking. Weldable steels such as X2CrNiMnMoNNb23-17-6-3 with ≈ 0.4 % N are used in cryostat tanks and in cooling systems for superconductive systems (Figure B.8.3). Because

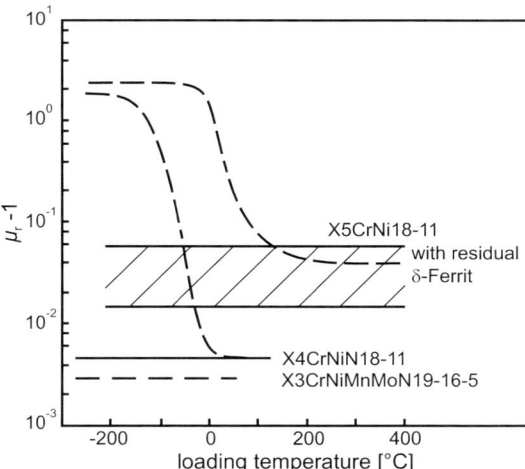

Fig. B.8.3 Relative magnetic permeability μ_r of austenitic steels (from W. Weßling and W. Heimann). Measured at room temperature after holding at the loading temperature (solid lines and scatter band) or after tensile testing at the same temperature (dashed lines). Low-temperature straining of steels containing 18 % Cr and 11 % Ni during the tensile test leads to the formation of ferromagnetic martensite. In contrast, a steel with 19 % Cr and 16 % Ni undergoes an antiferromagnetic transformation.

the Curie temperature increases above 20 % Ni (Figure A.4.22, p. 118), Mn and N are used to stabilise the paramagnetic or antiferromagnetic state at low temperatures.

Drill strings for deep-hole drilling have non-magnetisable sections to accommodate magnetic measuring instruments that are used to determine the position of the hole within the Earth's magnetic field. These up to 10 m-long drill collars are made of e.g. X5CrMnNiN21-23 with 0.85 % N.

Non-magnetisable cast irons have already been mentioned in Chapt. B.6 (p. 343). EN 13835 specifies grades containing flake and spheroidal graphite (GJLA-XNiMn13-7, GJSA-XNiMn13-7). This austenitic cast iron reaches $\mu_r \approx 1.02$ and is used e.g. for pressure-tight closures in turbogenerators, housings of electrical installations and insulating flanges.

B.8.4 Materials with a special thermal expansion

The coefficient of thermal expansion α at a given temperature corresponds to the difference in relative length produced by a temperature change of 1°C. In many cases, the mean coefficient of expansion between room temperature and the test temperature is of interest. Tab. B.8.3 gives values for ferrous alloys with special coefficients of thermal expansion.

As Figure A.4.23 (p. 120) shows, very different values of α can be obtained in the iron-nickel system. Figure B.8.4 compares Fe-Ni alloys with steels. As the test temperature increases, the thermal expansion of fcc Fe-Ni alloys is largely compensated by magnetostriction (Chapt. A.4, p. 120) as the Curie temperature is approached (Figure B.8.4 b). Fe-36 % Ni exhibits values of $\alpha < 2 \cdot 10^{-6} [°C^{-1}]$ at test temperatures from -100 to +100°C. The exchange of ≈ 5 % nickel for cobalt gives $\alpha \approx 0$ (Figure B.8.4 a). The Curie temperature can be tuned by varying the nickel content and alloying with cobalt or chromium. The coefficient of expansion can thus be made to match that of other materials. Tempering treatments after solution-annealing ensure resistance to ageing, i.e. dimensional stability of the alloys over a long period.

Table B.8.3 Alloys with special thermal expansion properties: ((from SEW 385 and Int. Nickel Co.).

Composition	Young's modulus at 20°C [GPa]	Mean coefficient of thermal expansion $[10^{-6}/°C]$ between 20°C and				
		100	200	300	400	600°C
Fe-36 % Ni	140	1.2	2.2	4.5	7.5	10.7
Fe-49 % Ni	157	9.1	9.1	9.1	8.9	10.7
Ni-42 % Fe-6 %Cr	147	6.9	7.2	8.3	10.1	12.4
Fe-29 % Ni-18 %Co	157	5.8	5.3	5.0	4.9	7.6
Fe-20 % Ni-6 %Mn	196	19.8	19.9	20	20.1	20.3

B.8.4 Materials with a special thermal expansion

The anomaly in the stiffness of Fe-Ni alloys compared to steels is based on magnetostriction (Figure B.8.4 c). Although Young's modulus increases on alloying with further elements, it is possible to obtain a material that is essentially temperature-independent at climatic temperatures (Fe-36 % Ni-12 % Cr or Fe-40 % Ni-10 % Mo). This can be improved by cold work hardening or precipitation hardening by alloying with aluminium, titanium or boron.

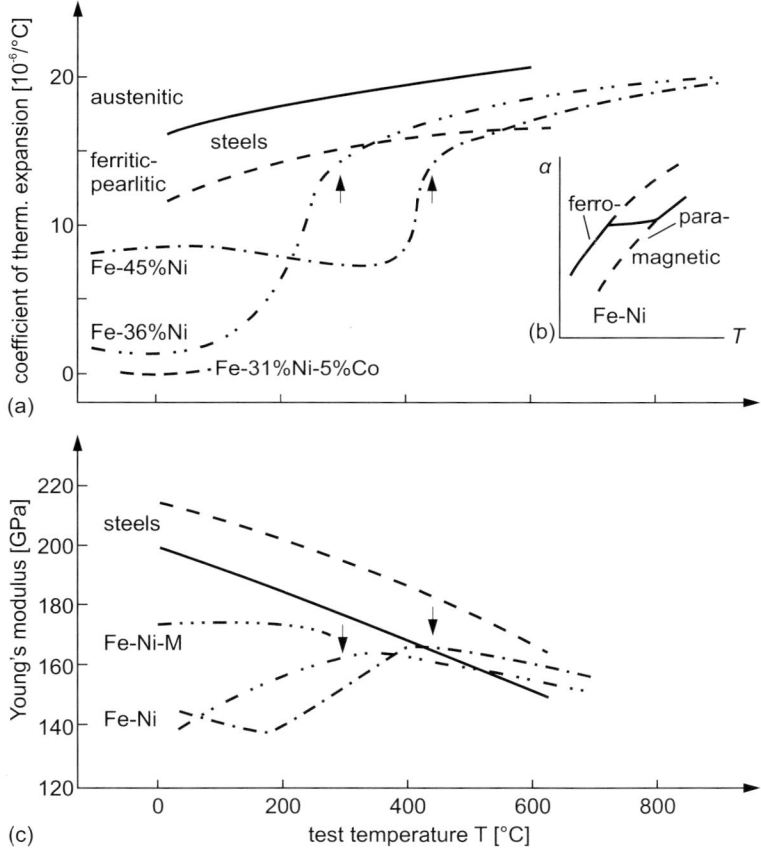

Fig. B.8.4 Comparison of the thermal expansion and stiffness of steels and iron-nickel alloys (from literature data). (a) Anomaly in the coefficient of thermal expansion α of Fe-Ni alloys below the Curie temperature (indicated by an arrow). (b) Expansion of fcc Fe-Ni alloys at the transition from the paramagnetic to the ferromagnetic state (magnetostriction, schematic). (c) Anomaly in the Young's modulus of Fe-Ni alloys. In Fe-Ni-M, M is chromium or molybdenum and the elements beryllium, aluminium and titanium that are responsible for precipitation hardening.

B.8 Functional materials

A low thermal expansion is required for primary measuring standards, tapes and rulers as well as clock pendulums, weighing beams and measuring instruments. Possible alloys include Fe-36 % Ni and Fe-31 % Ni-5 % Co. Electrical lead-through conductors (e.g. in light bulbs) are made of alloys such as Fe-49 % Ni or Ni-42 % Fe-6 % Cr and Fe-29 % Ni-18 % Co fused into soft and hard glasses, respectively. Because they have a similar coefficient of thermal expansion as the glass, the cooling stresses remain small and the seal airtight. Alloys with a low thermal expansion are mechanically bonded to austenitic steels or e.g. a Fe-20 % Ni-6 % Mn alloy by roll cladding to produce thermostatic bimetals that are used in thermometers and temperature controllers. Alloys with a constant Young's modulus are used for hairsprings in clocks and springs in spring balances. Other application fields include pressure gauge diaphragms, elements in load cells, mountings in precision instruments and vibrating forks with a particular natural frequency.

B.8.5 Materials with a shape memory

Certain alloys that undergo martensitic transformation exhibit a shape memory effect. For example, if a pipe socket is supercooled and then widened in the martensitic condition, it 'remembers' its previous shape during the reverse transformation on heating and shrinks onto the end of the inserted pipe. This one-way effect can be obtained within a selected narrow temperature range ΔT by tailoring the alloy composition. A two-way effect is obtained when the shape memory of the workpiece is 'trained' during several forward and reverse transformations induced by heating and cooling in the range ΔT, thus leading to reproducible shape changes. A force builds up if the shape change is constrained. This makes shape memory alloys suitable as e.g. actuators in control units: windows in greenhouses are opened when the temperature climbs to a preset value and then closed when the temperature drops. Some materials may also exhibit pseudoelasticity: as the stress increases, they initially deform elastically and then pseudoplastically, and they recover their initial shape on unloading. This rubber-like behaviour expresses itself in a large deformation for a small change of stress. These aforementioned effects can be represented in a stress-strain-temperature diagram (Figure B.8.5). The driving force of the shape memory effect is based on thermodynamic and mechanical factors. The forward and reverse martensite/austenite transformation is diffusionless. This means that although the maximum application temperatures are limited, the minimum ones are not. Small volume changes decrease the transformation work. Magnetostriction thus has a favourable effect and makes Fe-Ni-Co alloys particularly interesting materials. A mobile phase boundary between austenite and martensite is an important prerequisite for the shape memory effect. Austenite can be precipitation-hardened by alloying with aluminium and titanium. The precipitates lower the thickness of the martensite plates. After these thin plates have sheared, they provide elastically stored energy for the reverse transformation to austenite. Shape memory alloys of type

Fe-30 % Ni-10 % Co-4 % Ti are being studied in which the transformation range between the A_f and M_f temperatures can be shifted by changing the composition of the alloy. In this case, martensite has a cubic structure, whereas hexagonal ε -martensite forms in Fe-30 % Mn-1 % Si alloys. Important development objectives include: 1. obtaining a large strain hysteresis in two-way alloys with a narrow transformation temperature tailored to the application; 2. high mechanical strength to produce high actuating forces; 3. resistance to fatigue and reduction of the shape memory effect. In this respect, iron-base alloys are expected to perform better than the currently used Cu-Zn, Cu-Zn-Al and NiTi alloys.

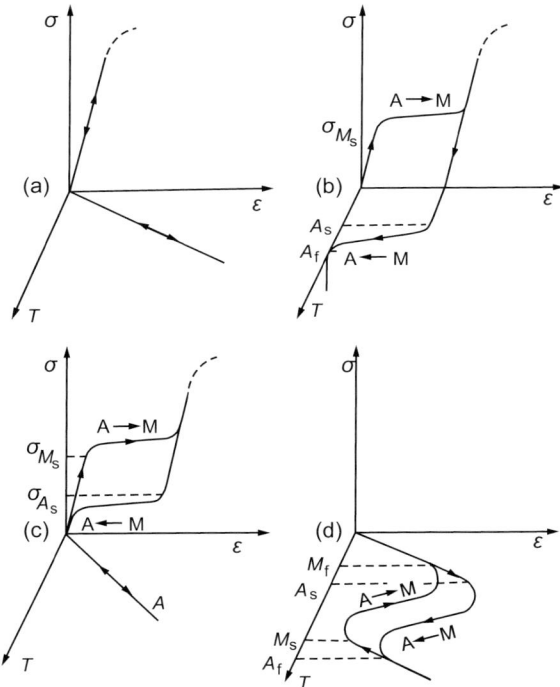

Fig. B.8.5 Types of shape memory effects: (a) Conventional behaviour, stress σ and temperature T act independently of each other on the strain ε. (b) One-way effect: a stress-dependent austenite (A) → martensite (M) transformation above the martensite stress σ_{Ms} together with a temperature-dependent reverse transformation between the austenite start and finish temperatures (A_s, A_f) produce one load cycle. (c) Pseudoelasticity: a transformation-dependent elongation starts above σ_{Ms} and its reversal starts below the austenite start stress σ_{As}, independently of the temperature. (d) Two-way effect: a temperature-dependent transformation-related strain hysteresis is obtained without an applied stress (from E. Hornbogen, N. Jost).

Examples of applications for shape memory materials are given in the literature. One-way: pipe connections, expansion rivets, plug-in connectors for electrical circuits, implants, clothing stiffeners. Two-way: valve control units, opening and lowering of car headlights, robot limbs, endoscope control units. Pseudoelasticity: gaskets, spectacle frames, highly shock-absorbing components, dental braces. Completely new applications are possible if the shape memory effect is not controlled by the temperature but by an electromagnetic field. This is currently being studied (cf. Chapt. A.2, p. 41).

B.8.6 Electrical resistance heating alloys

These materials are used to convert electric current into heat. DIN 17470 contains specifications for coiled spiral wires used in fan heaters, hair dryers and heating units for industrial furnaces and crucibles or as heating tapes in irons, toasters and industrial large-area heating systems. At a given load, wires are hotter than tapes owing to their smaller ratio between the radiant surface area and the heating cross-section. Heating elements usually have a ceramic support to prevent creep due to their own dead weight. Reactions with the supporting material and the surrounding air or other gases have to be avoided by means of a protective surface film. Similar considerations as for heat-resistant steels apply with respect to their resistance to high-temperature corrosion and creep (Chapt. B.6, p. 336). Austenitic materials use the surface film produced by chromium. They are more creep-resistant and are less susceptible to grain coarsening and embrittlement. A surface film of Al_2O_3 is formed on ferritic chromium steels containing $\approx 5\,\%$ aluminium (Tab. B.8.4). They are thus more resistant to scaling than austenitic steels. Aluminium also reduces embrittlement by the σ phase (Figure B.6.4 a, p. 341). Small amounts of rare earth metals and magnesium improve scale adhesion during temperature fluctuations.

Table B.8.4 Electrical resistance heating alloys (DIN 17470): A = austenitic alloy (reduced bendability between 500 and 900°C), F = ferritic alloy (cold brittleness due to α' phase after operating between 400 and 500°C or brittleness after exposure to temperatures > 1000°C due to grain coarsening), ρ = specific electrical resistance, $R_{p1/1000} = 1\,\%$ creep limit at 1000 h, T_{max} = maximum service temperature in air.

	Designation Composition	ρ [Ω mm²/m] 20°C	1200°C	$R_{p1/1000}$ 1100°C [MPa]	T_{max} [°C]
A	NiCr8020 Ni-20 %Cr	1.12	1.17	1.5	1200
A	NiCr3020 Fe-30 %Ni-20 %Cr	1.04	1.34	1.5	1100
F	CrAl25 5 Fe-25 %Cr-5 %Al	1.44	1.49	0.3	1300

B.8.6 Electrical resistance heating alloys

This is supported by the fact that common alloys do not undergo transformation. Their high alloy content leads to an elevated electrical resistance R, which gradually increases with the temperature. Nickel alloys exhibit a slight discontinuity due to a magnetic rearrangement below the Curie temperature. At a given current I, the thermal output is $I^2 R$. The temperature of the heating element increases and stabilises itself. Ferritic steels perform better in atmospheres containing sulphur or carbon. However, the austenites are better in atmospheres containing nitrogen and only a little oxygen. Both groups of alloys have little resistance to ammonia, chlorine-containing gases and heavy-metal vapours.

C
Appendix

C.1 Designation systems for steel and cast iron

C.1.1 Standardisation

Since the 2nd German edition of this book was published in 1993 there have been many changes in the standardisation of ferrous materials. This is not only due to technical advances, but is chiefly the result of changing over from national to international standards. Such harmonisation helped already the 3rd edition because the new designation systems for ferrous materials can be used in the text. Systematic designations enable unambiguous communications between material manufacturers and processors and also between design engineers and component users.

The first two editions referred to the relevant German DIN standards issued by the Deutsches Institut für Normung e.V. Berlin and published by the Beuth Verlag, Berlin, to Stahl-Eisen Materials Guidelines, Testing Guidelines and Application Lists (SEW, SEP, SEL) published by the Stahlinstitut (VDEh; German Iron and Steel Institute), Düsseldorf, to Werkstoff-Leistungsblätter der Deutschen Luftfahrt (Material Specification Sheets for the German aviation industry) and to various guidelines from the Verein Deutscher Ingenieure (VDI, Association of German Engineers) and Verein Deutscher Gießereifachleute (VDG, German Foundrymen Association).

In rare cases where there are no applicable European standards (EN) or international standards (ISO), we have used the above-mentioned set of German rules and regulations.

In the last decade there has been a rapid increase in the coverage of EN standards regulating ferrous materials issued by the European Committee for Standardization (CEN) in Brussels. In a few cases, the EN standard harmonises with the international standard, which is indicated by an EN ISO + number. Magnetic materials are regulated by the International Electrotechnical Commission (IEC + number). Some of the cited standards still have draft status. For this reason, the reader is requested to refer to the current status before applying a standard.

C.1.2 Designations for steels and cast steels

Steels and cast steels designated by European standard EN 10027 are referred to either by their material code or material number. In everyday dealings, material codes are more advantageous than material numbers because the combination of code letters and numbers in the former provide direct information on either important properties or the chemical composition of the steel. Designations using code names can be divided into two main groups:

Group 1

The code name for group 1 materials provides information on the use and on the mechanical and physical properties of the ferrous materials. This is why this type of designation is used for simple structural steels. According to EN 10027-1, the name is composed of principal symbols and additional symbols. The principal symbol includes a code letter for the steel group followed by the minimum yield strength in MPa for the smallest product thickness (Tab. C.1.1). Exceptions are groups Y and B, which give the minimum tensile strength, as well as group M, which gives the highest permissible magnetic loss. The code letters of cast steels are preceded by a G.

Table C.1.1 Designation system for group 1 materials according to EN 10027-1 and DIN V 17006-100 (selected additional symbols)

Main group			Additional symbols
Code letter		Code number	
S = structural steels		yield strength in MPa	notch impact energy T in °C
			27J 45J 60J RT
			JR KR LR 0
E = engineering steels		,,	J0 K0 L0 -10
			J1 K1 L1 -20
P = pressure vessel steels		,,	J2 K2 L2 -30
			J3 K3 L3 -40
L = steels for pipeline construction		,,	J4 K4 L4 -50
			J5 K5 L5
H = cold-rolled flat products in higher-strength drawing grades		,,	M = thermomech. rolled
			N = normalised, normalising rolled +Z = hot-dip Zn-coated
			Q = quenched and tempered
D = flat products made of mild steels intended for cold-working		,,	applies to fine-grain steels +ZE = electrolytically galvanised verzinkt
			A = precipitation-hardened
			B = bake hardening +A = hot-dip Al-coated
			P = alloyed with phosphorus
T = tinplate, very thin plate and strip		,,	X = dual phase +S = hot-dip Sn-coated
			Y = interstitial–free
			C = cold-workability requirement +H = hardenability requirement
B = concrete-reinforcing steels		tensile strength in MPa	D = intended for hot-dip coating
			W = weather-resistant
Y = pre-stressing steels		,,	+N = normalised
M = electrical steel sheet and strip		magnetic losses	G1 = cast in unkilled state
			G2 = cast in killed state +QT = quenched and tempered
(cast materials are indicated by a G preceding the code letter)			

The principal symbol is followed by an additional symbol giving information on the quality class of the steels (e.g. weldability, treatment condition or the minimum notch impact energy). The EN 10025 designation of the unalloyed structural steel known as St 52-3 under DIN is:

S355J0 C+N

S	steel for general steel construction
355	minimum yield strength = 355 MPa
J0	notch impact energy 27 J at 0°C
C	with special cold-workability
+N	normalised or normalising rolled

The new designation is particularly advantageous because the design engineer can see the more useful yield strength at a glance.

Group 2

The code names of group 2 are based on the chemical composition and thus on the former German DIN standard, although all the spaces are removed in the EN name to save space. The steels are divided into four subgroups that differ with respect to the content of alloying elements:

- unalloyed steels with a mean Mn content of $< 1\,\%$
- alloyed steels with a mean content of individual alloying elements below $5\,\%$ as well as unalloyed steels with a mean Mn content of $> 1\,\%$
- high-alloy steels
- high-speed tool steels

Alloyed and unalloyed steels differ with respect to the limiting contents specified in EN 10020. Ferrous materials that contain only carbon are regarded as unalloyed. High-alloy ferrous materials contain at least one alloying element with a content of $> 5\,\%$.

Unalloyed steels : Because the properties of unalloyed steels are essentially determined by their C content, the principal symbol is C followed by the hundredfold of the mean C content. An additional symbol, which can be appended to the principal symbol, gives further information on the use or on other elements. For example, the EN 10016-2 designation C70D stands for an unalloyed steel with 0.7 % C, a P and S content of $< 0.035\,\%$ and is suitable as a tool steel.

Alloyed steels : The principal symbol of this group of steels is a number that corresponds to the hundredfold of the mean carbon content. This is followed by the chemical symbols of alloying elements whose alloying content is given by numbers separated by a hyphen, whereby the first number refers to the first element, the second to the second element, etc. The contents of the various elements are multiplied by an element-specific factor to obtain whole numbers in the designation:

multiplier 4:   Cr, Co, Mn, Ni, Si, W
multiplier 10:   Al, Be, Cu, Mo, Nb, Pb, Ta, Ti, V, Zr
multiplier 100:   C, N, P, S, Ca
multiplier 1000: B

This type of designation can also be given an additional symbol to describe the condition or type of treatment (Tab. C.1.2).

Table C.1.2 Selected additional symbols for designations of group 2 steels

\multicolumn{2}{c}{Additional symbols}	
symbol	treatment
+A	soft-annealed
+AC	spheroidised annealed
+C	cold work-hardened
+CR	cold-rolled
+FP	treated to produce ferrite/pearlite
+M	thermomechanically rolled
+N	normalised and normalising rolled
+P	precipitation-hardened
+Q	quenched
+QT	quenched and tempered
+S	quenched and tempered for cold workability
+T	tempered
+U	untreated

The designation **50CrV4+QT** thus refers to an alloyed QT steel containing 0.5 % C, 1 % Cr and $< 1\,\%$ V in a QT condition.

High-alloy steels : High-alloy steels in which at least one element is present with a content of $> 5\,\%$ are characterised by a preceding X in the principal symbol. The X is followed by a number giving the hundredfold carbon content as well as percentages of alloying elements as whole numbers. According to EN 10088, the high-alloy hot-work tool steel X40CrMoV5-1 thus contains 0.4 % C, 5 % Cr, 1 % Mo and $< 1\,\%$ V.

High-speed tool steels : The fourth category of group 2 materials are also high-alloy steels and are represented by high-speed tool steels. The principal symbol comprises the letters HS (high speed) followed by numbers giving the contents of the elements tungsten, molybdenum, vanadium and cobalt, in that order, seperated by a hyphen. The designation of high-speed tool steel HS6-5-3 shows 6 % W, 5 % Mo and 3 % V, but not the 4 % Cr and 1.3 % C which it also contains.

In addition to designations with code names, steels can also be designated with material numbers according to EN 10027-2. The material numbers are composed of a number for the main material group followed by a full-stop, a steel group number and an ID number.

Main material group
0	pig iron, ferroalloys, cast iron
1	steel or cast steel
2	heavy metals except iron
3	light metals

Steel group number
00 and 90	unalloyed base steels
01 - 07 and 91 - 97	unalloyed quality steels
10 - 19	unalloyed stainless steels
08 and 09, 98 and 99	alloyed quality steels
20 - 29	tool steels
30 - 39	miscellaneous steels
40 - 49	chemical and heat-resistant steels
60	cast iron
50 - 80	structural, engineering and pressure-vessel steels

The above-mentioned steel X40CrMoV5-1 is a tool steel with the ID number 44 and is thus known as **1.2344**.

C.1.3 Designation of cast irons

A new designation system using symbols and material numbers was introduced in EN 1560.

The code name for cast irons is composed of up to six positions (Tab. C.1.3). The EN prefix is followed by the letter code GJ (for cast iron) followed by a letter to indicate the type of graphite and another, if necessary, for the microstructure. Grey cast iron with flake graphite is thus known as EN-GJL, grey cast iron with spheroidal graphite is EN-GJS. This is followed by a hyphen and a number characterising the mechanical properties or chemical composition. Information on the condition (e.g. H = heat-treated casting) can also be appended as code letters. The designation **EN-GJS-400-18-H** thus refers to a heat-treated spheroidal cast iron with a minimum yield strength of 400 MPa and an elongation at fracture of at least 18 %. Because the method of obtaining the test specimens is important if the mechanical properties are given, the following code letters can be appended:

C.1.3 Designation of cast irons

S   separately cast specimens
U   cast-on specimens
C   specimens cut from the casting

The designation of an austenitic cast iron with spheroidal graphite is e.g. GJSA-XNiCuCr15-6-2, which gives the percentages of characteristic alloying elements as whole numbers.

Table C.1.3 Designation system for cast-iron materials using symbols in accordance with EN 1560

1 obligatory prefix-silbe	2 obligatory type of metal		3 optional graphite-structure		4 optional micro- or macrostructure		5 obligatory a) or b) must be used				6 optional additional requirements	
		symbol		symbol		symbol	a) mechanical properties	symbol	b) chemical composition	symbol		symbol
EN–	cast iron	GJ	lamellar	L	austenite	A	aa) tensile strength, 3- or 4- digit number for the minimum value in MPa	e.g. 350	ba) letter symbol indicating the designation is based on the chemical composition	X	casting in as-cast condition	D
			spheroidal	S	ferrite	F	ab) minimum elongation in % as a 1- or 2-digit number after a hyphen	e.g. 19	bb) carbon content in %·100; however, only if the carbon content is significant	e.g. 300	heat-treated casting	H
			temper carbon	M	martensite	M	ac) 1 letter for the method of specimen preparation:		bc) chem. symbol of alloying elements	e.g. Cr		
			vermicular	V	pearlite	P	– separately cast specimen	S			suitable for production welding	W
			no graphite (white cast iron)	N	ledeburite	L	– cast-on specimen	U	percentages of alloying elements, each separated by a hyphen	e.g. 9-5-2	additional requirements specified by the customer	Z
			ledeburitic		quenched	Q	– specimen taken from a casting	C				
					quenched and tempered	T	ad) Brinell hardness, 1 letter and a 3-digit number for the hardness	e.g. H155				
			special structure specified in the respective material standard	Y	black	B	ae) notch impact energy with hyphen and 2 letters for the test temperature					
					white	W	– room temperature	RT				
							– low temperature	LT				

C.1.3 Designation of cast irons

Whereas wear-resistant cast irons are designated on the basis of their chemical composition, similarly to high-alloy steels, EN 12513 differentiates them according to their macrohardness, which is given after the letter combination EN-GJN (N for no graphite precipitates). Thus, white cast iron GX300CrNiSi9-5-2, known under the trade name of Ni-Hard IV, is now designated **EN-GJN-HV600**, which gives its minimum hardness.

Because the hardness criterion can be fulfilled by all high-chromium cast irons, they are given an additional symbol that indicates the approximate Cr content. The material previously designated as GX300CrMo27-1 by DIN 1695 is now designated as GJN-HV600 (XCr23) by EN 12513.

The first three positions of material number designations correspond to those of the code letter designations. This is followed by a number for the main characteristic (1 = tensile strength, 2 = hardness, 3 = chemical composition) as well as a two-digit ID number. The last single-digit number gives information on special requirements. Thus spheroidal cast iron EN-GJS-400-18 is now designated **EN-JS1022** whereby the last number '2' refers to a cast-on specimen.

C.2 A brief discourse on the history of iron

Iron is not a noble metal and its natural form is thus an ore rather than a native metal. Its history began with the reduction of its ores. Not only oxidic ores, such as haematite Fe_2O_3 and magnetite Fe_3O_4, can be used for ironmaking, but also carbonates (spars), hydroxides and sulphides that can be converted to oxides by heating in air (roasting). Archaeological finds indicate that the oxygen was successfully removed from iron oxide in a charcoal fire.

C.2.1 From a bloomery to a shaft furnace

A charcoal fire, banked up in a hillside depression with a diameter of 0.5 m or so, would have been fanned by the wind thus heating pieces of ore placed in it to 900 - 1000°C. Above the glowing embers, a mixture of CO and CO_2 gases formed that reacted with the iron oxide. Figure C.2.1 shows today's understanding of what could have been expected on mixing iron, oxygen and carbon. At the aforementioned temperatures, the Boudouard equilibrium (curve a-b) would push the CO content to $>95\,\%$ thus reducing the ore to slightly carburised sponge iron. The charcoal was not only a reducing agent, it also acted as a fuel and raised the temperature. The requisite fanning increased the CO_2 content; however, if this content exceeded 30 %, it stabilised wustite FeO and thus inhibited the formation of iron. The hotter and CO_2-richer part of the fire on the windward side shielded the steadier and CO-richer zones in which the ore was converted into pieces of sponge iron that still contained pieces of rock (gangue). These lumps of sponge iron were consolidated by hammering and forge welding to produce somewhat larger lumps that would have been large enough for a knife or an arrowhead. Ancient methods of directly reducing iron in the solid state would have taken place slightly to the left of curve c-d because the corresponding archaeological finds indicate practically carbon-free wrought iron.

The development of bellows allowed air to penetrate through higher beds of charcoal and ore, thus achieving even higher temperatures so that the gangue escaped as liquid slag. Figure C.2.1 does not take account of the interaction with slag or dilution of the gas phase with nitrogen, which slightly lowers the Boudouard curve. Nevertheless, this diagram is a good approximation of direct reduction in what was known as a bloomery furnace. Liquefaction of the gangue, which could be aided by adding flux materials that lower the melting point, afforded a cleaner wrought iron from the sponge iron. Subsequent carburisation under a coating of clay made the wrought iron hardenable. As the shaft above the fire continued to grow upwards, some of the freshly reduced sponge iron was carburised in CO-rich zones as far as a subeutectic melt that dripped downwards and coalesced, loosing some of its absorbed carbon in the CO_2-rich environment on the way. This process is known as fining. The resulting carbon-rich bloom was forged together with low-carbon steel to produce a composite steel with hardenable layers (wrought Damascus steel). Bloomery

C.2.1 From a bloomery to a shaft furnace

hearth furnaces were relocated from the windy hillsides to the valleys where water power was used to drive the bellows (and later on cylinder pumps) and forging hammers. Figure C.2.2 shows some of these stages of development plotted in the Fe-C phase diagram.

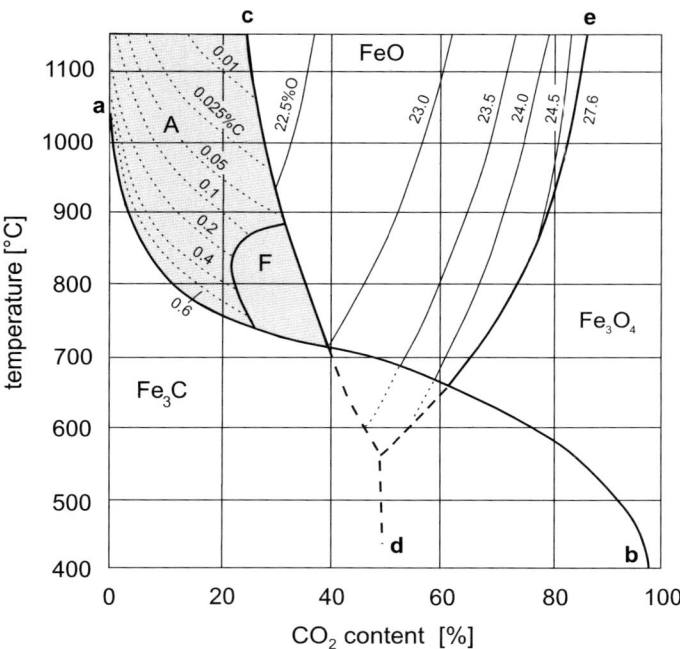

Fig. C.2.1 Reduction of iron oxide: The S-shaped curve a-b of the Boudouard equilibrium $2CO=CO_2+C$ describes the reduction of CO_2 by solid carbon as the temperature rises. The pitchfork curve c-d-e shows the reduction of iron oxide by an increasing amount of CO (mixture of $CO+CO_2$ gases at 1 bar, A = austenite, F = ferrite, from E. Schürmann).

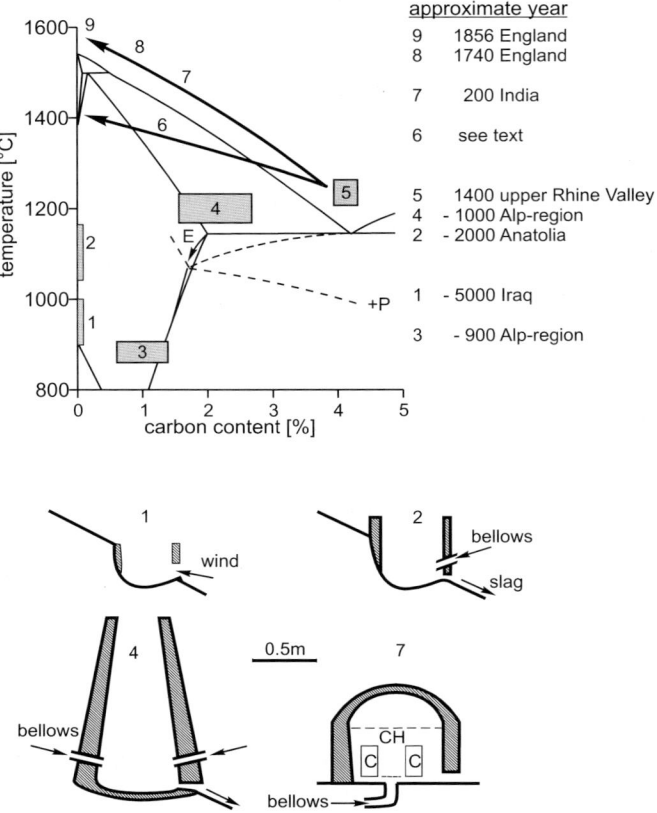

Fig. C.2.2 Stages in the development of iron-making plotted in the Fe-C phase diagram (Figure A.1.6, p.10) along with the corresponding furnace diagrams: (1) Bowl-shaped bloomery furnace on a hillside with a natural draught. (2) Bloomery hearth furnace with bellows and liquid slag. (3) Carburisation of wrought iron. (4) Partially liquid formation of iron in a shaft furnace. (5) Crude iron (pig iron) in a continuously charged blast furnace. (6) Processing of pig iron in a finery forge or puddle furnace and forging in a pasty condition. (7) Wootz crucible steel in a cupola furnace with crucibles (C) in charcoal (CH). (8) Crucible steel. (9) Mild steel (converter steel, open-hearth steel). Phosphorus-rich ores shift point E and lower the solidus temperature of crude iron, e.g. to 1050°C (dashed line). This means that some liquid is obtained at lower temperatures.

C.2.2 The spread of iron-making

Excavations of ancient furnaces and replicas of smelters are important sources of information because pieces of iron are liable to rust and are thus found increasingly rarely the older they are. In the following, we shall use Common Era dates with BCE dates in years represented by a preceding minus sign. Thus the oldest iron find from Samarra (Iraq) has been dated as -5000. Another 13 objects are dated as far back as -3000 and more than another 20 back to -2000. The first written mention of iron comes from the Sumer era around -2300. The accidental reduction of iron in fireplaces or forest fires appears to be less likely than its discovery in connection with reducing atmospheres in ceramic kilns or the smelting of iron-containing copper ores or fluxes. Some of the ancient iron artefacts contain $\approx 8\%$ Ni, which points to meteoritic iron. It might have been found as the metal or it may have originated from nickel-containing ores close to meteor impact sites. Agriculture began around -9600 at Göbekli Tepe near Urfa (Turkey), thus starting the transition to the Neolithic period. Sedentism appears to have fostered technological developments in this region because Northern Mesopotamia is currently regarded as being the cradle of iron smelting. From here, the first golden age of iron spread throughout Asia Minor under the Hittites from -1800 onwards. The first Central European iron find was found in Slovakia and is dated to the -17th Century. The first hardened artefact is dated to the -12th Century. Iron had spread from the Balkans to Scandinavia by about -1000. Between -750 and -500, the Celts of the Hallstatt culture brought the production of hardenable steel from the Balkans, over the Alps and into Northern France. This early Iron Age, named after Hallstatt in Austria, was followed by the late Iron Age from -500 to -100, named after La Téne in Switzerland. The know-how of the Celts was adopted by the Romans and used, amongst other things, to equip large armies. Starting from its cradle in the Near East, the Iron Age also spread northwards and eastwards via Persia and India to China and Japan. Near Lake Victoria in Africa, iron-making started either independently or it may have been initiated by the Phoenicians in the -3rd Century. The Spanish conquerors did not find any iron in America. The construction of the first iron smelters started in 1621 in North Carolina.

C.2.3 Cast iron and the fining process

Water power was harnessed in the 11th Century, and by the 12th Century it was already being used in iron-making. The wind pressure, shaft height and temperature increased so that it was possible to obtain a near-eutectic liquid crude iron at 1300°C in the 14th Century. This increased the iron extraction efficiency as well as the quantities produced. In contrast to the bloomery hearth, where it was difficult to introduce enough carbon into the wrought iron to make it hardenable, indirect reduction in a shaft furnace left a very large amount of carbon in the crude iron. One method of dealing with this was to cast the melt to produce near-net-shaped castings. Thus in the Upper Rhine region, cannon balls were being made of cast iron around the year 1400, and by 1500 there were more than 10 gunsmiths and master smelters who could also produce cannon barrels, furnaces, bells, chimney plates and line pipes. Based on well-known bronze casting techniques, pieces were made using the lost-wax method with undercuts and later on with cores as well. Clay or sandcasting was first carried out by remelting the crude iron in a cupola furnace and later, towards the end of the 14th Century, from continuously loaded blast furnaces. White solidification, seen in the fracture surface, was probably quickly exploited as white cast iron or chilled cast iron e.g. for stamp mills used to pound ore. In China there are large cast iron sculptures from the 7th to 10th Centuries.

The second method of processing crude iron produced by blast furnaces is in a finery forge. However, removing carbon increased the melting temperature. The period of almost five centuries between the late Middle Ages and Bessemer's invention of converter steel in 1856 was characterised by efforts to increase the temperature in order to process crude iron containing less carbon. The crude iron was first cast as a pig (whence its name). This pig iron was subsequently broken up, remelted in a finery forge and fined by adding iron oxide scale to produce a bloom, which was then "shingled" (hammered). The resulting steels were quenched from the forging temperature and broken up. The fractured pieces with a length of e.g. 250 mm were sorted according to the appearance of the fracture surface, i.e. according to their carbon content. The material was repeatedly forged, broken and sorted. Bundles (faggots) of matching bars were then forged together to produce wrought iron. A range of faggot steels could thus be produced from the differing carbon contents within one or more blooms.

The increasing production of cast iron and steel led to a shortage of charcoal in the 17th Century, in spite of careful forest management. Shaft furnaces had already grown to a height of 5 - 7 m, and hard coal was could not cope with the pressure exerted by the bed of material. Abraham Darby was able to increase the compressive strength and porosity of coal by coking to such a level that he produced the first coke-fired pig iron in England probably in 1709. Charcoal-fired finery forges were gradually replaced by coke-fired puddling furnaces, for which Henry Cort received an English patent in 1784. Large amounts of pasty steel were produced within the two-phase austenite/melt field by stirring (puddling) and then processed by hot-working. Cort started using the first steam-driven rolling mill for this in the same year. In 1740, Benjamin Huntsman was able to achieve a sufficiently high temperature to completely melt steel containing approx. 1 % C in graphite-containing crucibles. This meant that residual slag floated to the top. Owing to its cleanness, this crucible steel was in great demand as a tool steel until the beginning of the 20th Century; however, it was only available in limited quantities. *Wootz* or *bulat* crucible steel is regarded as being an ancient forerunner. It contained 1.4 to 1.8 % C and appeared in India in the 2nd Century. Stretch-forging of steel with impurities of good carbide-forming elements such as V and Mo produces effective banding onto which bands of hypereutectoid cementites precipitate during several annealing cycles around Ac_1. This banded microstructure is the basis of genuine Damascus steel, whose manufacture has also been proven in Turkmenistan around the year 1000.

A third method of processing pig iron was also used to a limited extent: dry fining, i.e. without a liquid phase. This is based on a report written in Paris by R.A.F. de Réaumur in 1722, who softened a white cast iron by tempering. Thin-walled *softened* castings with steel-like properties were already available in the 18th Century, before the advent of white heart malleable cast iron produced from 1827 in Switzerland.

C.2.4 Mild steel

New industries demanded steel: steam engines and boilers from 1769, steel ships from 1787, spinning and weaving machines in series from 1820, railways from 1829 in England and 1835 in Germany. The production of wrought iron from puddling furnaces reached its limits. In 1856, Henry Bessemer came up with the idea of blowing air through the crude iron melt in order to combust excess carbon in a short time. This oxidation process also raises the temperature to 1600°C, at which even low-carbon steels are completely molten. In 1878, S.G. Thomas and P.C. Gilchrist replaced the acidic silicate lining of the Bessemer converter with a basic lining of limestone and magnesite. In addition to the converter, the Siemens-Martin open-hearth furnace with regenerative firing was operating in 1864 with an acidic lining and in 1880 with a basic lining. Two types of converter and open-hearth furnace were thus able to produce large quantities of mild steel. In the meantime, blast furnaces had reached a height of 15 m and the flow of air was produced by steam power. By 1925, the annual global production had risen to around 100 million metric tons and in the following 50 years it rose to about 700 million tons. After 1950, the steel converters started using oxygen instead of air to avoid nitrogen-induced ageing (Chapt. B.1, p. 130). The transition from ingot casting to continuous casting began around 1960. Nowadays, the global annual production of ferrous materials is more than 1300 million tons. Although historians place the *Iron Age* in the first millennium BCE, it did not really get going until just before the end of the second millennium ACE.

The changeover from wrought iron to mild steel marked the start of alloying. Wrought iron contained small amounts of Mn and Si, occasionally Cu or Ni as well, from the respective ore. However, intentional alloying was not possible owing to incomplete melting. Alloying was not used for cast iron because the individual elements were not available as a ferroalloy at that time. This changed rapidly after the introduction of mild steel, as the following examples illustrate: manganese steel by R. Hadfield in 1888, case-hardenable nickel steel by Krupp in 1888, roller bearing steel by R. Striebeck in 1901, high-speed tool steel by F.W. Taylor in 1906 and stainless steel by Krupp in 1912.

Before it was displaced by mild steel, cast iron experienced a golden age, as shown e.g. by the iron bridge over the River Severn (1779) and the load-bearing cast iron structure of the Crystal Palace built to house the Great Exhibition in London (1851). A century later, cast iron enjoyed a new lease of life due to the discovery of spheroidal graphite. Spheroidising the graphite by treating the melt with magnesium was registered for a US patent in 1947. In the following year, a further patent was registered for what is now known as cast iron with vermicular graphite. The production of high-strength austempered ductile iron began in 1972. In the meantime, the global annual production of cast iron exceeds more than 60 million tons.

C.2.5 Ferrous materials

Cast iron was introduced in Central Europe more than 2000 years later than wrought iron as a consequence of the difficulties associated with achieving the necessary high temperatures. Another 450 years passed until mild steel was successfully produced, which was also characterised by a battle to raise the processing temperature. Crucible steel is regarded as the forerunner and its high quality was the incentive for displacing wrought iron. However, we must not lose sight of the fact that steam boilers, forging hammers, machine tools, locomotives and other investment goods made of wrought and cast iron brought about the Industrial Revolution.

During their 600 hundred years of common history, steel and cast iron were closely associated with respect to smelting; however, the blacksmith and the foundryman were worlds apart when it came to shaping these materials. This is still noticeable today: we speak of a steel industry and a foundry industry, which, however, includes more than just cast iron. In many countries, steel and cast iron are regulated by separate industrial associations and technical groups, each publishing their own journals and supporting different institutions. And there is very little interdisciplinary literature. This is why *ferrous materials* has not yet become an established term. The aim of this book is to change this situation: steel and iron should not be split into separate chapters, instead, each chapter should deal with both of them.

The history of ferrous materials will continue to develop in the future. The reason for this lies in the remarkable properties of iron, its basic element. Some of these properties were already recognised and prized by the ancients. Today's understanding allows us to name them: (a) Iron makes up more than 4 % of the Earth's crust, it is easily reduced with coke and is thus the most competitively priced metallic material with a high availability. (b) Iron has a high melting point as well as a high stiffness and creep resistance. (c) The cubic crystal structure gives iron its good plastic deformability, and both of its modifications provide good latitude with respect to alloying, heat treatment strategies and hardening. (d) The ferromagnetism of iron governs our use of electricity, and magnetic separation facilitates recycling of scrap. (e) Iron is a component of blood and can generally be classified as non-toxic.

Books on the history of iron

- Beck, L.: *Geschichte des Eisens*, Verlag Vieweg, Braunschweig, 1897.
- Johannsen O.: *Geschichte des Eisens*, Verlag Stahleisen, Düsseldorf, 1953.
- Smith C.S.: *A History of Metallography*, The MIT Press, Cambridge, Ma, 1988.
- Pleiner R.: *Iron in Archaeology*, Archeologicky ustav AV CR, Prague, 2000, (ISBN 80-86124-26-6, with about 950 citations).
- Berns H.: *Die Geschichte des Härtens*, published by Härterei Gerster, CH-4622 Egerkingen, Verlag Ditschi AG, CH-4601 Olten, 2002.
- Hayman R.: *Ironmaking: A History and Archaeology of the Iron Industry*, Tempus Publishing Ltd., Stroud Gloucestershire, UK. 2005
- Moore Swank J.: *History Of The Manufacture Of Iron In All Ages: And Particularly In The United States For Three Hundred Years 1585-1885*, Kessinger Publishing, Whitefish, MT, 2007

C.3 Bibliography for figures and tables

Figures

A.1.1b J. appl. Phys. (1965) p. 616. **A.1.1c** Z. Metallkd. (1969) p. 322. **A.1.9a** from Thermo-Calc, Users Guide, Royal Inst. of Technol., Stockholm. **A.1.9b** DEW-Techn. Ber. (1971) p. 147.
A.2.17 Atlas der Wärmebehandlung der Stähle Bd. 1, Verlag Stahleisen, Düsseldorf, 1961. **A.2.18** K. Röhrig, W. Fairhurst: ZTU - Schaubilder, Gießerei-Verlag, Düsseldorf, 1979. **A.2.24** Weld. Res. (1947) p. 601s, (1973) p. 281s, (1974) p. 273s.
A.3.3 Atlas der Wärmebehandlung der Stähle Bd. 3, Verlag Stahleisen, Düsseldorf, 1973. **A.3.6a** as Fig. A.2.17. **A.3.7a** Distortion in Tool Steels, Verlag ASM, Metals Park, 1959. **A.3.8** Dissertation Ruhr-Universität Bochum, see also VDI-Fortschr.-Ber., Series 5, No. 141, VDI-Verlag, Düsseldorf, 1988.
A.4.6 Dissertation Ruhr-Universität Bochum, see also Fortschr.- Ber. VDI-Z Series 5, No. 91, VDI-Verlag, Düsseldorf, 1985. **A.4.12** Z. Metallkde. (1968) p. 29. **A.4.13** Stahl und Eisen (1979) p. 841. **A.4.20** Chemical metallurgy of iron and steel, ISI Publ. 146, London, 1973, p. 395 and 419. **A.4.22** J. Phys. Ser. F (1981) p. 57. **A.4.23** Trav. Mem. Bur. Int. Poids et Mesures, (1927) p. 1.
B.1.3c Schweißen und Schneiden (1981) p. 363. **B.1.4** Archiv Eisenhüttenwes. (1975) p. 119. **B.1.5** Estel Ber. (1976) p. 139. **B.1.7** Konstruieren und Giessen 25 (2000) H.2 p. 23, 53. **B.1.8** Hochwertiges Gusseisen, Springer Verlag, Berlin, 1951 p. 585. **B.1.9a** Stahlguss und Gusslegierungen, Dt. Verlag für Grundstoffindustrie, Leipzig/Stuttgart, 1992. **B.1.9b** Gieß. Forsch. 37 (1985) H.1 p. 17. **B.1.10** Metals Handbook - 9th Edition, ASM, Metals Park Ohio, 1978, p. 40. **B.1.11** Konstruieren + Gießen 16 (1991) p. 7. **B.1.12** Konstruieren + Gießen 20 (1995) 2 p. 9. **B.1.14** Age Strengthening of Gray Cast Iron - Phase III, Technical Report, University of Missouri-Rolla, Metallurgical Engineering Dept., 2003.
B.2.2a Acta Met. (1981) p. 111. **B.2.3** Thyssen Edelstahl Techn. Ber. (1987) p. 89. **B.2.5** Proc. Int. Conf. on TRIP-Aided High Strength Ferrous Alloys, Gent, June 2002, p. 13-23. **B.2.6** steel research 73 (2002) p. 392. **B.2.7** steel research 75 (2004) p. 139. **B.2.10** Stahl u. Eisen (1987) p. 585. **B.2.11** Max Planck Institut für Eisenforschung, Annual Report, Düsseldorf, 2004, p. 105-108. **B.2.12** J. Phys. IV France 7 (1997) C5 p. 383. **B.2.13** Konstruktion (2005) p. IW6. **B.2.14c** Advanced Engineering Mat. 2 (2000) p. 261. **B.2.14d** steel research 68 (1997) p. 534. **B.2.14e** steel research 58 (1987) p. 369. **B.2.14f** ATZ Automobiltechn. Z. 100(1998) p. 918. **B.2.15** HTM (1968) p. 85. **B.2.16** Techn. Rundsch. Sulzer Forsch. Heft (1970) p. 1. **B.2.18** Stahl und Eisen (1973) p. 1164 and (1978) p. 157. **B.2.19** steel research (1986) p. 178. **B.2.20** Dissertation Ruhr-Universität Bochum, see also VDI-Fortschr. Ber., Series 18, No. 29, VDI-Verlag, Düsseldorf, 1986. **B.2.22** Metallurg. Trans. A (1977) p. 1025. **B.2.25** Wälzlagertechn. (1987)

p. 14. **B.2.26** Duktiles Gusseisen, Verlag Schiele & Schön, Berlin, 1996, p. 178.
B.2.27a as Fig. B.2.26, p. 179. **B.2.27b** Gießerei-Praxis 19 (1982) p. 203.
B.2.29 Gießereitechnik 19 (1973).

B.3.2 as Fig. A.3.3. **B.3.5** HTM 58 (2003) p. 191. **B.3.7** HTM 59 (2004) p. 25. **B.3.8** HTM 53 (1998) p. 359. **B.3.10b** HTM (1968) p. 296. **B.3.10d** HTM (1968) p. 101. **B.3.11** HTM 60 (2005) p. 233. **B.3.14** Atlas der Wärmebehandlung der Stähle, Verlag Stahleisen, Düsseldorf, 1972. **B.3.12** Dissertation Ruhr-Universität Bochum, see also VDI-Fortschr.-Ber., Series 5, No. 702, VDI-Verlag, Düsseldorf, 2005.

B.4.1 Hartlegierungen und Hartverbundwerkstoffe, Springer Verlag, Berlin Heidelberg, 1996, p. 24. **B.4.3** Dissertation Ruhr-Universität Bochum, s. a. Fortschr.-Ber. VDI Series 5, No. 340, VDI-Verlag, Düsseldorf 1994. **B.4.4** Hartlegierungen und Hartverbundwerkstoffe, Springer Verlag, Berlin/Heidelberg, 1996, p. 24.

B.5.1 AWT Seminar Werkzeugstähle, Berlin, 1984. **B.5.2** Arch. Eisenhüttenwes. (1958) p. 193. **B.5.4a** Arch. Eisenhüttenwes. 47 (1976) p. 391. **B.5.4b** HTM 29 (1974) p. 236. **B.5.5c** Metals and Mat. (1986) p. 421. **B.5.6** Process Developments and Applications for New PM-High Speed Steels, Proc. 5th Intern. Conf. on Tooling 1999, Inst. für Metallkd. Montanuniversität Leoben, Austria, p. 525. **B.5.7a** 3. Int. Conf. Fatigue, Charlottsville, USA, 1987. **B.5.7b** VDI-Z 127 (1985) p. 885. **B.5.8** Metals and Mat. (1986) p. 421. **B.5.9** samples from Dörrenberg Edelstahl. **B.5.11** AWT Seminar Werkzeugstähle, Berlin, 1984. **B.5.12** HTM 42 (1987) p. 25. **B.5.13** DEW-Techn. Ber. (1971) p. 178. **B.5.14** steel research 56 (1985) p. 433. **B.5.16** Radex Rundsch. (1989) p. 40.

B.6.1a Dissertation Universität Leiden, 1955. **B.6.1b,c** Archiv Eisenhüttenwes. (1937/38) p. 131. **B.6.3** Thyssen Edelst. Techn. Ber. (1986) p. 3. **B.6.4a** Archiv Eisenhüttenwes. (1963) p. 465. **B.6.4b** Dissertation RWTH Aachen, 1974. **B.6.4c** Dissertation RWTH Aachen, 1967. **B.6.4d** Nichtrostende Stähle, Verlag Stahleisen, Düsseldorf, 1989. **B.6.5** Thyssen Edelst. Techn. Ber. (1982) p. 153. **B.6.6** Archiv Eisenhüttenwes. (1938/39) p. 459. **B.6.7** Metal. Progr. (1949) p. 680 und 680B. **B.6.11a** Thyssen Edelst. Techn. Ber. (1982) p. 97. **B.6.11b** Thyssen Edelst. Techn. Ber. (1985) p. 3. **B.6.12** as Fig. B.6.4d. **B.6.14** Thyssen Edelst. Techn. Ber. (1989) p. 1. **B.6.15** HTM 38 (1983) p. 197 and 251. **B.6.17** Mat. Sci. Eng. (2006) p. 47. **B.6.19** Archiv Eisenhüttenwes. (1976) p. 241. **B.6.20** Proc. Int. Corrosion Forum, Boston, 1985. **B.6.21a** Konstruieren + Giessen 29 (2004) p. 2. **B.6.21b** Konstruieren + Giessen 26 (2001) p. 17.

B.7.4 Archiv Eisenhüttenwes. (1974) p. 263 a. 711. **B.7.6** Metallurg. and Mat. Trans 36A (2005) p. 333. **B.7.7** Iron and Steel (1961) p. 634. **B.7.8** Dissertation Ruhr-Universität Bochum, see also VDI-Fortschr. Ber. Series 5, No. 222, VDI-Verlag, Düsseldorf, 1991.

B.8.1a Stahl und Eisen (1987) p. 1119. **B.8.1b** steel research int. 76, (2005) p. 450. **B.8.3** Werstoffkunde Stahl Bd. 2, Verlag Stahleisen, Düsseldorf, 1985, p. 551. **B.8.4a,b** Physikalische Eigenschaften von Stählen,

Verlag Stahleisen, Düsseldorf, 1983 a. Int. Nickel Publ., 1966. **B.8.4c** Z. Metallk. (1943) p. 194. **B.8.5** Kontakt und Studium Bd. 259, Expert Verlag, Ehningen, 1988, p. 1.
C.2.1 Stahl u. Eisen 78 (1958) p. 1297.

Tables

A.1.2 Archiv Eisenhüttenwes. 50 (1979) p. 185. **A.2.1** Ausscheidungsatlas der Stähle, Verlag Stahleisen, Düsseldorf, 1983. **A.3.2** Material Science and Engineering A, Article in Press, doi:10.1016/j.msea.2006.11.181. **A.4.4** Materials and Corrosion 55 (2004) p. 341. **B.2.1** VDI-Ber. 428, VDI-Verlag, Düsseldorf, 1981, p.35. **B.8.3** Eisen-Nickel-Legierungen, document by Int. Nickel Deutschland GmbH, Düsseldorf, 1972.

Further Reading

- Totten G.E.: *Steel Heat Treatment Handbook, Metallurgy and Technologies, Equipment and Process Design*, Taylor & Francis, Boca Raton, 2007
- Bhadeshia, H.K.D.H and Honeycombe, R.W.K.. *Steels: Microstructure and Properties*. 3rd ed. Oxford, UK: Elsevier Ltd., 2006. ISBN 0-75068-084-9
- Krauss, George *Steels: Processing, Structure, and Performance* 1st ed. Materials Park, OH, USA: ASM International, 2005. ISBN 0-871-70817-5
- Durand Charre, Madeleine *Microstructure of Steels and Cast Iron* 1st ed. Berlin; Heidelberg; New York: Springer-Verlag 2004. ISBN 3-540-20963-8
- Llewellyn, D.T. and Hudd, R.C. *Steels: Metallurgy and Applications* 3rd ed. Oxford, UK: Butterworth-Heinemann 2000. ISBN 0-750-63757-9
- *ASM Handbook: Properties and Selection : Irons, Steels, and High Performance Alloys* 10th ed. Materials Park, OH, USA: ASM International, 1990. ISBN 0-871-70377-7
- Cahn, Robert W., Haasen, Peter, Kramer, Edward J. and Pickering, Frederick Br. (ed.) *Materials Science and Technology: A Comprehensive Treatment, Vol. 7: Constitution and Properties of Steels* 1st ed. Weinheim; New York; Basel, Cambridge: VCH 1992. ISBN 3-527-26820-0
- Stahlinstitut VdEh (ed.) *Steel A Handbook for Materials Research and Engineering, Vol. 1: Fundamentals* and *Vol. 2: Applications* Düsseldorf: Springer Verlag and Verlag Stahleisen 1992 ISBN 3-514-00377-7(Vol. 1) and 1993 ISBN 3-514-00378-5 (Vol. 2)

Keyword Index

A1, A2 ... temperatures   <u>29</u>
Abrasion   <u>106</u>, 251, 266, 282, 288, 299
Active screen   227, 233
Adhesion   <u>104</u>, 107, 127, 251, 283, 288
ADI   71, <u>208</u>
Ageing   <u>130</u>, 139, 158, 397
Alloying
  -, costs   191, 312
  -, effects   46
  -, elements   15, 44, 234, 397, 416
    -, austenite-stabilising   47, 323, 373
    -, corrosion-inhibiting   159, 310
    -, ferrite-stabilising   47, 323
    -, history   397
    -, in steel, fate of   14
    -, strengthening   94, 168
α'-embrittlement   313, 324, 339
Aluminising   137
Amorphous Zone   23
  -, layer   <u>286</u>, 288, 309
  -, ribbons   373
Anisotropy   <u>24</u>, 93, 131, 292
  -, in-plane   132
  -, perpendicular   <u>131</u>, 312
  -, size change   279, 303
Annealing   10, 34, <u>58</u>, 131, 154
Applications
  -, case-hardening steels   246
  -, cast iron   157, 290, 309, 398
  -, cold-work tool steels   288
  -, creep-resistant steels   361
  -, electrical resistance heating alloys   380

  -, fine-grain steels   179
  -, heat-resistant steels   339, 365
  -, high-speed tool steels   303
  -, high-strength steels   178, 192
  -, hot-work tool steels   64, <u>296</u>
  -, magnetically hard materials   374
  -, magnetically soft materials   369
  -, materials with
    -, shape memory   380
    -, special thermal expansion properties   378
  -, multi-phase steels   168, 180
  -, nitriding steels   232
  -, non-magnetisable steels   375
  -, pearlitic steels   184
  -, plastic mould steels   291
  -, precipitation-hardened ferritic-pearlitic steels, FP-PH steels   189
  -, stainless steels   331
  -, steels treated from the forging temperature   184
  -, surface-hardenable steels   67, 223
  -, thin films   287
  -, tools
    -, for processing materials   273
    -, for processing minerals   255, 259, <u>262</u>
  -, unalloyed structural steels   134
  -, weldable steels   175
Atomic
  -, diameter   5, 182
  -, volume   7

-, weight   7
Ausforming   202
Austempered Ductile Iron, (ADI)   208
Austempered Grey Iron, (AGI)   212
Austenite   9, 18, 47
  -, expanded   233
  -, grain size   241
  -, homogeneous, inhomogeneous   65
  -, micrograph   21
  -, reverse formation   201, 327
  -, stabilisation   70, 324
  -, transformation   29, 38, 41, 315
Austenitic steels   63
  -, creep-resistant   357
  -, electrical resistance heating steels   380
  -, heat resistant   337
  -, non-magnetisable   375
  -, stainless   322, 329
Auto-tempering   44, 186

Bainite   38, 45, 200
  -, cast iron   71, 208
  -, isothermal   204
  -, micrograph   44
  -, range   70, 208
  -, roller bearing steel   71, 204
  -, structural steel   168
Bake hardening   131, 172, 180
Banded microstructure   21, 64, 93
  -, structural steel   125
  -, tool steel   277
Bars, steel   139
Bauschinger effect   180
Bending strength   255, 280, 301
Bessemer steel   397
Blackplate   135
Blast furnace   394
Bloch walls   117, 374
Bloom   396
Bloomery hearth furnace   392
Blue brittleness   48, 199
Borides   15, 256, 356
Boriding   256
Boudouard equilibrium   225, 392
Bright steel   140, 196
Brittle fracture   80, 89, 99, 131

CADI   213

Carbide annealing   62
Carbides   15
  -, after cooling   33
  -, after tempering   48, 50
  -, arrangement   22
  -, micrographs   34, 44, 258, 282
  -, microhardness   103
  -, morphology   53
  -, types   48, 50, 52, 184
Carbon   9, 14
  -, activity   31, 144, 234, 313
  -, equivalent   31, 129, 138, 165, 176
  -, level   235
  -, steels   13, 135
Carbonitriding   238, 247
Carburisation   235
  -, heat-resistant   338, 381, 394
  -, process   242
Carrier gas   234, 238
Case-hardenable steels   240
  -, rolling bearings   206
Case-hardening   217, 234, 237, 244, 255
  -, depth (CHD)   217, 237, 246, 257
  -, in the oxide film   116
  -, in the surface layer   199, 245
  -, steels   72, 85, 234, 244
  -, gear components   246
  -, tools   256
  -, wear parts   257
Cast iron   31, 341, 360, 388
  -, austenitic   342
  -, chemically resistant   341
  -, ferritic   341
  -, grey   99, 144, 207
  -, with flake graphite   147
  -, with spheroidal graphite   150
  -, with vermicular graphite   152
  -, malleable   154
  -, black   154
  -, white   102, 156, 259, 344
Casting steel   142, 259, 319, 337, 384
Cementite   9, 11, 53
  -, decomposition   51
  -, precipitation   53
  -, cooling   30
  -, tempering   48
  -, quantity   10
  -, spheroidal   60
Charcoal   392

Chemical
-, compound   14
-, equipment   332
-, resistant materials   309
Chill casting   14, 33, 37, 260
Chromium cast iron   261, 265
Chromium equivalent   47, 322
Clean steels   197, 362
Cleanness   98, 204
-, hot cracking   179
-, ladle metallurgy   197
-, lifetime   205
-, non-metallic inclusions   28, 38, 138, 229, 317
-, structural steel   140
Cleavage   82
-, fracture   80, 83, 89, 100, 128
Coarse grain structure   87, 129, 242
-, chemically resistant   315
-, creep-resistant   351
-, magnetically soft materials   369
Coatings   72
-, corrosion protection   362
-, electroplated   107, 140, 175
-, enamelling   72, 137
-, hot-dip   72, 128, 136
-, organic   137
-, paint   72, 136
-, strip   135
-, thin films   285, 290, 304
-, wear   264, 267
Coefficient of friction   105
Coefficient of thermal expansion
-, cast iron   148
-, creep-resistant steels   353
-, iron-nickel   120
-, special materials   120, 376
Coercive force   369
Coherence   24, 51
Coherent particles   24, 51
Coiling temperature   173
Cold work hardening   212, 259
Cold workability   136, 313, 386
Cold-cracking susceptibility   166
Cold-rolled strip   136, 195, 198
Cold-work tool steels   274, 293
Comet tail   229
Composite casting   255, 269
Compound layer   224, 287

Compressive strength   100, 102, 283
Concrete reinforcing steels   135, 139, 385
Conductivity   85, 121
-, electrical   121
-, thermal   121, 148, 312, 353
Constitution   3, 134, 321
Contact fatigue   106, 251
Continuous casting   28, 37, 161, 397
Converter steel   394
Cooling
-, continuous   45
-, curve   40, 45, 66, 301
-, during hardening   65, 73
-, during rolling   14
-, in a vacuum furnace   238, 301
-, in the melt   12
-, rapid   36, 38, 48, 73, 344
-, slow   12, 26, 29, 30, 45, 355
Coordination number   5
Cord, steel   140, 184
Corrosion   109
-, fatigue   114
-, intergranular   49, 113, 310, 319
-, resistance   109, 127, 178, 311, 327
-, stainless steel   318
-, structural steel   127
-, transgranular   114
Crack propagation   87, 210, 245, 295
Cracking due to
-, applied loads   79, 88, 98, 190
-, carbides   86, 134, 204, 279
-, corrosion   114, 115, 179
-, fatigue   107, 204, 245
-, hardening   64, 73, 221, 304
-, hydrogen   58, 114, 178
-, martensite plates   44
-, nitriding   229
-, oxidation   75, 76, 116
-, segregation   86
-, surface hardening   206
-, thermal shocks   298
-, welding   266
Crash properties   170, 180
Creep   74, 89, 294, 349, 359
-, Coble   352
-, Nabarro-Herring   352
Creep rupture behaviour   338, 349, 354
-, creep-resistant steels   349

406   Keyword Index

-, embrittlement   358
-, hot-work tool steel   294
-, lifetime   363
-, testing   351
Creep-resistant materials   349
Creep-resistant steels   274, 294, 349
Crucible steel   394
Crude iron, pig iron   154, 394
Crumple zone   180
Crystal structure   6
Cubic lattice   5, 42
Curie temperature   117, 119
Current density/potential curve   112
  -, influence of chromium   310
  -, resistance   333
CVD coating   283
Cycle annealing   60, 219, 397

Damascus steel
  -, damascene steel   392
  -, genuine   397
Damping capacity   32, 100, 212
Decarburisation   75, 156, 185, 221, 223
Deep drawability   132
Deep drawing   133, 180, 275, 288
Defects   96, 279
Deformation   82
  -, martensite   312
  -, velocity   81, 85, 91
    -, strength   90
    -, transition temperature   83
Degassing   197
Degree of
  -, hardening   67, 198
  -, saturation   34, 147, 270
Delamination   106
Delayed fracture   89
Dendrites   26, 147, 277
Density   7, 106, 146, 159, 267
  -, current   112, 310
  -, dislocation   45, 51, 60, 94, 350
  -, ion   286
  -, magnetic flux   369
  -, power   246
Deoxidation   138, 292
Depth of hardening   217, 221, 291
  -, after nitriding   217, 225, 293
Designations
  -, of cast iron   388

-, of steel   383
Diffusion   16, 115, 235
  -, carburising   241, 338, 394
  -, layer   226, 230
  -, zone   229, 236
Direct hardening   240, 242
Direct reduction   392
Dislocations   23
  -, due to deformation   51
  -, in martensite   42
  -, pile-up   83, 85
Dispersion   165, 254, 277, 297
Distortion   73
  -, during case hardening   243
  -, during hardening   73, 189, 192, 278
  -, during nitriding   229
  -, reduction   58, 64, 301
DS   see Depth of hardening
Dual-phase
  -, microstructure   22, 125, 169, 320
  -, steels   169, 171, 181
Ductility   71, 83, 89
Duplex
  -, microstructure   21, 125, 182, 324
  -, steel   22, 322, 332, 345

Earing   132
ELC steels   319, 358, 361
Electrical resistance heating alloys   380
Electrical steel sheet   370, 371
Electron-beam hardening   217
Electroslag remelting, ESR   38, 296, 304
Elongation   70, 90, 130, 177, 183
Embrittlement due to
  -, creep rupture loading   353, 358
  -, grain boundary carbides   297
  -, heating to 300°C   48
  -, heating to 475°C   59, 313, 314
  -, heating to 500°C   50, 197
  -, hydrogen   114
  -, intermetallic phases   314
  -, irradiation   194
  -, low-temperature   84, 194
Endogas   234, 236
Energy
  -, conversion   370
  -, product   370, 373
Equilibrium   26, 36, 48, 115, 225, 393

Keyword Index 407

Erosion 108
expanded austenite 233
Extrapolation 364

Fatigue 205, 251, 295, 364, 379
 -, HCF 364
 -, LCF 364
 -, strength 82, 114, 212, 246
   -, rolling bearings 205
   -, springs 198, 199
   -, steel cord 140
   -, tools 281
Ferrite 9, 18, 47
 -, micrograph 21, 126
 -, net 22
 -, properties 311
Ferritic materials
 -, cast iron 99, 147, 151, 160, 341
 -, electrical resistance heating materials 380
 -, heat-resistant 337
 -, magnetically soft 373
 -, stainless 319
Ferritic-austenitic steels 324
Ferritic-martensitic steels 247, 328
Ferritic-pearlitic cast iron 146, 150, 212
Ferritic-pearlitic steels 127, 188
 -, precipitation hardenable 184, 188, 220, 233
Ferritising 62
Fibre 22, 91
Fine grain structure 182
 -, case hardening 241
 -, effect on T_T 185
 -, effect on the strength 168
 -, low-temperature toughness 194
 -, metal working 28, 165, 202
 -, produced by normalising 126
 -, quenching and tempering 69, 194
Fine-grain steels 85, 165, 177
Finery forge 395
Fining 392
Fish scales 138
Flaking
 -, avoidance 58, 197
Flame hardening 36, 72, 217, 223, 290
Flat products 134, 176, 385
Flow

-, curves 90, 196, 312
 -, stainless steel 312
 -, structural steel 196
-, lines 136
-, stress 91
Forging 141, 184, 187
FP-PH steels see Ferritic-pearlitic steels
Fracture 85
 -, intergranular 50, 87, 356
 -, resistance 190, 268
 -, strength 73, 90
 -, toughness 84, 89, 98, 201, 210
 -, transgranular 87
 -, work 84, 88, 98, 100, 182
Free-cutting steels 139
Friction 104
 -, internal 101, 132
Functional materials 369

G phase 52, 186
Galvanic corrosion 113
Gangue 392
Gas
 -, carburising 234
 -, nitriding 188, 224, 225, 228
 -, turbine 362
Glow discharge 227, 286
Grain boundaries 21, 23
 -, carbide 30
 -, segregation 50, 113, 197
 -, sliding 86, 351, 358
Grain growth 51, 73, 148, 279
Grains 21, 234
 -, elongated 21, 24, 25, 126
 -, equiaxial 21
Graphite
 -, formation 153
 -, degeneration 152
 -, formation 32, 145, 388
 -, interstitial 145, 157
 -, morphology 53
Graphitisation 70, 156, 260, 313, 341

Hadfield manganese steel 43, 259
Haematite 116, 392
Hallstatt Era 395
Hard phases
 -, cracking 86, 204

-, distribution   22, 107
-, hardness   103
-, type   52
-, wear resistance   252, 265
Hard steels   203
Hard-facing   255, 264
Hardenability   65, 190, 208, 242
-, costs   191
Hardening   42, <u>64</u>, 73, <u>273</u>, 274
-, capacity   221
-, from the forging temperature   184
-, of tools   254, 273
Hardening depth after nitriding   <u>217</u>, 225, 293
Hardening of the entire surface   217
Hardness
-, limit   217
-, penetration   221, 243, 273
Hardness of
-, microstructural building blocks   48, 103
-, minerals   103, 252
-, thin films   284
Heat resistance   312, 344
Heat treatment   57, 187
-, from the forging temperature   165, <u>184</u>
-, side-effects   73
Heat-affected zone   128
Heat-resistant steels   336
Heat-treatable cast iron   207
High-speed tool steels   274, 300
High-strength materials   165
High-strength steels   198
High-temperature corrosion   115, 309, 342, 380
-, effect of alloying   117
-, kinetics   115
HIP process   264, 267, 296
Homogenising   26, 63
Hot
-, cracking   27, 316
-, hardness   230, 279, 303
-, working   25, 165, 184, 277
    -, effect on the microstructure   28, 165
Hot-rolled strip   135, 143, 166, 170, 298
Hot-work tool steels   354
-, homogenising   64

-, secondary hardening   49
HSLA steels   165, 175, 180
Hydrogen
-, baking   58
-, damage   114, 178, 365
-, welding   129
Hydrogen-induced   114, 178, 336
Hydrolysis   128, 318, 336
Hypereutectoid steels   29, 67, 221
Hypoeutectoid steels   30
Hysteresis curve   369

Inclusions   28, 38
-, avoidance   190
-, fatigue   204, 205
-, fracture strength   96, 107
-, metal working   138, 317
-, polishing   229
Incoherence   24, 51
Induction hardening   72, 217, 221, 366
Induction, magnetic   217, 283
Ingot casting   37
Inoculation   147, 157
Instant fracture   <u>89</u>, 280
Intercritical microstructural tailoring   169
Interface   24, 117
-, energy   104
-, friction pairing   104, 107
-, phases   22, 24
Internal pressure   4
Internal zone   128
Interstices   5, 8, 16, 42
Interstitial elements   15, 326, 349
Interstitial-free steels   135, 138, 181
Interstition   <u>9</u>, <u>14</u>
Invar alloy   120
Ion implantation   72
Iron
-, Age   251, 395
-, α-, δ-, ε-, γ-Fe   5
-, annual production   150, 397
-, chemical resistance   109
-, constitution   5
-, crystal lattice   6
-, extraction   395
-, history   392
-, ore   9, 142, 392
-, properties   7, 94, 398

-, sponge   392
-, strengthening   94
Isoplethal section   17, 29, 31, 240, 276, 316

Jominy test   66, 68, 208

Kinetics
  -, high-temperature corrosion   115
  -, wet corrosion   111

Lüders elongation   90, 130, 180
La Téne Era   395
Ladle metallurgy   362
Lamellar fracture   132
Larson-Miller parameter   363
Laser applications
  -, hardening   217, <u>224</u>, 290, 372
  -, remelting   72
  -, welding   180
Latent heat   <u>4</u>, 39
Lattice
  -, bct   43
  -, fcc, bcc   6
  -, hcp   6, 44
  -, interstices   8
  -, parameters (constants)   5
  -, planes   82
  -, shearing   42
Lattice defects   <u>23</u>, 25
  -, reduction   57
  -, strengthening effect   57, 90, <u>94</u>
Laves phase   52, 313
Ledeburitic materials   233, 260, 278
Lehrer diagram   225
Lifetime   89, 113
  -, creep resistance   350, 363
  -, rolling bearings   205
  -, springs   199
  -, tool steels   281, 303
Loading   <u>79</u>, 88, 106
Localised corrosion   <u>113</u>, 317, 332
Long products   139
Low-pressure carburising   <u>237</u>, 239
Low-temperature steels   18, 194
  -, stainless   322, 342
Lower shelf   <u>84</u>, 96, 197, 268

Machinability   60, 157
-, cast iron   62
-, functional materials   320
-, structural steels   134, 317, 320
-, tool steels   292, 302
Machining   292, 305, 328
Magnesium treatment   32, 150, 398
Magnetic
  -, core loss   370, 372
  -, flux density   369
  -, permeability   369, 375
  -, permeability of free space   369
  -, properties   <u>117</u>, 370
Magnetically
  -, hard materials   373
  -, soft materials   117, 369
Magnetite   74, 116, 392
Magnetostriction   13, 376
Malleable cast iron   34, 63, 155
  -, black heart   154
  -, whiteheart   156
Maraging steels
  -, high-strength   201, 290
  -, stainless   49, 328
  -, tools   292
Martensite   38, <u>42</u>, 65
  -, ε-martensite   6, 45
  -, deformation-induced martensite   169, 182, 254
  -, hardness   103
  -, micrographs   43
  -, strength   95
Martensite phase steels   180
Martensitic forging steels   188
Matrix   <u>22</u>, 40, 99, 267
  -, potential   301
Mechanical properties   <u>100</u>, 193, 209, 211, 280, 343
Melting
  -, crude iron, pig iron   394
  -, steel   396, 398
Metal dusting   313
Metal working   28, 187
  -, influence of the microstructure   28
  -, normalising   166, 177, 386
  -, thermomechanical   186, 202
Metastable
  -, solidification   33
  -, system   10, 35
Meteoritic iron   395

M_f, M_s temperatures   39, 47, 222, 302
Microalloyed steels   165, 184
Microstructure   <u>21</u>
-, anisotropic   25, 93, 126
-, banding   21
-, effect on T_T 85
-, effect on the strength   91
-, microhardness of phases   103
-, morphology   21, 42
-, near-equilibrium   26
-, non-equilibrium   36
-, structure   21
Mild steel   394
Mismatches   24
Morphology   42, 53
Mould filling capacity   157
Multi-component system   17, 276, 316
Multi-phase steels   168, 180
Multiaxiality   <u>79</u>, 88, 96, 97

Néel temperature   119
Network microstructure   22, 255, 303
Nickel equivalent   <u>47</u>, 316, 357
Nickel white cast iron   260
Nickel-base alloys   294, 340
Nihard   102
Nitrided layer   226, 231, 287
Nitrides   15, 225
-, distortion   229
-, microhardness   103
-, precipitation   <u>46</u>, 52, 165, 319
-, surface layer   225
-, tools   286
Nitriding   72, <u>224</u>
-, distortion   229
-, in a plasma   227
-, low-temperature   233
-, microhardness curve   228
-, potential   226
-, steels   <u>224</u>, 292
-, tools   286
Nitrocarburisation   225, 233
Nitrogen-alloyed steels   46, 292, 328
-, creep-resistant   361, 365
-, production   328
-, rolling bearings   335
-, stainless   317, 322
-, tempering   48
-, tools   335

Non-magnetisable
-, materials   375
-, steels   119, 370, <u>375</u>
Non-metallic inclusions   *see Inclusions*
Normalising   <u>62</u>
-, creep-resistant steels   352
-, high-strength steels   178
-, structural steels   126, 138, 166
-, surface-hardenable steels   219
Notch effect   <u>81</u>
-, fracture   85, 100
-, fracture strength   92
-, internal   82, 99
-, multiaxiality   <u>79</u>, 88, 96
-, on T_T 84
Nucleation
-, creep   353
-, heterogeneous   26
-, lattice defect   30, 48, <u>50</u>, 68
-, supercooling   36, 46
-, transformation   62

ODS alloy   363
Open hearth steel   397
Ostwald ripening   48, <u>60</u>, 353
Overageing   201
Overcarburising   <u>235</u>, 246
Oxidation   75, 341, *see also Scale formation*
-, case hardness   237
-, electrical steel sheet   371
-, internal   236, 371
Oxide film   <u>116</u>, 148
Oxygen probe   236
Oxynitriding   225, 233

Passivation   <u>112</u>, 182, 321, 332
Patenting   <u>63</u>, 140
Pearlite   <u>30</u>
-, continuously cooled   39, 40
-, isothermal growth   46, 200
-, micrograph   <u>30</u>, 126, 258
-, normalised   62
-, properties   96, 101, 127, 154
-, wear resistance   127, 140, 149
Pearlitic
-, cast iron   149, 153, 221, 276
-, forged steels   189
-, steels   140, 184

-, white cast iron   34, 258, 259
Permanently magnetic
    see Magnetically
Permeability   see Magnetic
Petrochemistry   365
Phase diagram
    -, Fe   4
    -, Fe-13 % Cr-0.2 % C-N   244
    -, Fe-13 % Cr-C   276
    -, Fe-18 % Cr-18 % Mn-C-N   330
    -, Fe-30 % (CrNi)   316
    -, Fe-C   10
    -, Fe-C-X   18
    -, Fe-Cr   18
    -, Fe-Cr-4 % Ni   18
    -, Fe-Cr-C   17
    -, Fe-Fe$_3$C   10
    -, Fe-N   231
    -, Fe-Ni   18
    -, Fe-Si-C   31
    -, HSS   17
Phases   3, 23, 82
    -, boundary   23
    -, coarse   25
    -, complex-phase steels   172
    -, dual-phase steels   171
    -, hard   107, 252
    -, intermetallic   22, 52
    -, morphology   3, 25
    -, multi-phase steels   168
    -, precipitation in   50
        -, creep-resistant steels   359
        -, high-strength steels   202, 328
        -, stainless steels   314
Phosphide eutectic   148, 222
Phosphorus alloyed steels   181
Pipes   141, 176, 203, 207, 333
Pitting
    -, corrosion   113, 233, 247, 318
    -, resistance equivalent number,
        PREN   317, 324
Plasma
    -, carburising   238
    -, nitriding   227, 286, 299
Plastic mould steels   292
Plate
    -, carbide   85, 95, 126
    -, cementite   69
    -, martensite   43

-, materials   132, 143, 168, 371
-, thick   138
-, thin   135, 290
-, very thin (blackplate)   135
-, wear   261
Polycrystals   24
Porous zone   229
Potential   110, 112
Powder metallurgy   28, 374
    -, microstructure   273, 277
    -, structural steels   142
    -, tools   267, 275, 296, 303
Precipitates   15, 24, 32, 52, 85
    -, coherence   24, 188
    -, embrittling   207, 322
    -, graphite   145, 156
    -, shape   23
    -, types   14, 23
Precipitation hardening (Ageing)
    -, creep-resistant   322
    -, high-strength   211
    -, microalloyed   188, 207
    -, resistant to scaling   314
    -, stainless   311, 314
    -, tools   274, 279, 296
Preheating   129, 256, 298
Pressure metallurgy   206, 329
Primary grain size   37
Profiles   57, 139
Progressive hardening   206, 217, 218
Properties of steel
    -, chemical   72, 109
    -, magnetic   41, 117
    -, mechanical   79, 91, 96, 100
    -, physical   79, 117
    -, tribological   103
Puddling furnace   394
Purging   197

Quasi-isotropy   24
Quenching and tempering   69, 188, 207
    -, steels   190, 193
        -, creep-resistant   352

r Value   132
Rails   135, 139, 184
Rapid solidification   9, 28, 300, 373
Recovery
    -, during creep   350

-, in martensite   48, 60, 254
Recrystallisation
  -, austenite   37, 165
  -, inhibition   165
  -, martensite   48
  -, metal working   28
  -, secondary   313
Reduction
  -, Fe ore   9, 142, <u>393</u>
  -, wet corrosion   109
Reheat
  -, cracking   356
  -, hardening   <u>243</u>, 257
Reheating   48
Residual stresses   114, 372
  -, after bainite formation   204
  -, after hardening   41, 72, 204
  -, after shot peening   199
  -, after welding   58, 356
  -, in the scale film   116
  -, in the surface layer   229, 244
  -, relief of   58, 74, 202, 263
Resistance
  -, to corrosion   109, 127, 178, 311, 327
  -, to pressurised hydrogen   365
  -, to wear   127
Retained austenite   43, 67
  -, case hardening steel   209
  -, cast iron   70, 209, 260
  -, micrograph   43, 47
  -, multi-phase steels   171
  -, rolling bearing steel   67, 204
  -, tools   254, 257, 301
  -, transformation   173, 206, 254, 279
  -, wear resistance   279
Rivets   380
Roasting   392
Rolling bearing steel   204, 329
Rolling mills   167
Roughness   <u>104</u>
Ruling thickness   141, 191

S phase   233
Salt bath nitrocarburising   227, 228
Sampling   102
Saturation polarisation   369
Scale formation   74–76, 115, 338
Scale-resistant steels   309

-, electrical resistance heating steels   380
Schaeffler diagram   47, 315
Scratch energy   253
Secondary graphite   53
Secondary hardening   49
  -, high-strength   202, 207
  -, nitriding   228, 233
  -, rolling bearings   329
  -, stainless   206
  -, tool steels   273, 279, 287
Secondary-hardening alloy carbides   50, 274, 297, 354
Segregation   <u>24</u>, 26
  -, ageing   130
  -, causes   27
  -, corrosion   113
  -, cracking   304
  -, local   28
  -, reduction   38, 63, 64
  -, temper embrittlement   197, 356
Self-quenching   128, 217, 223
Semi-hot metal working   186, 233
Separations   167
Shaft furnace   392
Shallow hardness penetration   74, 289
Shape change during hardening   <u>73</u>, 223, 232
Shape memory   45, <u>378</u>
Shaping   <u>37</u>, 57, 140, 190, 247, 251
Shear
  -, cutting   273, 289
  -, fracture   86, 100
  -, lip   88
Shot peening   72, 199, 245
Shrink holes   28, 38, 102
Side-effects   72
  -, thermal   73
  -, thermochemical   74
Sigma phase   52, 313, 315, 339
Signal conversion   369
Sintered
  -, MMC   258, 269
  -, steel
    -, applications   195, 247
    -, production   142
Size
  -, change   51, 68, <u>73</u>, 204, 279, 376
  -, stability   73, 204

Skeleton   22, 74, 102
Skin-passing   136, 137
Sliding friction   104, 108
Slip   42, 82, 352
Soft annealing
   -, of cast iron   62
   -, of steel   59, 242
Soft martensitic steels   325, 335
Soft skin   75, 223
Solid-solution   9
   -, hardening   94, 147, 312, 350
Solidification   9, 12, 26, 31
Solubility   9, 15
Solution-annealing   63, 130, 202, 259
Solution-nitriding   239, 247
Specific surface layer thickness   217
Spheroidising annealing   60, 387
Spinel   75, 117, 309
Spinodal decomposition   182, 373
Spot welding   176
Spray-forming   273, 303
Spring steels   70, 185, 198
Stabilisation   320
Stable Fe-C system   9
Stacking faults   44, 90, 182, 259
Stainless steels   240, 317
   -, austenitic   322
   -, ferritic   319
   -, high-N steels   328
   -, martensitic   325
Standard potentials   110
Standardisation   383
Static friction   104
Steadite   see Phosphide eutectic
Steam
   -, power plant   360
   -, turbine   361
Steels
   -, designations   384
   -, history   392
   -, post-treatment   190, 197
Stiffness   101, 160, 377
Stirring   28, 396
Strain rate   84, 183
Strain-hardening exponent   132, 312, 315
Strength   90, 150, 313
Strengthening   94
   -, cast iron   100

   -, mechanisms   94
Stress
   -, corrosion cracking   114, 335
   -, reduction   59
   -, stainless steels   311, 332, 335
   -, structural steels   179
   -, intensity   80, 89
   -, relief annealing   58, 187
   -, relief cracking   356
   -, state   79–81, 99
Stress-induced martensite   212, 279
Strip   134, 195, 198
   -, thick   138
   -, thin   135, 290
   -, very thin (blackplate)   135
Structural steels   26, 125, 352, 384
   -, for full heat treatment   190
Substitution   14
Sulphides   15, 32
Sulphides morphology, influence on
   -, anisotropy   132
   -, hydrogen embrittlement   178
   -, melting   197
   -, metal workability   247
   -, toughness   177
Supercooling   36, 46
Superferrite   316, 333
Surface film
   -, chemically resistant   111, 182, 309, 340, 357, 380
   -, wear-resistant   286
Surface layer   33, 220, 228, 237
   -, decarburised   198, 223
   -, hardening   71, 217, 219
   -, remelting   71
   -, treatment   217, 239
   -, white solidification   33, 161
Surface rolling   72, 212, 325
Surface-hardenable materials   220, 228, 241
Susceptibility   310, 322, 357

$t_{8/5}$-time   38, 45, 65, 207
Tailored
   -, blanks   161, 180
   -, castings   161
   -, strips   161, 181
   -, tubes   161, 180
Temper

-, carbon   34, 146
-, diagram   68, 273, 301
-, embrittlement   50, 69, 197, 208, 356
-, parameters   69
Tempering   48, 68, 188, 263, 329
-, auto-tempering   44, 186
-, multiple   302
-, partial   199
-, soft martensitic steels   329
-, tools   273, 279, 301
Tensile test   90
Test methods   96
Test temperature, influence on the
-, creep   354, 364
-, hardness   92
-, strength   91, 349, 354
-, thermal expansion   377
-, toughness   84
Tetragonal lattice distortion   42, 43
Texture   24, 25
-, deep-drawing sheet   95
-, electrical steel sheet   371
-, strengthening   133
Thermal
-, activation   84, 96, 349
-, conductivity   121, 148, 312, 353
-, expansion   7, 120, 376
-, shock cracks   153, 295
-, treatment of the surface layer   72, 217
-, side-effects   73
Thermomechanical metal working   37, 60, 165, 177, 186, 202
Thermostatic bimetals   120, 378
Thick plate   138
Thin
-, plate   136, 171
-, strip   135, 290
Thomas steel   397
Tool steels   265, 273, 388
Toughness   84, 88, 313
-, fine-grain steels   177
-, low-temperature steels   194
-, stainless steels   322
-, upper/lower shelf   84
Tramp elements   4, 25, 144, 416
Transformation   10
-, bainitic   53, 172, 208

-, change in volume   3, 73
-, coupled   12
-, diffusion-dependent, diffusionless   42
-, discontinuous   12, 30
-, eutectic   12
-, eutectoid   12, 29
-, isothermal   36, 39, 46, 70
-, lamellar   30
-, latent heat   36, 39
-, martensitic   39, 47, 67, 102, 210, 375
-, non-equilibrium   26, 36
-, on cooling   12
-, on heating   65
Transformation-free steels   18, 311
Transition temperature   83, 95
-, QT steels   193, 197
-, cast iron   151, 211
-, fine-grain steels   165, 168
-, stainless steels   322
Tribochemical reaction   105, 108
Tribological system   108
Tribology   103
TRIP steels   181, 183
TRIPLEX steels   181, 183
TTA diagram   65, 129
-, 42CrMo4   65
-, 50CrMo4   219
-, surface-hardening   217
TTT diagram   129
-, 42CrMo4   39
-, 56NiCrMoV7   45
-, HS6-5-2   301
-, S355   129
-, case-hardening steel   244
-, cast iron with spheroidal graphite   40
-, continuous   301
-, effect of alloying   47
-, isothermal   46
-, multi-phase steels   171
-, welding   129
Twin boundaries   21
Twinning   8, 42, 182
TWIP steels   181, 183

Unalloyed steels   30, 48, 125, 135, 220, 386

Uniform corrosion 112, 318, 331
Unit cell 6
Upper shelf 84, 96, 132, 151, 268

Vacancy 23
Vacuum
 -, carburising 237
 -, hardening 238, 301
 -, sintering 269
Valves 344, 365
Vertical forging 273
Volume change 42, 73

Wall thickness, effect of 147
Water-tempered steels 192, 256
Wear 106
 -, ADI 212
 -, duplex 325
 -, ferrite-pearlite 127
 -, mechanisms 106
 -, resistance 127
 -, types 108
Wear-resistant ferrous materials 258

Weather-resistant steels 128, 135, 385
Weldability
 -, casting alloys 161
 -, high-alloys 256, 292, 314
 -, low-alloys 128, 138, 165
Weldable structural steels 26
Welding 129, 140, 161, 166, 175, 355
Wet corrosion 109, 310, 317
 -, kinetics 111, 112
 -, local 113
 -, thermodynamics 110
White cast iron 102, 120, 154, 259
Wire 247
 -, quenched and tempered 185
 -, unalloyed 139, 195
Wrought iron 392
Wustite 74, 115, 356, 392

Yield strength 90
 -, influence of T, $\dot{\varepsilon}$ 92, 349
Young's modulus 7, 82, 100, 120

Zinc-coated steel 111, 136, 180

List of alloying and tramp elements

Aluminium 15
ageing 131, 136, 172, 352 ◇ case-hardening steels 241 ◇ creep-resistant austenite 294, 338 ◇ deoxidation 134, 263 ◇ electrical resistance heating steels 340 ◇ fine grain 165 ◇ free-cutting steel 134 ◇ functional materials 370 ◇ heat-resistant steels 309, 337 ◇ high-speed tool steels 303 ◇ inoculation 150 ◇ lightweight steels 182 ◇ maraging steels 200 ◇ nitride 241, 284, 339 ◇ nitriding steel 220, 228 ◇ oxide 182, 263, 268 ◇ protective film 137 ◇ rail steel 184 ◇ rolled texture 131 ◇ stainless steels 321, 324 ◇ γ' phase 359 ◇ σ phase 314

Antimony 15
graphite shape, influence on 145 ◇ temper embrittlement 50, 197, 208, 362

Arsenic 15
temper embrittlement 50, 197, 208, 362

Boron 15
borides 15, 251, 256 ◇ boriding 256 ◇ case-hardening steel 243, 257 ◇ cold-heading steels 196 ◇ creep resistance 357 ◇ creep-resistant steels 356 ◇ hardenability 179, 190, 195, 256 ◇ magnetically soft materials 373 ◇ quenching and tempering steels 192, 196 ◇ temper embrittlement 192

Carbon 15
carbides 52, 103 ◇ carburisation 234 ◇ graphite 32 ◇ hardness increase 66, 273 ◇ reduction 393 ◇ important in most ferrous materials

Chromium 15
carbides 50, 103, 107, 144 ◇ case-hardening steel 242 ◇ cold-work tool steels 277, 278 ◇ creep-resistant casting alloys 148 ◇ creep-resistant steels 352 ◇ electrical resistance heating steels 121, 380 ◇ equivalent 47 ◇ ferrite stabilisation 18, 47 ◇ hardenability 190, 204, 242, 256, 273, 277, 302 ◇ heat-resistant cast iron 343 ◇ heat-resistant steels 309, 337 ◇ high-speed tool steels 302 ◇ hot-work tool steels 294 ◇ nitriding steel 228, 233 ◇ passivation 112, 310 ◇ pearlitic steels 184 ◇ quenching and tempering steels 94, 193 ◇ rolling bearing steels 204, 329 ◇ secondary hardness 50, 273, 301 ◇ stainless steels 309, 310, 321 ◇ weather-resistant structural steels 127 ◇ white cast iron 259 ◇ σ phase 313, 316, 357, 360

Cobalt 15
Co-base alloys 366 ◇ creep resistance 294, 302, 357 ◇ high-speed tool steels 300 ◇ hot-work tool steels 296 ◇ Invar effect 120 ◇ maraging steels 49, 200, 292 ◇ materials with special magnetic properties 370 ◇ materials with special thermal expansion properties 376 ◇ Ni-base alloys 294 ◇ precipitation hardening 202 ◇ saturation polarisation 373

Copper 15
austenite stabilisation 323 ⋄ cold heading 323 ⋄ corrosion-resistant 318, 324 ⋄ high-strength 336 ⋄ high-strength steels 188, 207 ⋄ in cast iron 144, 151, 263, 207, 309, 342 ⋄ precipitation 50, 328 ⋄ stainless steels 332 ⋄ weather-resistant structural steels 127

Hydrogen 15
corrosion 109, 179, 336 ⋄ cracking 114, 178 ⋄ electroplating 58, 128 ⋄ embrittlement 58, 114, 178 ⋄ enamelling 138 ⋄ flaking 197 ⋄ thin films 286 ⋄ welding 58

Lead 15
graphite shape, influence on 145 ⋄ impurities 137 ⋄ machinability 134

Manganese 15
austenite stabilisation 169, 321, 330 ⋄ cast iron 32, 144 ⋄ creep-resistant steels 352 ⋄ evaporation 238 ⋄ free-cutting steels 134, 135 ⋄ hardenability 190, 204, 263 ⋄ non-magnetic steels and casting alloys 336, 342, 370, 375 ⋄ stainless steels 330, 336 ⋄ sulphide 14, 132 ⋄ temper embrittlement 192, 362 ⋄ tools for mineral processing 44 ⋄ weldable steels 126

Molybdenum 15
carbide 263 ⋄ cast iron 70 ⋄ corrosion resistance 317, 322 ⋄ creep resistance 298, 352, 358 ⋄ hardenability 207, 242, 263 ⋄ secondary hardness 50, 301 ⋄ temper embrittlement 192

Nickel 15
austenitic cast iron 309 ⋄ austenite stabilisation 70, 320, 324 ⋄ austenitic steels 322 ⋄ case-hardening steels 206, 220, 397 ⋄ cast iron 151, 208 ⋄ equivalent 47, 322 ⋄ free electrons 85, 331 ⋄ functional materials 370, 378 ⋄ hardenability 190, 208, 256, 263 ⋄ hot cracking 316 ⋄ low-temperature toughness 194, 322, 342 ⋄ quenching and tempering steels 190, 257 ⋄ maraging steels 49, 201 ⋄ meteoritic iron 395 ⋄ non-magnetisable steels and cast iron 342, 376 ⋄ soft-martensitic steels 336 ⋄ transition temperature 322 ⋄ white cast iron 261

Niobium 15
carbides 103, 107, 165, 266 ⋄ fine grain 168, 184 ⋄ intergranular corrosion 319, 357 ⋄ nitrides 165, 361 ⋄ precipitation hardening of pearlite 184 ⋄ secondary hardness 298 ⋄ wear resistance 107

Nitrogen 15
ageing 52, 130, 158 ⋄ creep embrittlement 358 ⋄ grain stability 165 ⋄ inert gas 269 ⋄ interstitial-free 138 ⋄ laser treatment 160 ⋄ martensite 46 ⋄ nitrides 46, 52 ⋄ nitriding 224 ⋄ stainless steels 206, 314, 322, 330

Phosphorus 15
cast iron 127, 151, 394 ◇ steels 127 ◇ temper embrittlement 50, 197, 208, 356

Silicon 15
ADI 209 ◇ bainite/austenite 71, 169 ◇ carbon activity 75, 144 ◇ carbon equivalent 31 ◇ cast iron 35, 51, 144, 152 ◇ decarburisation 75, 338 ◇ degree of saturation 34, 343 ◇ ferrite stabilisation 18, 47, 85 ◇ hardenability 190 ◇ heat-resistant cast iron 75, 341 ◇ heat-resistant steels 117, 309, 337, 310 ◇ high-strength steels 50, 200 ◇ magnetically hard materials 373 ◇ magnetically soft materials 119 ◇ multiphase steel 169 ◇ scale formation 76 ◇ soft annealing 59 ◇ solidification 31 ◇ spring steel 200 ◇ stainless cast iron 341 ◇ stainless steels 323, 332 ◇ transition temperature 85, 151 ◇ temper embrittlement 192, 201, 362 ◇ tool steels 233 ◇ white cast iron 33

Sulphur 15
banded microstructure 132 ◇ cast iron 144 ◇ cracking 132 ◇ machinability 135, 241, 292, 302, 320, 328 ◇ sulphides (Mn, Ti) 132 ◇ toughness 177

Tellurium 15
machinability 134

Tin 15
cast iron 35, 144, 160, 341 ◇ temper embrittlement 197, 362 ◇ tinplate 136

Titanium 15
carbides 103, 266, 305 ◇ case-hardening steels 243 ◇ creep-resistant steels 359 ◇ fine grain 168 ◇ functional materials 378 ◇ graphite shape, influence on 145 ◇ heat-resistant steels 309 ◇ intergranular corrosion 319 ◇ interstitial-free 136 ◇ machinability 302, 305, 317, 373 ◇ nitrides 103, 204, 284 ◇ oxidation resistance 309 ◇ precipitation hardening 328, 359, 377 ◇ stainless steels 319, 328 ◇ thin films 284 ◇ wear resistance 266 ◇ welding 165

Tungsten 15
carbide 267, 294, 300 ◇ creep-resistance 357 ◇ creep-resistant steels 357, 358 ◇ ferritic-austenitic steels 324 ◇ high-speed tool steels 251 ◇ hot-work tools steels 299 ◇ secondary hardness 50, 301

Vanadium 15
carbide 103, 107, 165, 266 ◇ fine grain 168, 184 ◇ hardenability 190 ◇ nitride 165, 361 ◇ pearlite refinement 96 ◇ precipitation-hardened ferritic-pearlitic steels, FP-PH-steels 184 ◇ precipitation-hardening of pearlite 184 ◇ secondary hardness 273, 301 ◇ tool steels 273 ◇ wear resistance 107, 266

Printing: Krips bv, Meppel, The Netherlands
Binding: Stürtz, Würzburg, Germany